INSTITUT INTERNATIONAL DE CHIMIE-SOLVAY

PREMIER CONSEIL DE CHIMIE. - BRUXELLES, 21-27 AVRIL 1922.

BENJAMIN COUPRIE, PHOTO. 28, AVENUE LOUISE, BRUXELLES

G. CHAVANNE O. DONY-HÉNAULT F. SWARTS CH. MAUGUIN E. HERZEN L. FLAMACHE E. HANNON AUG. PICCARD

M. DELÉPINE E. BIILMANN H. WUYTS T.-M. LOWRY G. URBAIN J. PERRIN F.-M. JAEGER A. DEBIERNE H. RUPE A. BERTHOUD R.-H. PICKARD

CH. MOUREU F.-W. ASTON Sir W.-H. BRAGG H.-E. ARMSTRONG Sir W. POPE E. SOLVAY A. HALLER S. ARRHÉNIUS F. SODDY

100 Years of Solvay Conferences

26th International Solvay Conference on Chemistry

Chemistry Challenges of the 21st Century

Hotel Plaza, Brussels, Belgium 17 – 19 October 2022

Scientific Editors

K. Wüthrich

Scripps Research, La Jolla, USA & ETH Zürich, Switzerland

B. L. Feringa

University of Groningen, The Netherlands

International Solvay Institutes Editors

L. Rongy

Université libre de Bruxelles, Belgium

A. De Wit

Université libre de Bruxelles, Belgium

World Scientific

NEW JERSEY • LONDON • SINGAPORE • BEIJING • SHANGHAI • HONG KONG • TAIPEI • CHENNAI

Published by

World Scientific Publishing Co. Pte. Ltd.
5 Toh Tuck Link, Singapore 596224
USA office: 27 Warren Street, Suite 401-402, Hackensack, NJ 07601
UK office: 57 Shelton Street, Covent Garden, London WC2H 9HE

Library of Congress Control Number: 2023920413

British Library Cataloguing-in-Publication Data
A catalogue record for this book is available from the British Library.

CHEMISTRY CHALLENGES OF THE 21ST CENTURY
Proceedings of the 100th Anniversary of International Solvay Conferences on Chemistry

ISBN 978-981-12-8231-7 (hardcover)
ISBN 978-981-12-8232-4 (ebook for institutions)
ISBN 978-981-12-8233-1 (ebook for individuals)

For any available supplementary material, please visit
https://www.worldscientific.com/worldscibooks/10.1142/13569#t=suppl

Desk Editor: Nur Syarfeena Binte Mohd Fauzi

Typeset by Stallion Press
Email: enquiries@stallionpress.com

https://doi.org/10.1142/9789811282324_fmatter

Contents

https://doi.org/10.1142/9789811282324_fmatter

I. The International Solvay Institutes

Board of Directors

Pierre Gurdjian
Chairman of the Board of Directors of the ULB

Baron Daniel Janssen
Former Chairman of the Board of Directors of the Solvay Group

Honorary Members

Franz Bingen
Emeritus Professor VUB
Former Vice-president and Treasurer of the Solvay Institutes

Philippe Busquin
Minister of State
Former European Commissioner for Research

Eddy Van Gelder
Former Chairman of the Board of Directors of the VUB

Jean-Louis Vanherweghem
Former Chairman of the Board of Directors of the ULB

Lode Wyns
Former Vice-rector for Research VUB
Former Deputy Director for Chemistry of the Solvay Institutes

Guest Members

Gert Desmet
Professor VUB and Deputy Director for Chemistry

Anne De Wit
Professor ULB and Scientific Secretary of the International Solvay Committee for Chemistry

Marc Henneaux
Professor ULB
Director

Franklin Lambert
Emeritus Professor VUB

Alexander Sevrin
Professor VUB and Deputy Director for Physics and Scientific Secretary of the International Solvay Committee for Physics

Marina Solvay
Chairwoman of the Archives Committee of the Solvay Institutes

Karel Velle
Secretary of the Royal Flemish Academy for Science and the Arts of Belgium

Didier Viviers
Secretary of the Royal Academy for Science and the Arts of Belgium

https://doi.org/10.1142/9789811282324_fmatter

II. International Solvay Scientific Committee for Chemistry

Professor Kurt Wüthrich (Chair)
Scripps Research, La Jolla, USA and ETH Zurich, Switzerland
2002 Nobel Laureate

Professor Joanna Aizenberg
Harvard University, Cambridge, USA

Professor Thomas Cech
University of Colorado, Boulder, USA
1989 Nobel Laureate

Professor Gerhard Ertl
Fritz-Haber-Institut der Max-Planck-Gesellschaft Berlin, Germany
2007 Nobel Laureate

Professor Ben Feringa
University of Groningen, The Netherlands
2016 Nobel Laureate

Professor Stefan Hell
Max Planck Institute, Germany
2014 Nobel Laureate

Professor JoAnne Stubbe
Massachusetts Institute of Technology, Cambridge, USA

Professor Bert Weckhuysen,
Utrecht University, The Netherlands

Professor George M. Whitesides
Harvard University, Cambridge, USA

Professor Anne De Wit
Université libre de Bruxelles, Belgium, Scientific Secretary

https://doi.org/10.1142/9789811282324_fmatter

III. Acknowledgements

The organization of the 26th Solvay conference has been made possible thanks to the generous support of the Solvay Family, the Solvay Group, the Université libre de Bruxelles, the Vrije Universiteit Brussel, the Belgian National Lottery, the FWO, the FRS-FNRS, the Brussels Capitale Region and the Fédération Wallonie-Bruxelles.

https://doi.org/10.1142/9789811282324_fmatter

IV. Participants

De Wit Anne
ULB, Brussels, Belgium

Ebbesen Thomas
University of Strasbourg, France

Feringa Ben
University of Groningen, The Netherlands

Fire Andrew
Stanford University, USA

Flitsch Sabine
University of Manchester, UK

Fujita Makoto
University of Tokyo, Japan

Grey Clare
Cambridge University, UK

Giuseppone Nicolas
University Strasbourg, France

Goldberg Karen
University of Pennsylvania, USA

Haile Sossina
Northwestern University, USA

Hilvert Donald
ETH Zurich, Switzerland

Höbartner Claudia
Julius-Maximilians-U. of Würzburg, Germany

Joyce Gerald
The Salk Institute for Biological Studies, USA

Kanan Matthew
Stanford University, USA

Katsonis Nathalie
University of Groningen, The Netherlands

Kirkegaard Karla
Standford University, USA

Knowles Robert
Princeton University, USA

Krishnan Yamuna
University of Chicago, USA

Liao James
Acedemia Sinica, Taiwan

Lutz Stefan
Codexis, Inc., USA

MacMillan David
Princeton University, USA

Manthiram Karthish
Caltech, USA

Meijer Bert
Eindhoven, The Netherlands

Mirkin Chad
Northwestern University, USA

Narayan Alison
University of Michigan, USA

Nocera Daniel
Harvard University, USA

Osuna Sílvia
University of Girona, Spain

Palese Peter
Icahn School of Medicine at Mount Sinai, USA

Poliakoff Martyn
University Park Nottingham, UK

Rongy Laurence
ULB, Brussels, Belgium

Sauer Joachim
Humboldt University of Berlin, Germany

Snaith Henry
University of Oxford Clarendon Laboratory, UK

Sutherland John
MRC Laboratory of Molecular Biology, UK

Szostak Jack
University of Chicago, USA

Thomas Arne
TU Berlin, Germany

Turberfield Andrew
University of Oxford, UK

Turner Nicholas
Manchester Institute of Biotechnology, UK

von Heijne Gunnar
Stockholm University, Sweden

Weckhuysen Bert
Utrecht University, The Netherlands

Wüthrich Kurt
Scripps Research, La Jolla, USA & ETH Zürich, Switzerland

Yaghi Omar
UC Berkeley, USA

Zou Xiaodong
Stockholm University, Sweden

Züttel Andreas
EPFL, Switzerland

https://doi.org/10.1142/9789811282324_fmatter

V. Auditors

Aprile Carmela (UNamur, Belgium)
Back Olivier (Solvay SA)
Ballet Steven (VUB, Brussels, Belgium)
Baron Gino (VUB, Brussels, Belgium)
Debroye Elke (KULeuven, Belgium)
Elias Benjamin (UCLouvain, Belgium)
Geerts Yves (ULB, Brussels, Belgium)
Govaerts Cédric (ULB, Brussels, Belgium)
Gueydan Cyril (ULB, Brussels, Belgium)
Hennecke Ulrich (VUB, Brussels, Belgium)
Kapelyushko Valeriy (Solvay SA)
Luhmer Michel (ULB, Brussels, Belgium)
Maestro Patrick (Solvay SA)
Marion Philippe (Solvay SA)
Martens Chloé (ULB, Brussels, Belgium)
Mastroianni Sergio (Solvay SA)
Meegan Jonathan (Solvay SA)
Remaut Han (VUB, Brussels, Belgium)
Rocci Alessio (VUB, Brussels, Belgium)
Surin Mathieu (UMons, Belgium)
Theunissen Cédric (ULB, Brussels, Belgium)
Tielens Frederik (VUB, Brussels, Belgium)
Valkenier Hennie (VUB, Brussels, Belgium)
Van Assche Guy (VUB, Brussels, Belgium)
Vorobieva Anastassia (VUB & VIB, Brussels, Belgium)

https://doi.org/10.1142/9789811282324_fmatter

26th Solvay Conference on Chemistry — Brussels, 17–19 October 2022

HOTEL PLAZA — 17 OCT 2022

Dear Jean-Marie,
Dear Members of the Solvay Family,
Dear Colleagues,
Dear Friends,

It is my great honour and pleasure to open the 26th Solvay Conference on Chemistry. Its theme is "Chemistry Challenges of the 21st Century".

We are celebrating this year the 100th anniversary of the first Solvay Conference on Chemistry, which took place in Brussels in 1922.

The title of the 1922 conference was "5 questions d'actualité". What were these 5 topics which were drawing the research of chemists in those days? These were:

- "Isotopes and radioactivity" with reports by Soddy, Aston Perrin and Urbain
- "Analysis of molecular structure by X-rays" with a report by Bragg
- "Molecular structure and optical activity" with reports by Pope and Lowry
- "Valence "with a report by Mauguin
- "Chemical mobility" with a report by Job.

How much knowledge has been gained since then! Perhaps the best way to illustrate the expansion of the scope of chemistry in the century that separates us from the first conference is to give the list of all the Solvay conferences on chemistry,

1. 1922 « Cinq Questions d'Actualité »
2. 1925 « Structure et Activité Chimique »
3. 1928 « Questions d'Actualité »
4. 1931 « Constitution et Configuration des Molécules Organiques »
5. 1934 « L'Oxygène, ses réactions chimiques et biologiques »
6. 1937 « Les Vitamines et les Hormones »
7. 1947 « Les Isotopes »
8. 1950 « Le Mécanisme de l'Oxydation »
9. 1953 « Les Protéines »
10. 1956 « Quelques Problèmes de Chimie Minérale »
11. 1959 « Les Nucléoprotéines »
12. 1962 « Transfert d'Energie dans les Gaz »
13. 1965 « Reactivity of the Photoexited Organic Molecule »
14. 1969 « Phase Transitions »
15. 1970 « Electrostatic Interactions and Structure of Water »
16. 1976 « Molecular Movements and Chemical Reactivity as conditioned by Membranes, Enzymes and other Molecules »
17. 1980 « Aspects of Chemical Evolution »
18. 1983 « Design and Synthesis of Organic Molecules Based on Molecular Recognition »
19. 1987 « Surface Science »
20. 1995 « Chemical Reactions and their Control on the Femtosecond Time Scale »
21. 2007 « From Noncovalent Assemblies to Molecular Machines »
22. 2010 « Quantum Effects in Chemistry and Biology »
23. 2013 « New Chemistry and New Opportunities from the Expanding Protein Universe »
24. 2016 « Catalysis in Chemistry and Biology »
25. 2019 « Computational Modeling: From Chemistry to Materials to Biology »

This list shows great diversity in the themes, ranging from physical chemistry to biochemistry. The theme of the centenary conference that starts today is voluntarily broad and reflects the huge span of topics in which chemistry plays a central role.

***

There are many interesting scientific aspects connected with the history of the Solvay conferences on chemistry.

However, while the history of the Solvay conferences on physics is well documented, with many books written about it, the history of the conferences on chemistry is still to be written to a large extent.

Only a few conferences have drawn some attention of the historians of science and much more remains to be done.

Thanks to the remarkable support of the Solvay family — and thank you again, Marina and Jean-Marie for supporting this initiative -, we have started the systematic classification and study of the chemistry archives of the Institutes. They constitute a mine of information about the development of chemistry in the 20th century, which we must bring to light!

You will have a first taste of the work being done in two lectures on the history of the 1922 conference, one this morning, and another one at the banquet tomorrow.

***

Before we start, let me recall the format of the Solvay Conferences. Ernest Solvay wanted to put in contact the best scientists from the world by inviting them to meet periodically in small number in order to discuss in depth questions chosen by them before the meeting.

These are thus conferences by invitation-only, with a limited number of participants. The presentations are short in order to leave a lot of time to the discussions. Scientific interactions are privileged over talks.

For the discussions to be fruitful, at least two conditions must be met: (1) First, participants should attend the full conference. (2) Second, an extremely careful preparation is clearly needed. We all know that spontaneous discussions only work if there is preparation!

I would like to express the gratitude of the International Solvay Institutes to all the scientists behind this careful preparation: the two chairs of the conference, Professors Kurt Wüthrich (who is also the chair of the International Solvay Scientific Committee for Chemistry) and Ben Feringa, the session chairs and also, the local scientific team led by our colleague Anne De Wit, who has dedicated an enormous amount of her time to make sure that the conference could proceed as smoothly as possible, something that has not been easy given the many last-minute cancellations due to the sanitary crisis.

***

Professor Kurt Wüthrich will now also say a few welcome words. Kurt has been helping us for about 12 years now as Chair of the International Committee for Chemistry and has been directly involved in the organization of 4 Solvay conferences. It is a great pleasure for me to give him the floor.

***

Thank you very much for your attention.
Marc Henneaux

https://doi.org/10.1142/9789811282324_fmatter

Preface by Professor Kurt Wüthrich, chair of the 26th "Centenary" Solvay Conference on Chemistry

"Chemistry Councils" organized under the auspices of Ernest Solvay, of which the first was held in 1913, paved the avenue that led to the first Solvay Conference in Chemistry in 1922. This early history of the Solvay Conferences on Chemistry was reviewed in the November/December 2013 issue of *Chemistry International*, pages 4–11, under the titles "A Look Back at Ernest Solvay" and "The Solvay Chemistry Councils". Following the inaugural 1922 Conference, Solvay Chemistry Conferences were held in regular three-year intervals until 1937, and again after the Second World War from 1947 to 1965. From 1969 to 1995, seven conferences were held at irregular intervals, and after an interlude of twelve years the regular three-year schedule was adopted again in 2007, so that we had the 26th Solvay Conference on Chemistry in 2022. Remarkably, the conferences up to 1962 were held in French, and English has been the conference language only since 1965.

The themes of the first three conferences, "Cinq Questions d'Actualité", "Structure et Activité Chimique" and "Questions d'Actualité" were very general, but since 1931 the individual Conferences were focused on special areas of chemistry. During much of the ensuing history, biology and biochemistry were prominently represented in the scientific Conference programs. Examples are 1937 "Les Vitamines et les Hormones", 1953

"Les Protéines", 1959 "Les Nucléoprotéines", 1976 "Molecular Movements and Chemical Reactivity as Conditioned by Membranes, Enzymes and other Molecules", 2013 "New Chemistry and New Opportunities from the Expanding Protein Universe", and the word "Biology" was also part of the themes of the 2016 and 2019 Conferences. In the centenary meeting we returned to a general theme, similar to the first three Conferences, addressing burning societal issues of the 21st century. We investigated possible roles of chemistry in dealing with environmental issues and climate change, emphasizing sustainability for the future of chemistry, research on new materials and searches for novel renewable energies. Considering the Covid-19 pandemic during the years preceding the Centenary Conference, we addressed the important role of chemistry in the defense against newly emerging pathogens.

The early history of modern chemistry relates to a small number of outstanding individuals, and this was also reflected in the early format of the Solvay Conferences. Each one of the scientific sessions contained in a Conference was introduced by a single "Rapporteur", who presented the current status of the session theme. However, over the years the number of scientists working in an ever-increasing variety of chemistry specializations increased to the point where the individual view of a "Rapporteur" had to be replaced by presentations of diverse views by a group of scientists, providing a broader foundation. It is quite unique to the Solvay Conferences that the remaining part of each session is devoted to an open discussion among all 40 to 50 invited participants, which is recorded as part of the Proceedings.

For the Centenary Conference this format of the preceding three Conferences was maintained, and all invited participants had formal assignments in the scientific program. In advance of the Conference, the chairpersons and the panelists of each session contributed short articles presenting their current views on the theme of their session. These papers were sent to all participants during the week before the Conference. In Brussels, each of the six scientific sessions was then opened with a general introduction by the chairperson and short presentations of five to seven panelists. The transcripts of the ensuing general discussions among all invited participants, as edited by "Auditors" and then further checked by the discussion contributors for correct scientific content, are the core of the Proceedings. For each session, they are preceded by the introductory reports of the session chair and the panelists. The first Session on "Catalysis for Sustainable Chemistry" had originally been organized by the late Robert H. Grubbs, who we dearly missed in Brussels. David MacMillan had kindly accepted to take over the chair of this Session, adding his personal touch to its program. The Session 2, "From Molecules to Dynamic Supramolecular Systems" was organized and vividly chaired by Ben Feringa. Omar M. Yaghi presented a highly exciting Session 3 on "Reticular Chemistry and New Materials». In Session 4, "New Chemistry for Renewable Energy", Daniel Nocera had assembled a program that presented this important aspect of the future of humankind from carefully selected different viewing angles. In Session 5, Sabine Flitsch addressed another important direction of sustainable chemistry, with the theme "Directed Protein Evolution for Green Chemistry". The final Session 6, organized by Gerald Joyce, "RNA Chemistry to Explore our Origins and Fight Viral Pandemics" combined two additional themes of high interest and urgency. It was a special privilege for me to enjoy the help of so many in the preparation of this broad scientific program for the Centenary Solvay Conference on Chemistry. I want to especially acknowledge the co-Chair of the conference, Prof. Ben Feringa, and the Chairpersons of the individual Sessions for their support.

Today's Centenary Conference and the scientific activities during the past one-hundred years were only possible with the continuous unwavering support of the Solvay family, to whom we owe deep respect and gratitude. Mr. Jean-Marie Solvay, President of the Solvay Institutes, Mrs. Marie-Claude Solvay de la Hulpe and Mrs. Marina Solvay showed their close personal interest and kind support by attending parts of the scientific program and the social events of the meeting. We greatly appreciate the generous support by the Solvay Institutes represented by Prof. Marc Henneaux. Special thanks go to Prof. Anne De Wit and Prof. Laurence Rongy, who organized the recording and transcription of the scientific discussions by a group of local scientists (the "Auditors"). Dominique

Bogaerts and Isabelle Van Geet from the Solvay Institutes made our lives easy and enjoyable with their commitment and dedicated assistance; we owe them our heartfelt thanks.

2023-03-21
Kurt Wüthrich
Chair of the Conference

https://doi.org/10.1142/9789811282324_fmatter

Tribute to Professor K. Wüthrich

Dear Professor,

You made your first appearance at our conferences in 1983.

The title of your presentation was "Glucagon Conformation in different Environments: Implication for Molecular Recognition".

At the time, the structure of glucagon had already been identified by X-ray crystallography. However YOU discovered through nuclear magnetic resonance spectroscopy that in solution in water its structure was totally different. By investigating the environment and their interactions of the protein with solvent molecules such as water. You could see the glucagon in 3D, 29 amino acids that end in a helicoidal structure.

When you presented your structure, Robert Huber who studied proteins with X-ray crystallography said "This is not possible, one can not identify a structure in solution in water because the molecule moves all the time. You can not photograph with precision an object in rapid movement".

You were quite upset!

So you left the lab and went skiing for two years in Wengen. You became a ski instructor. You even trained with the ski team.

Then Hoechst gave simultaneously to you and to Huber one of their products to study. In a few months you had identified its structure in solution with NMR spectroscopy.

On the other hand, Huber and his assistant took a longer time with their mono crystals images only to discover that they had found the same structure as yours.

In 1997, Europe was going through the Mad Cow Crisis, and you described the structure of the protein "Prion".

This protein had a long tail, it was new!

When you use the NMR spectroscopy method you are able to work in solution. You do not need mono crystals and you are closer to physiological conditions.

You were awarded the Nobel prize in 2002 for the "development of nuclear magnetic resonance spectroscopy for determining the three-dimensional structure of biological macromolecules in solution".

And then you made the Institutes the great honor to join the International Chemistry Committee in 2005.

In 2011, you became Chairman of the committee.

You have been thriving at bringing together brilliant minds to discuss their views and ideas on topics in the same way started by Lorentz in 1911.

We have appreciated your work and the commitment you gave to this task.

Thanks to you the Solvay International conferences of Chemistry have lived up to their reputation and excellence.

Many thanks for all what you have done for the Solvay conferences.

Marina Solvay

Image courtesy of David MacMillan

Session 1

Catalysis for Sustainable Chemistry

https://doi.org/10.1142/9789811282324_0001

Sustainable Synthesis via Novel Activation Modes: Organocatalysis and Photoredox Catalysis

Colin A. Gould and David W. C. MacMillan

Department of Chemistry, Princeton University, Princeton, NJ 08540, USA

Catalysis for sustainable chemistry

Catalysis impacts society broadly, enabling advancements in diverse areas that range from renewable energy to drug discovery. Indeed, around 35% of gross domestic product (GDP) globally relies on catalysis [1], and over 75% of all industrial chemical transformations and 90% of newly developed processes utilize catalysts [2]. Due to the ubiquitous and essential nature of catalysis, the development of more efficient and environmentally friendly catalytic processes represents a key step towards a more sustainable future.

Within chemical industry, the pharmaceutical and fine chemical sector is a particularly crucial area for the implementation of sustainable catalysis. Pharmaceutical and fine chemical synthesis is currently reliant on non-sustainable technologies, such as precious metal catalysis [3], and this sector generates over an order of magnitude more waste per product (kg/kg) than other areas of the chemical industry (e.g., bulk chemical synthesis or oil refining) [4]. Over the past several decades, significant progress towards the sustainable synthesis of pharmaceuticals and fine chemicals has been achieved by replacing non-sustainable chemistry, reducing waste, and increasing efficiency [5]. For instance, technologies such as base-metal catalysis [6], biocatalysis [7], and organocatalysis [8] have been developed that can be substituted for precious metals in some chemical transformations. As an example, in the manufacturing process of the type 2 diabetes medication Sitagliptin, scientists at Merck developed a biocatalytic transamination reaction to replace a rhodium-catalyzed asymmetric hydrogenation, resulting in a substantial increase in productivity (53% increase in kg/L per day) and yield (10–13% increase) and a reduction in waste (19% decrease), while simultaneously eliminating the need for precious metals in this process [9]. Clearly, the development of more sustainable technologies can render the pharmaceutical and fine chemical sector more environmentally friendly, while simultaneously increasing productivity and efficiency.

In the following sections, I will briefly summarize my lab's contributions towards sustainable catalysis, focusing on the areas of organocatalysis [10] and photoredox catalysis [11, 12], and discuss future directions in these areas.

My recent research contributions

My lab focuses on the development of novel activation pathways that enable new chemical transformations, including work in the fields of organocatalysis [10] and photoredox catalysis [11, 12]. In particular, synthetic methodologies developed by my lab utilize native functional groups such as carboxylic acids [13, 14], alcohols [15], and even C–H bonds [16–18] as versatile cross-coupling handles to enable the expedited synthesis of complex molecules from abundant, bench-stable precursors. Such methodologies have been rapidly adopted by the fine chemical and pharmaceutical industry, enabling more efficient and sustainable syntheses of target compounds.

For example, the fragrance company Firmenich currently synthesizes a target compound — Lily of the Valley (Fig. 1) — through a single-step

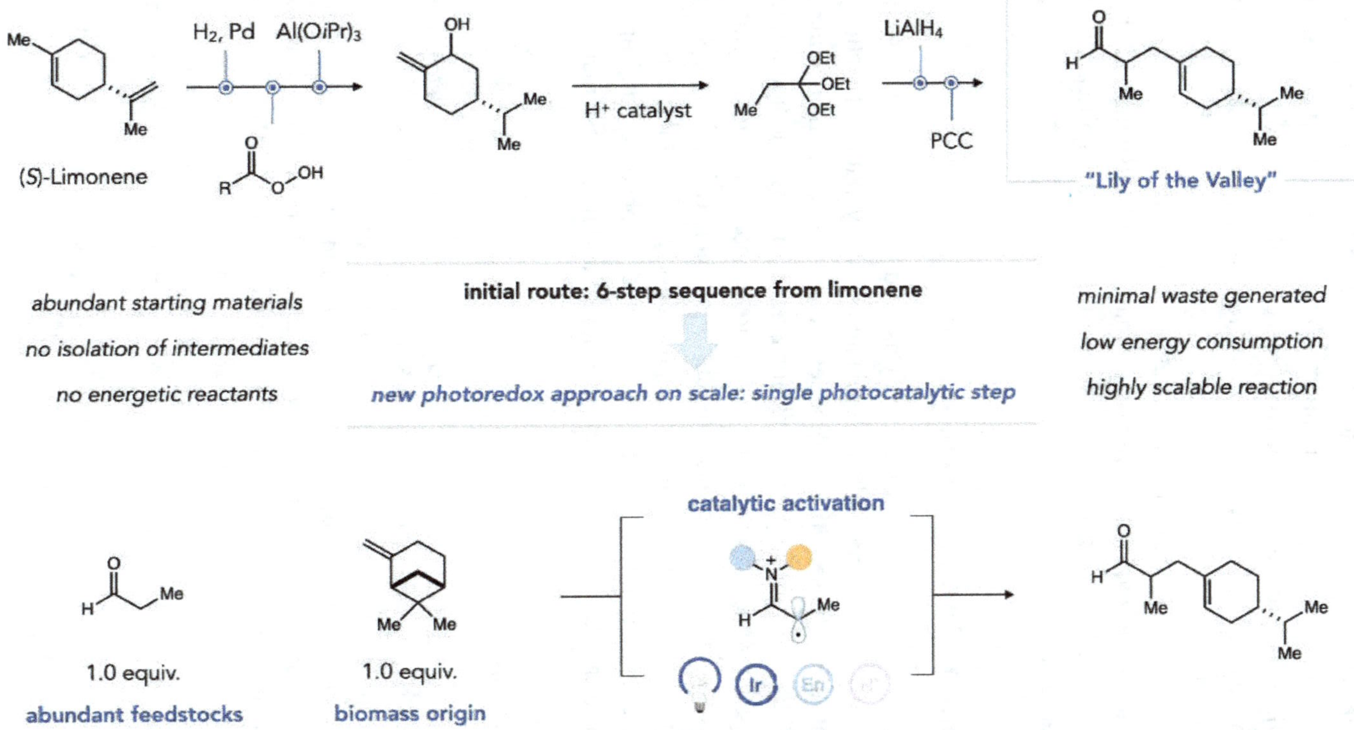

Fig. 1. Expedited synthesis of Lily of the Valley through the merger of organocatalysis and photoredox catalysis.

chemical reaction that is promoted by the combination of an organocatalyst and a photocatalyst [19], an approach that was first developed in my lab [20]. This organophotocatalytic process allows the initial six-step route to Lily of the Valley to be reduced to a single step, starting from a biomass-derived precursor, (−)-β-pinene, and a feedstock chemical, propanal. The optimized route eliminates toxic or energetic reagents (i.e., peroxy acids and pyridinium chlorochromate), while also minimizing the waste generated during the synthesis, resulting in a significantly more sustainable process overall.

Outlook to future developments

Since their development, organocatalysis, and photoredox catalysis have been rapidly adopted by both academia and industry, but most methodologies that utilize these activation modes are designed for small-scale processes (i.e., <1 gram output). Translating more of these methodologies to an industrial scale is thus an essential goal, and, excitingly, significant progress is currently being made in this direction [21, 22]. I am optimistic about the long-term adoption of these activation modes at the commercial scale by chemical industry, and I believe that this constitutes an important route towards making catalysis more efficient and more sustainable.

References

[1] Z. Ma and F. Zaera, *Encyclopedia of Inorganic Chemistry — Heterogeneous Catalysis by Metals*, John Wiley & Sons, Ltd., p. 1 (2005).

[2] S. Bhaduri and D. Mukesh, *Homogeneous Catalysis: Mechanisms and Industrial Applications*, John Wiley & Sons, Ltd., p. 1 (2014).

[3] A. O. King and N. Yasuda, *Organometallics in Process Chemistry: Topics in Organometallic Chemistry Vol. 6*, Springer, p. 205 (2004).

[4] R. A. Sheldon, *J. R. Soc. Interface* **13**, 20160087 (2016).

[5] R. Ciriminna and M. Pagliaro, *Org. Process Res. Dev.* **17**, 1479 (2013).

[6] J. D. Hayler, D. K. Leahy, and E. M. Simmons, *Organometallics* **38**, 36 (2019).

[7] J. P. Adams, M. J. B. Brown, A. Diaz-Rodriguez, R. C. Lloyd, and G.-D. Roiban, *Adv. Synth. Catal.* **361**, 2421 (2019).

[8] B. Han, X.-H. He, Y.-Q. Liu, G. He, P. Cheng, *et al.*, *Chem. Soc. Rev.* **50**, 1522 (2021).

[9] C. K. Saville, J. M. Janey, E. C. Mundorff, J. C. Moore, S. Tam, *et al.*, *Science* **329**, 305 (2010).

[10] D. W. C. MacMillan, *Nature* **455**, 304 (2008).

[11] M. H. Shaw, J. Twilton, and D. W. C. MacMillan, *J. Org. Chem.* **81**, 6898 (2016).

[12] A. Y. Chan, I. B. Perry, N. B. Bissonnette, B. F. Buksh, G. A. Edwards, *et al.*, *Chem. Rev.* **122**, 1485 (2022).

[13] Z. Zuo, D. T. Ahneman, L. Chu, J. A. Terrett, A. G. Doyle, *et al.*, *Science* **345**, 437 (2014).

[14] C. P. Johnston, R. T. Smith, S. Allmendinger, and D. W. C. MacMillan, *Nature* **536**, 322 (2016).

[15] Z. Dong and D. W. C. MacMillan, *Nature* **598**, 451 (2021).

[16] J. D. Cuthbertson and D. W. C. MacMillan, *Nature* **519**, 74 (2015).

[17] M. H. Shaw, V. W. Shurtleff, J. A. Terrett, J. D. Cuthbertson, and D. W. C. MacMillan, *Science* **352**, 1304 (2016).

[18] I. B. Perry, T. F. Brewer, P. J. Sarver. D. M. Schultz, D. A. DiRocco, *et al.*, *Nature* **560**, 70 (2018).

[19] C. Chapuis, International Patent PCT/EP2017/083377 (2018).

[20] J. A. Terrett, M. D. Clift, and D. W. C. MacMillan, *J. Am. Chem. Soc.* **136**, 6858 (2014).

[21] F. Xu, *Organocatalysis for Asymmetric Synthesis: From Lab to Factory*, John Wiley & Sons, Ltd., p. 1 (2013).

[22] K. C. Harper, E.-X. Zhang, Z.-Q. Li, T. Grieme, T. B. Towne, *et al.*, *Org. Process Res. Dev.* **26**, 404 (2022).

https://doi.org/10.1142/9789811282324_0002

Designing Catalysts for Selective Aerobic Oxidations of Hydrocarbons

Karen I. Goldberg

Department of Chemistry, University of Pennsylvania, Philadelphia, PA 19146, USA

Catalysis for sustainable chemistry: Catalytic systems for the direct aerobic oxidation of alkanes

Despite the growing deployment of renewable sources of energy, society's dependence on fossil fuels is predicted to continue for decades into the future [1]. Our heavy reliance on crude oil for energy and as a feedstock for our chemicals severely limits our efforts to address climate goals. The large amount of natural gas that is currently being wasted by flaring (direct conversion to carbon dioxide with no energy capture) is also alarming. Natural gas, a by-product in oil drilling, is often flared because no economically viable means exist for transporting it or converting it on-site to a liquid chemical or fuel; more than 144 billion m^3 of natural gas were flared worldwide in 2021 [2]. Methane, the largest component of natural gas, has greater warming potential than carbon dioxide and is a significant contributor to the temperature increase experienced since the industrial revolution. Notably, fossil fuel production is only one source of methane emission, with other large anthropogenic sources being landfills and enteric fermentation.

A two-step conversion of methane to methanol (a liquid product, which can be more economically transported) has been operational since the middle of the last century. However, this commercial process is highly energy intensive and must be carried out on a large scale to be economically viable. It is an indirect process in which methane is steam reformed over a heterogeneous catalyst at high temperatures (ca. 700–1000°C) to produce carbon monoxide and hydrogen gas followed by recombination of the carbon monoxide and hydrogen over a different heterogeneous catalyst to produce methanol. Developing technologies for direct methane to methanol conversion under mild conditions which could operate on a smaller scale could significantly impact climate change progression. Similarly, mild aerobic alkane dehydrogenation processes would allow for the low-temperature conversion of the light alkanes in natural gas (ethane and propane) to olefins.

The remarkable ability of homogeneous transition metal compounds to selectively activate strong alkane C–H bonds under mild conditions makes them particularly attractive to target as catalyst candidates for the direct conversion of natural gas to valuable chemicals and fuels. Homogenous compounds are already known to selectively activate methane and convert it to methanol or derivatives thereof [10–11]. However, the oxidants used in these reported processes are expensive and/or hazardous. For a commercially viable process, molecular oxygen as the oxidant will almost certainly be required. Air or oxygen is the ideal oxidant because it is abundant, cost-effective, and environmentally benign, generating only water or hydrogen peroxide (which typically decomposes to water) as byproducts. Notably, almost all large-scale commercial organic oxidations use O_2, either in the form of air or separated from air as the oxidant [12].

Mechanism of the reactions of homogeneous metal complexes with oxygen

The first examples of C–H bond activation of alkanes, including methane [13], by homogeneous metal complexes were reported some

50 years ago, and the mechanisms of these reactions have since been studied in great detail [3–11] However, selective activation of the C–H bond is only one step required in the direct aerobic partial oxidation of alkanes, and most of the studies, on this step and on the subsequent steps, have been carried out in oxygen-free environments. Because most organometallic chemists rigorously avoid oxygen, working in glove-boxes and on Schlenk/vacuum lines, the field has very limited information about how organometallic compounds react with oxygen and further how the key C–H activation step itself can be carried out effectively in the presence of oxygen. Thus, a significant challenge for the development of commercially viable aerobic oxidations of methane, other light alkanes, and hydrocarbons in general is to learn how organometallic complexes react with oxygen.

With the goal of selective aerobic hydrocarbon oxidation in mind, our team has been actively investigating the reactions of late transition metal complexes with molecular oxygen [14–19]. Important with respect to aerobic methane oxidation, we have demonstrated the insertion of molecular oxygen into metal–methyl bonds as well the aerobic oxidation of metal–methyl complexes leading to methanol products and have carried out extensive mechanistic studies of these reactions.

In addition to learning how metal complexes react with oxygen, we also need to learn how to promote C–H activation effectively in the presence of oxygen. It appears that the key to an oxygen-compatible C–H activation reaction may be the mechanism of the C–H bond cleavage. For example, the low valent electron-rich metal centres that are required for an oxidative addition mechanism of C–H activation often react directly with oxygen, water, and/or the functionalized alcohol products of aerobic oxidation. In contrast, systems that activate alkane C–H bonds via mechanisms in which the metal maintains its oxidation state, such as metal–ligand cooperation (MLC) or concerted metalation–deprotonation (CMD), are less likely to be affected by oxygen. Shown in Fig. 1 are two proposed cycles for aerobic oxidation of alkanes as mediated by a transition metal centre; the one on the left is for methane to methanol and the cycle on the right is for alkanes to olefins. Both are shown using CMD-type C–H activation mechanisms, but MLC pathways can also be envisioned. The

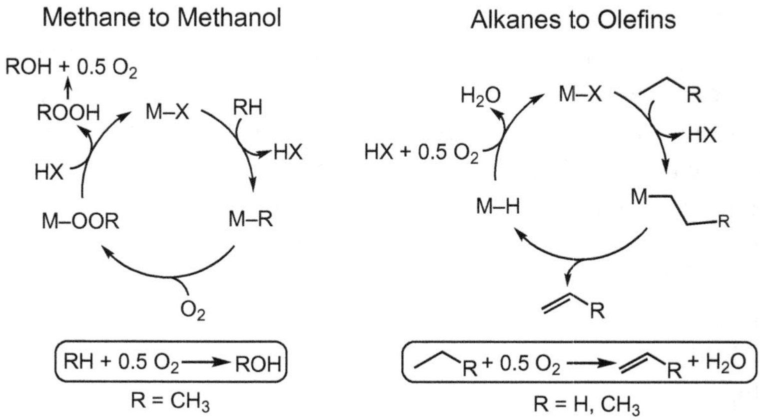

Fig. 1. Two proposed catalytic cycles for aerobic oxidation of methane and light alkanes to methanol and olefins, respectively.

individual reaction steps shown in these cycles each have literature precedent, and studies of the mechanisms of these individual steps are continuing to provide insight into the best metals, ligands, and reaction conditions. These factors for each reaction must then be balanced to optimally design a single metal–ligand system that can catalyze the full cycle with high activity and selectivity.

The outlook for this sustainability catalysis challenge

Direct alkane functionalization was first termed a "holy grail" in 1995 [3]. We are not there yet, but tremendous progress has been made. Global interest in sustainability and climate change have heightened the importance of carrying out this transformation on commercial scale using oxygen as the oxidant. From catalytic studies to detailed mechanistic understanding, many of the stumbling blocks have been identified and strategies to remove them or design around them are being deployed. In terms of sustainability and the health of the planet and humanity, developing catalytic systems for the aerobic oxidation of alkanes will remain a high-priority goal.

Acknowledgements

The United States (US) National Science Foundation (NSF, CHE-1856547) is acknowledged for support of this work.

References

[1] U.S. Energy Information Administration, *International Energy Outlook 2021*, October 6, 2021.
[2] The World Bank, *Global Gas Flaring Reduction Partnership (GGFR)*, https://www.worldbank.org/en/programs/gasflaringreduction/global-flaring-data (Accessed September 20, 2022).
[3] B. A. Arndtsen, R. G. Bergman, T. A. Mobley, and T. H. Peterson, *Acc. Chem. Res.* **28**, 154 (1995).
[4] A. E. Shilov and G. B. Shul'pin, *Chem. Rev.* **97**, 2879 (1997).
[5] S. S. Stahl, J. A. Labinger, and J. E. Bercaw, *Angew. Chem. Int. Ed.* **37**, 2180 (1998).
[6] J. A. Labinger and J. E. Bercaw, *Nature* **417**, 507 (2002).
[7] U. Fekl and K. I. Goldberg, *Adv. Inorg. Chem.* **54**, 259 (2003).
[8] R. A. Periana, G. Bhalla, W. J. Tenn III, K. J. Young, X. Y. Liu, *et al.*, *J. Mol. Catal. A: Chem.* **220**, 7 (2004).
[9] M. Lersch and M. Tilset, *Chem. Rev.* **105**, 2471 (2005).
[10] J. A. Labinger and J. E. Bercaw, *Chem. Rev.* **117**, 8483 (2017).
[11] K. I. Goldberg and A. S. Goldman, *Acc. Chem. Res.* **50**, 620 (2017).
[12] F. Cavani and J. H. Teles, *ChemSusChem* **2**, 508 (2009).
[13] N. F. Gol'dshleger, M. B. Tyabin, A. E. Shilov, A. A. Shteinman, *et al.*, *Russ. J. Phys. Chem.* **43**, 122 (1969).
[14] L. Boisvert and K. I. Goldberg, *Acc. Chem. Res.* **45**, 899 (2012).
[15] M. L. Scheuermann and K. I. Goldberg, *Chemistry* **20**, 14556 (2014).
[16] K. E. Allen, D. M. Heinekey, A. S. Goldman, K. I. Goldberg, *et al.*, *Organometallics* **33**, 1337 (2014).
[17] J. D. Prantner, W. Kaminsky, and K. I. Goldberg, *Organometallics* **33**, 3227 (2014).
[18] K. A. Smoll, W. Kaminsky, and K. I. Goldberg, *Organometallics* **36**, 1213 (2017).
[19] H. E. Zeitler, W. Kaminsky, and K. I. Goldberg, *Organometallics* **37**, 3644 (2018).

https://doi.org/10.1142/9789811282324_0003

Reducing the Environmental Impact of Solvents for Catalytic Reactions

Martyn Poliakoff and Michael W. George

School of Chemistry, University of Nottingham, University Park, Nottingham NG7 2RD, UK

The present state of research on catalysis for sustainable chemistry

Ever since the birth of Green Chemistry, there has been a focus on increasing the sustainability of chemistry by reducing waste. Catalysis has since made impressive advances in maximizing the selectivity of reactions and minimizing the waste from stoichiometric reagents [1]. More recently, there have been further improvements by the use of photo- and electrochemistry to carry out "reagentless reactions" [2]. Nevertheless, large volumes of waste in pharmaceutical, speciality and fine chemicals manufacture originate in the solvents used in the processes and for the purification of the products [3]. Not only can the solvents sometimes be toxic compounds, for example, chlorinated organics, but most are also hydrocarbon-based and pose flammability and environmental hazards.

Even the use of water as a solvent can be problematic as it is costly to clean traces of organic substrates from the water before it can be safely discharged into the environment. Therefore, there remains a pressing need for reducing solvent waste to create more sustainable chemical manufacture via catalysis. Here, we suggest one possible way forward, linked to efforts for mitigating climate change.

Our recent research contributions to catalysis for sustainable chemistry

We have pioneered the use of supercritical carbon dioxide ($scCO_2$) as a solvent for chemical reactions and devised methodologies for high-pressure continuous catalytic reactions on laboratory and larger scales. CO_2 becomes supercritical above its critical point of 30.9°C and 73.8 bar. $scCO_2$ has relatively low solvent power (similar to that of an alkane) and, in general, the solubility of a compound increases with pressure. However, $scCO_2$ has several potential advantages as a solvent: (i) CO_2 is very inexpensive compared to most solvents; (ii) solute separation merely requires release of the pressure and purification of the used CO_2 is simpler than purification of typical organic solvents; (iii) CO_2 is miscible with gaseous O_2 and is completely non-flammable, offering safety advantages for aerobic oxidation reactions [4]; (iv) CO_2 is also miscible with H_2 making it highly suitable as a solvent for hydrogenation. Therefore, our initial work focused on hydrogenation of relatively simple organic compounds with H_2 using a Pd heterogeneous catalyst, with the reactions eventually being scaled up to a 1000-ton p.a. commercial plant [5], which was technically highly successful but was rendered uneconomic by steeply rising energy costs for recompressing the CO_2 (see the outlook). However, we demonstrated that $scCO_2$ can offer additional advantages such as in the hydrogenation of levulinic acid to form γ-valerolactone where the high-pressure phase behaviour can be manipulated to separate pure product (Fig. 1) from the co-product H_2O with less energy than would be required for conventional separation by distillation [6].

We have also combined some of the advantages of photochemistry [2] and $scCO_2$ to demonstrate that $scCO_2$ can be used for reactions with photo-generated singlet oxygen ($^1O_2$). The reactions of $^1O_2$ are potentially

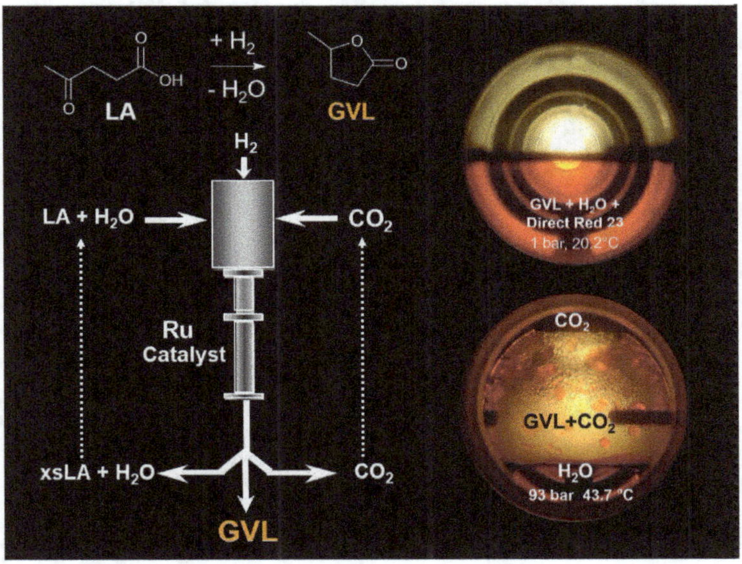

Fig. 1. Schematic of the hydrogenation of levulinic acid (LA) over a ruthenium catalyst to form γ-valerolactone (GVL) in $scCO_2$ using H_2O as the co-solvent and exploiting the pressure-induced phase separation of GVL and H_2O to obtain pure GVL at the output of the reactor. The photos on the right show this phase separation, with a water-soluble dye, Direct Red 23, to highlight the separation of the water more clearly. For more details, see [6].

attractive because of their high atom economy, but, until now, there has been a reluctance to use them on a large scale because of the unstable nature of the intermediate products and the need to minimize the flammability of the solvent. Both of these problems can be addressed by the use of $scCO_2$ in high-pressure flow photo-reactors, which have the additional advantage that $^1O_2$ has a long lifetime in $scCO_2$ [7]. Although our reactors are still relatively small, we have produced a variety of products using $^1O_2$ in $scCO_2$ on scales that are at least an order of magnitude greater than have been previously reported [8]. One example, which involves both $^1O_2$ and an unusual dual-function heterogeneous catalyst, is the semi-synthesis of the anti-malarial drug artemisinin (Fig. 2) using toluene as a co-solvent and eliminating the need for toxic CF_3CO_2H, which is conventionally used as the acid catalyst in this reaction [9].

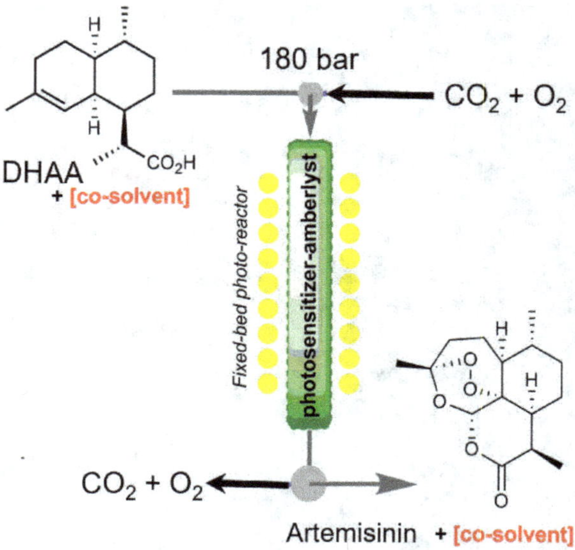

Fig. 2. The photo-oxidation of dihydroartemisinic acid, DHAA, by photo-generated singlet O_2 in $scCO_2$ with a toluene co-solvent to form Artemisinin with the use of a bi-functional catalyst, the photosensitizer tetraphenylporphyrin, immobilized on the acid catalyst Amberlyst 15. Adapted from [9].

We have also shown how such $scCO_2$ flow reactors can easily be automated and combined with AI for self-optimizing catalytic reactions [10].

In addition, we have considered the wider environmental impact of making chemicals and have been inspired by the ideas of Chemical Leasing [11], which proposes that chemical manufactures should recognize that most customers buy chemicals for the effect that those chemicals produce (e.g., coating a surface and treating an illness). This goes against the conventional business model that is based upon the assumption that you will earn more from selling more products. The supplier does not sell quantities of Chemical Leasing, but the supplier sells the performance i.e., the function of the chemical. Thus, by analogy with the electronics industry [12], we proposed Moore's Law for Chemistry, namely that *sustainable chemists should strive to reduce the amount of a chemical needed to produce a given effect by a factor of two, for example within a 5 year cycle, and this process should be repeated for a number of cycles* [13]. We build on this concept in the outlook.

Outlook: Future developments of research on catalysis for sustainable chemistry

CO_2 is cheap and abundant and $scCO_2$ is a solvent with considerable potential for catalytic reactions. At the same time, $scCO_2$ is beset by the problem of high energy costs for the compression of CO_2. We believe that increasing concerns about climate change offer a real opportunity to resolve this problem. The opportunity lies at the core of some philosophical aspects of Green Chemistry, namely the need to expand the focus from the design of chemical routes and the selection of feedstocks to the wider problems of manufacturing, business models, and supply chains [14]. Our proposal involves Carbon Capture and Storage (CCS) which is increasingly recognized as a key technology, at least in the short term, for decarbonizing not only the energy sector but also other industries (e.g., cement-making or steam reforming to make so-called "blue" ammonia). However, a major barrier to implementation worldwide is that CCS introduces significant additional costs compared to releasing CO_2 into the atmosphere. Therefore, there is a real need for strategies which add value to the captured CO_2 by transforming it into a revenue-earning asset. Thus, one needs to address not only the environmental but also the economic pillar of the triple bottom line [15], a difference between sustainable and green chemistry.

The opening of the 1000-ton p.a. commercial plant [5] in 2002 mentioned above coincided with a large rise in energy prices, a problem which continues to reoccur and impact chemical and other manufacturing. This meant that the energy cost for compressing CO_2 rendered the plant uneconomic to operate. Since then, we have worked extensively in CCS and realized that the key is to combine CCS and $scCO_2$ solvents [16, 17]. That would make a huge difference because all of the compression costs will be met, as they are already an integral part of CCS. The novelty arises from linking significant potential revenue to what is currently a costly implementation of the net-zero agenda via CCS. It transforms captured

CO_2 from a worthless waste to a valuable resource without the need to modify the CO_2 itself. This is achieved by exploiting the compression energy stored in captured CO_2 thereby valorizing CO_2 by a physical process that is already inherent to the CCS technology. It tackles one of the key aspects within the circular economy of how to reduce solvent waste — one of the unsolved issues to drive sustainability noting the amount of CO_2 needed to be used for chemical industry is incredibly small compared to the vast amount that needs to be captured. Of course, the captured CO_2 is likely to contain impurities, but we have demonstrated [17] that the predicted levels of impurities for CCS would be acceptable for the type of hydrogenation reaction that we have previously carried out on an industrial scale [5]. Most CCS systems are still at the development stage, so now is the right moment to take a system's view and combine CCS plants with sustainable chemical production by catalysis in $scCO_2$.

Moore's Law for Chemistry (see above) [13] focuses on the effect of chemicals and, in many ways, catalysts are the "ultimate" effect of chemicals. They facilitate transformations that chemists would otherwise struggle to carry out efficiently. Therefore, we propose that, if catalysis is really to help ensure our future supply of chemicals in a sustainable way, we need to adopt and implement Moore's Law for Chemistry. This will need both existing and new catalysts together with future developments to reduce not only the amount of catalysts required but also the quantity of chemicals needed to make those catalysts. That is, we need to reduce over a 5-year cycle by 50% the chemical footprint of a process including the catalysts that we use in the future. This means not only reducing the amount of catalyst needed to promote a given transformation but also halving the amount of chemicals required to make those catalysts. And we need to repeat that 50% reduction over several cycles so that catalysis as well as the whole process can sustainably fulfil the chemical needs of future generations.

We have presented two innovative ideas, namely (i) harnessing CCS to provide a source of CO_2 solvents which can be fed back into the CCS process after use for easy recycling and (ii) applying Moore's Law for Chemistry to catalysts and catalytic processes. Individually, each of these ideas could substantially increase the sustainability of chemical manufacture in the future. Linking them together could be transformative, but there is still a need to apply to $scCO_2$ some of the innovations in reactors and processes that we, and others, have already made for continuous chemistry [2] so as to create scalable supercritical photochemical processes. Chemical and other manufacture is now in the throes of a 4th industrial revolution driven by digitization [18], which makes the application of Moore's law to catalytic processes a realistically achievable goal for sustainable chemistry exploiting the ever-increasing application of AI to manufacturing.

Acknowledgements

We thank the University of Nottingham and all of the agencies, charities, and companies who have funded our research. We are grateful to our collaborators, colleagues, co-workers, students, and technicians for their contributions to our research.

References

[1] R. A. Sheldon, *Green Chem.* **19**, 18 (2017).
[2] M. Poliakoff and M. W. George, *Phil. Trans. R. Soc. A* **378**, 20190260 (2019).
[3] P. J. Dunn, S. Galvin, and K. Hettenbach, *Green Chem.* **6**, 43 (2004).
[4] G. R. Akien, A. O. Chapman, P. Licence, and M. Poliakoff, *Green Chem.* **12**, 310 (2010).
[5] P. Licence, M. Sokolova, S. K. Ross, and M. Poliakoff, *Green Chem.* **5**, 99 (2003) .
[6] R. A. Bourne, J. G. Stevens, J. Ke, and M. Poliakoff, *ChemComm.* 4632 (2007).
[7] R. A. Bourne, X. Han, M. Poliakoff, and M. W. George, *Angew. Chem. Intl Ed.* **48**, 5322 (2009).
[8] L. Q. Wu, B. L. Abreu, A. J. Blake, L. J. Taylor, W. Lewis, *et al.*, *Org. Proc. Res. Dev.* **25**, 1873 (2021).
[9] Z. Amara, J. F. B. Bellamy, R. Horvath, S. J. Miller, *et al.*, *Nat. Chem.* **7**, 489 (2015).
[10] R. A. Skilton, R. A. Bourne, Z. Amara, R. Horvath, J. Jin, *et al.*, *Nat. Chem.* **7**, 1 (2015).

[11] https://chemicalleasing.com (Accessed September 19 2022).

[12] D. C. Brock, *Understanding Moore's Law*, Chemical Heritage Foundation, Philadelphia (2006).

[13] M. Poliakoff and P. Licence, M. W. George, *Angew. Chem. Int. Ed.* **57**, 12590 (2018).

[14] An_Agenda_to_Mainstream_Green_Chemistry.pdf (greenchemistryandcommerce.org) (2015) (Accessed Sept 19 2022).

[15] A. Scerri and P. James, *Int. J. Soc. Res. Methodol.* **13**, 41 (2010).

[16] *Big Ideas for the Future*, RCUK, London, p. 45, (2011).

[17] J. G. Stevens, P. Gomez, R. A. Bourne, M. W. George, *et al.*, *Green Chem.* **13**, 2727 (2011).

[18] P. Bellini, D. Cenni, N. Mitolo, P. Nesi, G. Pantaleo, *et al., J. Ind. Inf. Integration* **26**, 100276 (2022).

https://doi.org/10.1142/9789811282324_0004

Electrification and Decarbonization of Chemical Synthesis

Karthish Manthiram

Division of Chemistry and Chemical Engineering, California Institute of Technology, Pasadena, CA 91125, USA

Synthetic paradigms, such as total synthesis, biosynthesis, and semi-synthesis, have driven significant progress towards synthetic routes with fewer steps. My group is developing a synthetic paradigm in which organic molecules are synthesized and functionalized with CO_2, N_2, and H_2O, using renewable electricity at ambient conditions (Fig. 1). In some cases, the carbon, hydrogen, nitrogen, and oxygen atoms derived from just these three precursors (CO_2, N_2, and H_2O) can be converted into useful products; in other cases, these three precursors can be used to sustainably functionalize petroleum-based and bio-based feedstocks. This paradigm makes use of increasingly available and cheap sources of renewable electricity, providing energy for sustainably driving synthetic steps at mild conditions; this electrification of chemical synthesis, along with the use of CO_2 as a precursor, provides a pathway towards decarbonizing chemical synthesis [1]. Given that the energy and feedstocks required are ubiquitous, the processes are conducive to distributed production of chemicals, close to where they are needed.

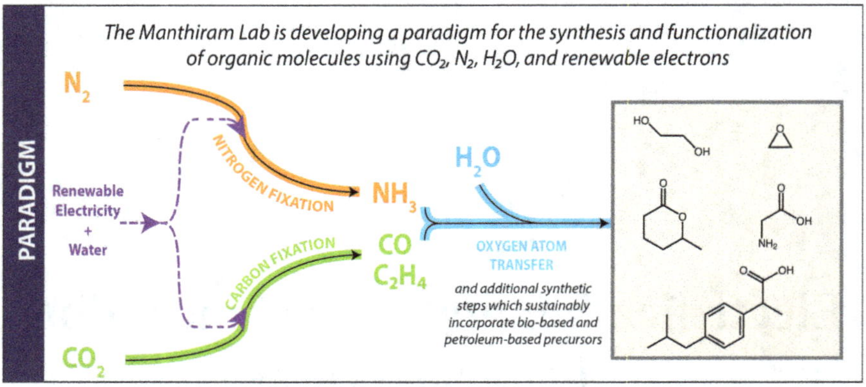

Fig. 1. My group is developing a paradigm for the synthesis and functionalization of organic molecules using CO_2, N_2, H_2O, and renewable electrons.

Ammonia synthesis at ambient conditions

The Haber–Bosch process, which involves reacting nitrogen and hydrogen gas at 150–250 bar and 400–500°C, has served as the dominant method of fixing atmospheric nitrogen to produce ammonia for over a century; ammonia serves as the gateway molecule through which nitrogen functionality is ultimately introduced into diverse molecules. The use of high temperature and pressure limits the modularity of the Haber–Bosch process, requiring centralized ammonia synthesis at large scales, which does not match the distributed availability of renewable electricity, such as solar and wind. Electrochemical polarization can allow for overcoming both the use of temperature for improving kinetics and pressure for shifting the equilibrium [1], as well as replacing methane as the hydrogen source with water. My lab has advanced the use of a one-pot lithium-mediated process for ammonia synthesis which operates at ambient conditions (Figs. 2(a) and 2(b)) [2]. Lithium is attractive due to its intrinsic chemical reactivity with nitrogen, which leads to the breaking of the nitrogen–nitrogen triple bond through the formation of lithium nitride. The lithium nitride can react with protons to generate ammonia and lithium ions; the catalytic cycle is closed by electrochemically reducing lithium ions to produce lithium metal. Through mechanistic studies, we have put forth a coupled kinetic-transport model that quantitatively predicts

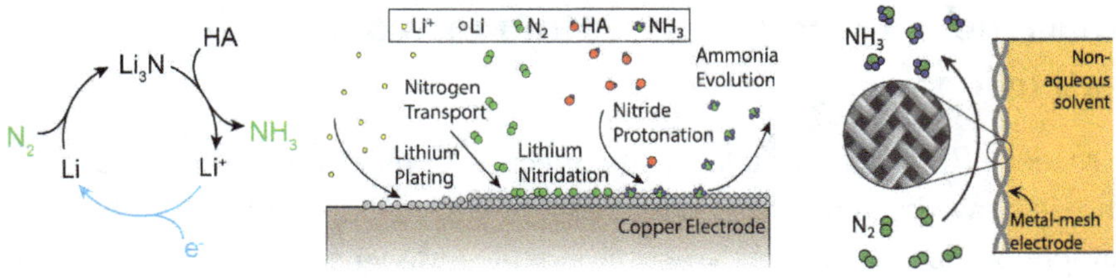

(a) (b) (c)

Fig. 2. (a) Lithium-mediated reaction network for conversion of N_2 to NH_3 under ambient conditions, involving a proton donor (HA), which is typically ethanol. (b) Schematic of surface processes which occur continuously on the surface of the cathode. (c) Metal mesh-based cell architecture developed in my lab which enhances transport of sparingly soluble gases in non-aqueous solvents.

reactivity, enabling the identification of more selective reaction conditions [2]. This model led to the first recognition of transport limitations in lithium-mediated nitrogen fixation, due to the poor solubility of nitrogen in the electrolyte. Typically, such transport limitations would be overcome through the use of a gas diffusion electrode; however, a long-standing challenge in electrochemical architectures has been that commonly used gas diffusion electrodes based on carbon fibres are easily flooded by non-aqueous solvents due to favourable surface interactions. My lab has developed non-aqueous gas diffusion electrodes based on metal meshes (Fig. 2(c)), which have less favourable surface interactions with the non-aqueous electrolyte, avoiding flooding and leading to effective contacting of the gas and liquid reactants. This method may be broadly applicable for non-aqueous electrosynthesis involving sparingly soluble reagents [3]. Our approach has led to record Faradaic efficiencies of up to 45% for one-pot ammonia synthesis at higher rates than those previously achieved at ambient conditions in the literature [3]. Cryogenic electron microscopy, typically used in structural biology, has been applied to image the reactive interface between lithium metal and the electrolyte, providing an unprecedented atomic-scale view of how electrolyte composition controls the transport of critical reagents through the solid-electrolyte interface. We are developing methods by which the ammonia made through our electrochemical route can then be electrochemically activated [4] for use in downstream synthesis.

Electrochemical carboxylation for carbon chain extension

Broader efforts in the field on the electrochemical utilization of carbon dioxide have been focused on the reduction of carbon dioxide to relatively simple products, such as carbon monoxide, ethylene, or formic acid. It is important to establish a broader toolkit through which carbon dioxide can be used to synthesize and functionalize more complex organic molecules. To this end, my group has been advancing electrochemical routes for carboxylation.

We have developed an electrochemical route for the carboxylation of benzylic C–N bonds using carbon dioxide [5], which can be used to sustainably produce surfactants and pharmaceuticals, such as ibuprofen. This procedure does not require stoichiometric metals, external reducing agents, or sacrificial anodes, which are needed in conventional methods, making column chromatography unnecessary for product purification.

We have extended this methodology to aliphatic halides, dramatically expanding the substrate scope [6]. Importantly, the addition of inorganic salts such as magnesium chloride to the electrolyte allows for the stabilization of the carboxylate product over other competing side products, in part by suppressing the nucleophilicity of the carboxylate product. In ongoing work, we have established design principles for solvents involved in carboxylation chemistry, moving beyond the typical screening-based selection of solvents in organic electrosynthesis.

Electrochemical oxygen atom transfer from water

Epoxidation reactions are a subset of oxygen atom transfer reactions in which an olefin is partially oxidized. An epoxidation agent, such as meta-chloroperbenzoic acid or tert-butylhydroperoxide, reacts with the olefin to generate an epoxide and a stoichiometric side product, in this case, the corresponding benzoic acid and alcohol, respectively. These epoxidation agents are highly hazardous, and the resulting side products must be separated and dealt with. Other epoxidation methods, such as the direct oxidation of olefins with oxygen, can only be used with a small number of olefins, such as ethylene, and lead to large CO_2 footprints arising from over-oxidation; in fact, the production of epoxides is one of the top five contributors to CO_2 emissions from chemical manufacturing worldwide.

My group has developed a reaction in which *water* is a sustainable O-atom source for olefin epoxidation at ambient conditions; this reaction generates no stoichiometric waste products, involves no hazardous reagents, and operates at ambient conditions [7]. My lab elucidated the mechanism through which a hypothesized metal-oxo species generated from the activation of water at a manganese oxide nanoparticle electrocatalyst drives epoxidation. While supporting high rates of epoxidation, our process simultaneously generates hydrogen at the cathode. In a typical water electrolyzer, the oxygen generated at the anode is simply vented; our approach provides a new paradigm through which the oxygen can instead be used for high-volume oxidative functionalization reactions. We have recently developed next-generation catalysts for this reaction, involving single atoms of iridium deposited on manganese oxide, which are more selective for electrochemical olefin epoxidation, as the iridium promotes the generation of electrophilic oxygens on the surface [8].

In addition to electrocatalytically generating electrophilic oxygen atoms from water for use in epoxidation, my group has pursued the generation of nucleophilic oxygens from water for use in reactions, such as lactonization [9]. Specifically, we have found that platinum can activate water for the conversion of cyclic ketones into branched lactones, providing unique selectivity patterns compared to the conventional Baeyer–Villiger reaction. In ongoing work, we are establishing how water's thermodynamic non-ideality in a blended electrolyte controls rates of oxygen atom transfer.

Outlook to future developments of research on Catalysis for Sustainable Chemistry

The research efforts described above lay a foundation for the effective utilization of CO_2, N_2, and H_2O in chemical synthesis. There are many opportunities for continuing to build on this foundation. For instance, ongoing efforts in my lab are focused on harnessing: (1) CO produced from CO_2 reduction for electrochemical hydroformylation of olefins, (2) NH_3 produced from N_2 reduction for electrochemical reductive amination, and (3) water as an oxygen atom source for C–H activation. These reactions and others will continue to expand the realm of chemical synthesis and functionalization which can be conducted with ubiquitous precursors. These new reactions need to be integrated into multi-step syntheses through which one can make more complex monomers or building blocks. For instance, one could also envision routes through which amino acids could be generated abiotically from just carbon dioxide, nitrogen, and water, using the foundational methods developed in my group.

As the complexity of the target molecules increases, it will be important to implement enzyme-inspired design principles which allow for the recognition of molecules at an electrode surface in order to direct selective oxidations and reductions at particular sites. This will be especially important for molecules that present sites with similar reactivity or for electrode reactions that otherwise have poor functional group tolerance. The field will also need to develop methods of discriminating molecules on the basis of their transport to an electrode surface as a means of imparting selectivity in reactions involving multi-component mixtures. Substrates can be discriminated in their transport based on size or charge; in a way, this will be a means of integrating separations with electrode reactions. Altogether, these approaches will provide greater fidelity over synthetic steps driven directly at electrode surfaces.

Acknowledgements

K.M. gratefully acknowledges support of work on oxygen atom transfer by the U.S. Department of Energy (DOE) Office of Science, Office of Basic Energy Sciences, Catalysis Science Program, under Award No. DE-SC0020999, work on ammonia synthesis supported by the National Science Foundation under grant no. 2204756, and efforts on carboxylation supported by the National Science Foundation under grant no. 2204757.

References

[1] Z. J. Schiffer and K. Manthiram, *Joule* **1**, 10 (2017).

[2] N. Lazouski, Z. J. Schiffer, K. Williams, and K. Manthiram, *Joule* **3**, 1127 (2019).

[3] N. Lazouski, M. Chung, K. Williams, M. L. Gala, and K. Manthiram, *Nat. Catal.* **3**, 463 (2020).

[4] Z. J. Schiffer, N. Lazouski, N. Corbin, and K. Manthiram, *J. Phys. Chem. C* **123**, 9713 (2019).

[5] D. Yang, M. Zhu, Z. J. Schiffer, K. Williams, X. Song, X. Liu, and K. Manthiram, *ACS Catal.* **9**, 4699 (2019).

[6] N. Corbin, D. Yang, N. Lazouski, K. Steinberg, and K. Manthiram, *Chem. Sci.* **37**, 12365 (2021).

[7] K. Jin, J. H. Maalouf, N. Lazouski, N. Corbin, D. Yang, and K. Manthiram, *J. Am. Chem. Soc.* **141**, 6413 (2019).

[8] M. Chung, K. Jin, J. S. Zeng, T. Ton, and K. Manthiram, *J. Am. Chem. Soc.* doi: 10.1021/jacs.2c04711.

[9] J. H. Maalouf, K. Jin, D. Yang, A. Limaye, and K. Manthiram, *ACS Catal.* **10**, 5750 (2020).

https://doi.org/10.1142/9789811282324_0005

Organic Synthesis Away from Equilibrium

Robert R. Knowles

Department of Chemistry, Princeton University, Princeton, NJ 08544, USA

An emerging area of study in catalysis and sustainable chemistry

The great majority of organic transformations are thermal processes which occur on a single potential energy surface. Due to microscopic reversibility, thermal processes are limited to transforming higher-energy starting materials into lower-energy products. A significant goal for the field of synthetic organic chemistry in the years to come is to devise strategies that decouple reaction direction from reaction thermochemistry and to employ these strategies to develop reactions and processes that proceed against a thermodynamic bias and thus cannot be achieved using any conventional methods.

Towards this goal, significant advances have been made through the use of photoredox catalysis. One of the most interesting and enabling aspects of photoredox chemistry is the ability of these methods to drive organic reactions "uphill" in opposition to a thermodynamic bias.

Drawing parallels to solar fuels chemistry, photoexcitation events provide an exogenous driving force that offsets the otherwise unfavourable energetics of the bond-forming and bond-breaking steps, allowing access to products that are formally higher in energy than the starting materials. However, for synthetic applications, these reaction manifolds also exhibit unique kinetic benefits. Specifically, as all the reaction pathways proceeding from the initial excited state are thermodynamically favourable, the partitioning between potential pathways is kinetically controlled and fully decoupled from the ground-state thermochemistry of the resulting products — even for systems that are dynamic. These attributes create an exciting prospect for catalysis where chemists can, in principle, use molecular catalysts to kinetically select any accessible product state in a dynamic system irrespective of the system's ground-state thermochemical preferences. The applications of these ideas in synthesis have begun to be explored and developed in recent years. Important contributions have been made by a number of groups, including Thorsten Bach, Alison Wendlandt, David MacMillan, Aiwen Lei, and Ryan Gilmour [1–5]. In the following sections, I will highlight some of my lab's contributions to this area and thoughts on where these advances may lead in the future.

My recent research contributions

In recent years, our lab has reported a number of methods that make use of excited-state redox events to drive reactions away from equilibrium (Fig. 1). Our first report in this area was on the development of intermolecular hydroamination of unactivated olefins with secondary alkyl amines [6]. Under visible light irradiation, an excited state chromophore oxidizes an amine substrate to its corresponding aminium radical cation, which then readily reacts with alkene reaction partners. This protocol constructs valuable tertiary alkyl amine products with complete anti-Markovnikov regioselectivity. Notably, many of these hydroamination products are less thermodynamically favoured than their corresponding olefin and amine starting materials, which makes them inaccessible through conventional ground-state catalysis. Recently, we have extended this protocol to the selective synthesis of secondary amines via direct intermolecular

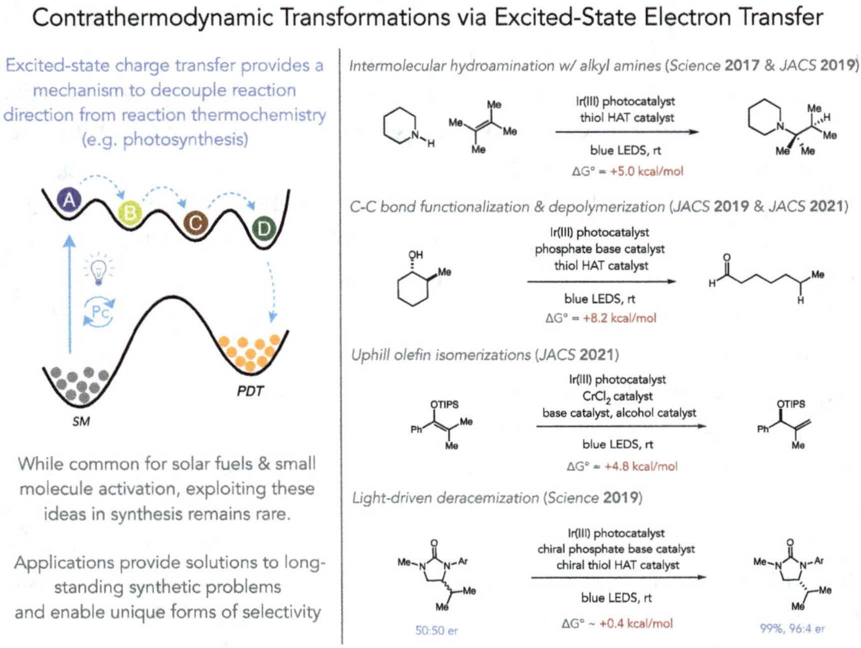

Fig. 1. Examples of light-driven contrathermodynamic transformations in synthesis enabled by excited-state electron transfer.

hydroamination of primary alkyl amines and simple olefins with minimal generation of over-alkylated tertiary amine products [7].

We have also described a catalytic protocol for C–C bond cleavage with simple aliphatic alcohols that mediate redox-neutral isomerizations of cyclic aliphatic alcohols to linear carbonyl compounds [8]. In these reactions, the alcohol O–H bonds are homolytically activated via excited-state proton-coupled electron transfer. The resulting alkoxy radical intermediates undergo intramolecular C–C β-scission to provide fragmented carbonyl products, which are generally higher in energy than their corresponding alcohol-starting materials. We have further advanced this chemistry for the development of photocatalytic methods for the depolymerization of native lignin and high molecular weight hydroxylated polyolefins [9, 10].

We have also applied the ideas above for the contra-thermodynamic positional isomerization of internal olefins into terminal olefins [11].

In this work, stepwise oxidation and deprotonation of a more substituted and more thermodynamically stable olefin substrate is mediated by an excited-state oxidant and a Brønsted base to afford an allylic radical. This radical is captured by a Cr(II) co-catalyst to furnish an allylchromium(III) intermediate that undergoes *in situ* protodemetalation with methanol, affording an isomerized and less thermodynamically stable alkene product with high regioselectivity. The higher oxidation potential of the less substituted olefin isomer renders it inert to further reaction with the excited-state oxidant, enabling it to accumulate in solution over the course of the reaction. A similar light-driven method that proceeds by a distinct mechanism was also reported by Wendlandt and co-workers [12].

Lastly, we have applied these ideas in out-of-equilibrium catalysis to asymmetric synthesis. In particular, we have developed, in collaboration with Professor Scott Miller, a light-driven deracemization reaction wherein a racemic urea substrate undergoes spontaneous optical enrichment (up to 94% ee and 95% yield) simply upon visible light irradiation in the presence of three distinct molecular catalysts [13]. Excellent enantioselectivity is achieved for the homolysis and reformation of a stereogenic C–H bond through orchestrated movements of an excited-state electron, proton, and hydrogen atom. Remarkably, this system establishes a stable, non-equilibrium distribution of urea enantiomers without any use of stoichiometric chemical reagents. It is notable that a number of other light-driven deracemization technologies have been reported in recent years, including seminal contributions from the group of Bach [1].

Outlook

These approaches will create new possibilities and efficiencies in synthesis. The results above show that a variety of canonical, redox-neutral reaction types (additions, eliminations, isomerizations, etc.) can be driven away from their equilibrium positions using excited-state electron transfer-based mechanisms. This framework can potentially be extended to a large number of reaction classes, providing a platform to drive known transformations in reverse, introduce dynamic behaviour into otherwise static systems, and access reaction products that cannot be achieved using conventional thermal methods. Moreover, the use of visible-light excitation obviates the need to use stoichiometric chemical reagents to provide a reaction's driving force, thus limiting the production of waste and increasing the sustainability of the developed technologies.

A particularly exciting direction for this area extends the ideas presented in the deracemization work discussed above. Specifically, as these technologies advance, one could imagine a new paradigm where stereocomplex target molecules could be made as stereo-random mixtures and then edited post-synthetically to adjust all the stereochemical relationships [2, 3, 13] Moreover, this could be achieved with catalyst-controlled selectivities that are not subject to thermodynamic constraints. While such an idea would have seemed distant or perhaps impossible several years ago, it is increasingly clear that such strategies are tractable. If successful, this approach points to a fundamentally distinct way of thinking about stereoselective synthesis with potential impacts at the level of retrosynthetic analysis. Taken together, the prospects for this area appear bright and it will be exciting to see what progress can be made in the years to come.

Acknowledgements

The projects described above were supported either by the National Institutes of Health (NIGMS R35 GM134893) or through BioLEC, an Energy Frontier Research Center funded by the U.S. Department of Energy, Office of Science, Basic Energy Sciences under Award No. DE-SC0019370.

References

[1] A. Holz-Hobelmeir, A. Bauer, A. V. Silva, S. M. Huber, C. Bannwarth, and T. Bach, *Nature* **564**, 240 (2018).
[2] Y. Wang, H. M. Carter, and A. E. Wendlandt, *Nature* **578**, 403 (2020).
[3] C. J. Oswood and D. W. C. MacMillan, *J. Am. Chem. Soc.* **144**, 93 (2022).
[4] H. Wang, X. Gao, Z. Lv, T. Abdelilah, and A. Lei, *Chem. Rev.* **119**, 6769 (2019).
[5] J. B. Metternich and R. Gilmour, *J. Am. Chem. Soc.* **137**, 11254 (2015).

[6] A. J. Musacchio, B. C. Lainhart, X. Zhang, S. G. Naguib, T. C. Sherwood, and R. R. Knowles, *Science* **355**, 727 (2017).

[7] D. C. Miller, J. M. Ganley, A. J. Musacchio, T. C. Sherwood, W. R. Ewing, and R. R. Knowles, *J. Am. Chem. Soc.* **141**, 16590 (2021).

[8] E. Ota, H. Wang, N. L. Frye, and R. R. Knowles, *J. Am. Chem. Soc.* **141**, 1457 (2019).

[9] S. T. Nguyen, E. McLoughlin, J. H. Cox, B. P. Fors, and R. R. Knowles, *J. Am. Chem. Soc.* **143**, 12268 (2021).

[10] S. T. Nguyen, P. R. M. Murray, and R. R. Knowles, *ACS Catalysis,* **10**, 800 (2020).

[11] K. Zhao and R. R. Knowles, *J. Am. Chem. Soc.* **144**, 137 (2022).

[12] G. Occhialini, V. Palani, and A. Wendlandt, *J. Am. Chem. Soc.* **144**, 145 (2022).

[13] N. Y. Shin, J. M. Ryss, X. Zhang, S. J. Miller, and R. R. Knowles, *Science* **366**, 364 (2019).

https://doi.org/10.1142/9789811282324_0006

Heterogeneous Catalysis: Where Chemistry Meets with Material Science

Avelino Corma

Instituto de Tecnología Química, Universitat Politècnica de València-Consejo Superior de Investigaciones Científicas, Avenida de los Naranjos s/n, 46022 Valencia, Spain

Present status and objectives

The main objective in catalysis is to reach the *ab initio* design of the catalyst starting from the reaction mechanism and introducing the single or multiple active sites accordingly. This already occurs in the case of enzymes and in many cases of homogeneous transition metal complexes and organocatalysts where the electronic properties of the single sites could be adapted by properly selecting the ligands and functional groups. However, the presence of single isolated sites is not always enough to achieve very high selectivities, and we have learned from enzymes that reactant selection, directed adsorption and transition-state stabilization by means of weak interactions, can be determinant for the catalytic process. While all this can be better rationalized in the case of molecular homogeneous catalysts, it becomes more difficult to be achieved in the case of solid catalysts.

It is then a challenge and a clear objective for the *ab initio* preparation of solid catalysts, to go beyond catalysts' design by accumulated knowledge, trial and error, and judicious interpretation of results. It will be then mandatory to develop general fundamental surface reactivity models at the molecular level. This necessarily implies the symbiosis of physical chemistry, material science, organic and inorganic chemistry, together with the development of surface characterization techniques able to follow the surface chemical process under operation conditions, in order to establish structure–reactivity relationships.

Recent contributions

For the design of well-defined single isolated sites, we and others [1–5] have generated them on framework (and extraframework) positions of high surface micro- and mesoporous crystalline materials. The advantages of these materials not only rely on the possibility to stabilize, in specific structural positions, the single (or multiple) sites but also to make available pores and cavities of the size of the molecules that can select the reactants, while introducing confinement effects within the pores and cavities, i.e., the presence of weak (dispersion) forces that can be determinant for catalyst selectivity. To this respect, we recently show that by using adapted mimics of the reactions transition states as organic structure directing agents, it is possible to crystallize microporous structures with well-defined single isolated active sites, while the templated cavities stabilize the reaction transition state during the preselected reaction [6].

Single metal atoms and small clusters (Pt, Pd, Au, and Ni) within solutions have shown to be catalytically active for a number of reactions. However, they agglomerate with time on stream and become catalytically inactive. Then, the stabilization of these single atoms on different supports was achieved and Au, Pt, Co, and others were atomically isolated on the surface of CeO_2, functionalized carbon and iron oxide [8–10]. Those single atoms show much higher activity, for some reactions, than supported metal nanoparticles. In other cases, subnanometric metal clusters [10, 11] show enhanced activity and selectivity. The stabilization, geometric, and electronic properties of single atoms as well as subnanometric monometallic and bimetallic nanoclusters could, in principle, be tailored through well-defined bonding with the support. Since stability requires either encapsulation and/or strong ionic or covalent interactions with the support, this can introduce charge transfer between the support and the isolated metal atoms or clusters, and the final electronic properties will determine their adsorption and catalytic properties.

The potential of single-atom and nanocluster catalysts is being explored in thermo-, photo-, and electrochemical catalytic processes with the possible objective of substituting noble with non-noble metal atoms [12].

An additional scientific objective is to define and synthesize low-nuclearity clusters with well-defined and homogeneous number of atoms in monometallic and bimetallic entities. This has been recently achieved within the structure of microporous materials, in where low-nuclearity sites formed by two, three, four, and six Pd, Pt, and Pt-Sn atoms have been obtained and the catalytic benefits are shown [13–15].

The developments on single atom centres (SACs) and low nuclearity clusters, especially bimetallic clusters, have required and have impulsed developments in analytical characterization techniques, such as transmission electron microscopy, SAXS, as well as theoretical calculation techniques.

Future developments

In Fig. 1, the progress in single-atom catalysis is presented as taken from [15].

Future developments in the *ab initio* synthesis of solid catalysts for specific reactions will require new and improved synthesis methods to achieve higher loadings of single atom centres and control of nuclearity of the centres. Efforts should also be directed to measure and quantify the

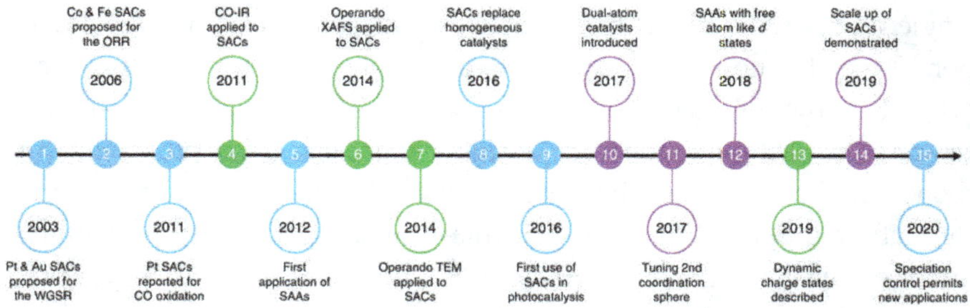

Fig. 1.

surface-accessible SACs and their behaviour and evolution during reaction conditions. When SACs are located in crystalline microporous structures, it will be of much interest to select, by synthesis, the T atom in the framework position to be located. Their activity and selectivity can be different because of the bonding constraints and the dimensions and topology of the channel or cavity to which they will be exposed.

In the case of clusters of low nuclearity, it is required to develop synthesis techniques to form the bimetallic particles stable and homogeneously distributed with the adequate metal ratio. Moreover, characterization techniques to follow the evolution of the bimetallic particles during the reaction will have to be improved.

It should also be discussed the generality or not of the preparation and application of high-entropy alloy particles for catalysis, i.e., synthesis control, reproducibility, characterization, and evolution during the reaction. Note that in a knowledge-based system, it should be possible to predesign "in silicon" the single-atom and mono- and bi-metallic catalysts with the desired structural and electronic properties.

Finally, we have still much work to do for synthesizing chiral solid catalysts able to carry out enantioselective synthesis.

Acknowledgements

The European Union is acknowledged by ERC-AdG-2014-671093 — SynCatMatch.

References

[1] J. M. Thomas, R. Dewi, and W. Lewis, *Angewandte Chemie Int. Ed.* **44**, 6456 (2005).

[2] M. G. Clerici, G. Bellussi, and U. Romano, *J. Catal.* **129**, 159 (1991).

[3] A. Corma, L. T. Nemeth, M. Rentz, and S. Valencia, *Nature* **412**, 423 (2001).

[4] A. Corma, M. T. Navarro, and J. Pérez-Pariente, *J. Chem. Soc. Chem. Commun.* 147 (1994).

[5] X. Liang, S. Ji, Y. Cheng, and D. Wang, *iScience* **25**, 104177 (2022).

[6] E. M. Gallego, M. T. Portilla, A. Leon-Escamilla, M. Boronat, M. Moliner, and A. Corma, *Science* **355**, 1051 (2017).

[7] Q. Fu, H. Satlsburg, and M. Flytzani-Stephanopoulos, *Science* **301**, 935 (2003).

[8] R. Bashyan and P. A. Zelenay, *Nature* **443**, 63 (2006).

[9] B. Quiao *et al.*, *Nat. Chem.* **3**, 634 (2011).

[10] U. Heiz, F. Vanolli, A. Sanchez, and W. D. Schneider, *J. Am. Chem. Soc.* **120**, 37, 9668 (1998).

[11] M. Boronat, A. Leyva, and A. Corma, *Acc. Chem. Res.* **47**, 834 (2014).

[12] L. Liu and A. Corma, *Chem. Rev.* **118**, 4981 (2018).

[13] L. Liu *et al.*, *Nat. Mater.* **18**, 866 (2019).

[14] M. Viciano-Chumillas, *et al., Acc. Chem. Res.* **53**, 520 (2020).

[15] S. Mitchel and J. Pérez-Ramirez. *Nat. Commun.* **11**, 4 (2020).

https://doi.org/10.1142/9789811282324_0007

Discussions of Session 1 — Catalysis for Sustainable Chemistry

Chair: David MacMillan
Auditors: C. Theunissen*, F. Tielens†

*Laboratoire de Chimie Organique, Service de Chimie et PhysicoChimie Organiques, Université libre de Bruxelles (ULB), Avenue F. D. Roosevelt 50, CP160/06, 1050 Brussels, Belgium
†Algemene Chemie (ALGC) — Materials Modelling Group — Vrije Universiteit Brussel (VUB), 1050 Brussels, Belgium

David MacMillan:
Thank you everyone for your participation this morning. I was asked by Kurt Wüthrich to make a quick note of something which I think is important. Originally, this first session was supposed to be chaired by Bob Grubbs, who unfortunately passed away in December of 2021. He was a person who attended this meeting many times. He was a wonderful guy, and he was the person who actually came up with this concept of sustainable catalysis. That was the reason we led off with it and I just wanted to remind everyone of that and what a really wonderful and fantastic guy Bob was. So now I think we'll get started on the discussion. Maybe we could lead off with some questions to put to the panel and also to the group. The first one I wanted to start off with, when we're talking about sustainable catalysis, is about performing chemical processes on scale.

For societal impact, scale is often very important but there is also a very large community of people who perform these catalytic processes in academia that are not on scale. They might never need to be on scale. I guess the first question I would have is: should we care about sustainable catalysis only for scale or is this something we should care about across the board? I'll open with that question.

Karthish Manthiram:
When selecting targets, our group at least is very motivated by knowing the carbon footprint behind those targets. That can be clarifying in terms of knowing what we want to pursue in terms of decarbonisation. That being said, these targets that are commodities are going to be exceptionally difficult to commercialize, because you're going up against products that have virtually no margins in their sales. These are really low margins, highly competitive markets. For that reason, I think that developing these more sustainable chemistries is enormously important in the context of chemicals that have margins. Often, this is done in the pharmaceutical industry first, and then we need finding ways to work towards the commodities. It's been really exciting to see how many steps have been electrified in the last few years in the pharmaceutical industry. I think it's really encouraging to see that we're on a pathway now where this knowledge of reactor design, of scalability for pharmaceuticals, and proving out these new concepts, exists, even when the initial cost could be greater. That will then create a pathway towards ultimately implementing these new technologies for chemicals that are made at scale. In some ways, there's actually a blueprint for doing this already. If we look at other electrified technologies, whether they are solar panels or batteries, there's a history of starting with small scale niche implementations. If you look at solar panels, first and earliest implementations are for space technology. It wasn't really having a clear impact here on Earth, but one can see that, over a period of many decades, the technology worked its way down in price to go from something that was a niche application in space to be something that could have impact here on Earth for decarbonisation. The same is true for batteries: earliest developments happened in the 1970s at Exxon. At the time, they couldn't really see the path to commercializing that as an energy technology so, instead, it went to camcorders first, with

Sony commercializing this in 1991. It has taken another 30 years for it to get to the tipping point where we can see the impact that it will have to use lithium-ion batteries in electric vehicles and perhaps even for grid scale storage. So there is this sort of roadmap that we see with other electrified technologies. First, a niche with smaller scale application, whether it's solar cells or batteries. I think the same will be true for electric synthesis, starting at small scale for niche chemicals, and then working our way up towards things that are commodities that will have the decarbonisation impact that we want to see.

Martyn Poliakoff:
I'd like to address your question about small scale catalysis. I think it's really important to try and make small scale catalysis more sustainable because academic chemistry is carried out very unsustainably in fume hoods that pump large amounts of heat out of labs, making chemistry buildings by far the most expensive buildings to run on any campus. It also uses huge amounts of solvent compared to an industrial process which has been optimized to use less solvent. Although the total global impact of this is probably not huge, the potential impact on our profession as chemists is enormous. In times of financial stress, which is around the whole world at the moment, if chemistry departments are seen to be very expensive to run, there is a huge incentive on universities to reduce the size of their chemistry departments, perhaps completely close them to save on running costs. So I think it's really important that chemists learn to work sustainably from the outset. That also trains our students to think sustainably rather than this being something that they might think about if they eventually go and work in industry.

Ben Feringa:
Coming back to your question about scalability, I got this question several times. I mentioned this example of photo-oxidation (singlet oxygen photocatalysts) and people from industry said to us: can you make this an industrial process? Because the petrochemical industry gets worried and they say "we are not accustomed to use light so can you ever do it on scale?" This is one of the reasons why we teamed up with the engineering schools and the students together built this photoreactor that

I showed — and Martyn (Poliakoff) has such reactors in his own lab — in order to show them that, using LEDs that you can buy for a penny in the supermarket, you can use electricity and light and you can do this in flow and make a kilogram a day. You don't have to build a factory or scale it up to ton quantities. I don't see it as our duty, but to show them that indeed, you can build a photoreactor. It reminded me the reason why we did this. When I worked after my PhD at Shell, they were working in a parallel department on the conversion of methane to diesel. Why did they do that? Because in the world, how much methane is burned/flared off? So, they developed, in the fume hood, a reactor of the size we have now with photochemistry and at the end they could show that the diesel came out. The gas went in and the diesel came out and they produced one liter a day. It took 25 years. Now in Qatar, they produce 260.000 barrels of gas-to-liquids a day. They produce the cleanest diesel in the world. It might take a while but, in my opinion, I think, our duty is not to do all this stepwise and large-scale innovation, that is the task for industry. We don't even have the money and the facilities, but we need to show them and to demonstrate that in the future, it will be profitable. It might take a few decades before the mindset has changed. This was my experience, also in industry when I still was a child.

Karen Goldberg:

One aspect of sustainability is this issue of precious metals versus non-precious metals. First of all, it's important to know that a lot of chemistry is more challenging with more Earth-abundant metals and that you can form stronger bonds with some of the precious metals. It's important to be able to demonstrate chemistry on precious metal systems, and then later, it can be transferred to non-precious metals. It's also important to think about where that process is going so that when you use an Earth-abundant or a precious metal, what happens when you send it out into the world. For example, in a catalytic converter, you take a precious metal that's going out into the world and you're not getting that metal back. That's a problem. But if you had a chemical process that's running with a precious metal that is staying within the system and that is able to carry out many turnovers, it has not to muck up the system or decompose. You don't have to turn the reactor off to run the process, clean it all out and go again.

So, there are different purposes for different metals, and we need to think about that in the design of different systems.

Daniel Nocera:

There's some fundamental core principles that are at the laboratory scale and large scale and, especially for sustainability, it's your energy input. That's really critical when we finally get to scale and start looking at cost. Even if you're doing small scale things, it's good to have in mind what your real energy input is for the entire reaction. That means really trying to control high photo-efficiencies or faradaic efficiencies, which I might be doing at little lab scale. If I have a large energy input, that's going to be deadly to large scale. So that's just one. Rob (Knowles), you might want to talk about a life cycle analysis, like we were saying, when you start putting the entire energy input in. I think our field kind of ignores that. Right now, we look at reactant to product and, in between, there are massive inefficiencies that are going to be hard to scale. But maybe Rob, you could mention what we were talking about for a life cycle analysis, for instance.

Robert Knowles:

One of the real challenges in all of the photoredox types of chemistries are the low quantum efficiencies. Oftentimes, these have quantum efficiencies less than 1%. You can excite molecules and they can engage in very efficient charge transfer reactions but then those charged separated states just recombine. This is also a problem in solar cells and other types of photo-driven applications. It turns out in a lot of the photoredox chemistry, Dan (Nocera) has studied this in his lab, that 80 to 90% of the charge separated states just recombine which is an enormous loss of quantum efficiency. That's absolutely a challenge that has to be overcome for these things to be scaled, but it's also a fundamental science problem to be worked on and figured out. New catalysts and systems could be designed to overcome those limitations. I think as these fields, that previously haven't had a lot of interactions, come together, these problems start to become elucidated. I'm optimistic at least I think that there's prospects for being able to solve these problems as we all move forward together. And we learn from folks from the solar fuels communities for how to think about and quantify these issues in terms of energy and try to make our processes more sustainable and efficient.

Karen Goldberg:
I just want to also add along these lines that we're trying to do transformations that haven't been done before and done selectively. If you think back at the Haber-Bosch process, it was brought up many times today. When that process was first developed, that took a really long time just to come up with the catalyst and then to come up with it on scale. The contribution of the chemical engineer in that whole process was enormous, as we couldn't run reactions under those pressures before. I think it's a complete process and we don't want to tie our hands by saying that we can only think about sustainable processes as we're working towards developing something that, if it can be done on scale, could be sustainable.

Karthish Manthiram:
It's quite likely that we're going to increase the energy footprint of the chemical industry by making it more sustainable. That's something which I think can sound counterintuitive at first, right? It may seem as this should be harmful to the planet but because we're starting with precursors that are at the very bottom of the free energy landscape, it's going to require more energy to be able to go from CO_2, nitrogen, water and other more sustainable inputs to the chemicals that we desire. That's going to be a big change but we should still aspire to minimize that increase in energy footprint as much as we can in the way that Dan (Nocera) articulated just a moment ago.

Sossina Haile:
One comment about this idea of scalability is that we recognized that, in many cases, we need to make the fundamental breakthroughs that could allow large scale processes. If we understand that well enough, that will allow that pathway. The other piece is to think about processes which benefit simply from being small. That would be certainly electrochemical systems and also some of these interesting flow dynamic systems where the economy of scale comes from manufacturing many of these. You have the opportunity to replace small parts and not have large legacy

infrastructure, which now cannot be improved. That's just a different way of thinking about scalability.

Chad Mirkin:
I think you cannot look at this in a box. I think Karthish's (Manthiram) analysis is spot on. I came out of a group that talked about artificial photosynthesis in the late 80s and we're still talking about it in the same way we talked about in the 80s, in most senses. You have to map where we need to go to really understand how to get there and what the most effective way is. This idea of picking who you're going to go after is really critical in this moving into pharma. Electrification, that's something that could impact them in a significant way. The bar is a lot lower, they're used to making bets. To me, that's really strategic. But I wanted to ask a question: you lead with the Haber process? It seems so completely counter to your argument, I think we often do that. And I think it's a mistake. Because at the end of the day, the chemical industry is an old, I shouldn't say this being at Solvay, somewhat stodgy industry fixed in their ways. Very conservative in terms of approach. That means the bar is really high to make them do things differently. And so you have to validate in areas where people will take a lot more risk and then move to those other areas. I think that's ultimately the path and something we shouldn't lose sight of.

Karthish Manthiram:
I think the point you bring up Chad (Mirkin) is essential. In many ways, some of the research that we're working on is kind of protected by basic science funding. We're able to look at the 2030s time horizons, but the startups that were spinning out are the ones that looked specifically at targets that are viable today. So there is this sort of tension in a way between those two, and we hope in some ways to let the academic research continue to look at long term targets, while still spinning out ones that are more easily implementable today.

Chad Mirkin:
Just one more comment because you brought up the idea that we're going to be more energy intensive, which is true. But a lot of that is going to be clean energy, if we do it right. The other thing is that when you go the

electrolysis route, the idea of a plant becomes mobile. The old view of the plant is that we build something on seven square miles. Some of these plants are incredible, you go to BASF or Dow in Texas, they're really impressive engineering feats. They're not mobile entities. That's what a lot of this does, it allows you to take chemical production to a site closer to where energy resources are, to basically delocalize it.

Karthish Manthiram:
One thing to add is that even though those electrons might be clean, which is essential, the cost of those electrons, especially when storage is accounted for, can be pretty steep. So there's a techno-economic reason, even if the energy is clean, to still try to minimize the energy input as much as one reasonably can. On the point of modularity, I think that what Sossina (Haile) brought up as well, is going to be hugely impactful. One kind of interesting trend to see is, if you look at wind turbines today, even though these began as relatively small implementations, it is astonishing how large wind turbines are getting. If you look at the blades of the most recent wind turbines that are implemented by GE, these blades are the length of three football fields. This technology, that was really tolerated as being small and modular, is still modular. They're making a lot of these and putting them out, but they're huge, even an implementation of an individual unit, and they are getting bigger and bigger to get to the smallest dollars per kilowatt hour of any power source that we have today. So, we have to revise how we believe in modularity, and in small scale production and be open to the fact that even these electrochemical technologies might become relatively large relative to what we're doing today.

Karen Goldberg:
I wanted to comment on this idea of costing more in energy and if it is clean energy. We have to balance that against what we're paying and what we're going to continue to pay in terms of the damage of climate change. And I think once you do that, this increased cost and energy is going to save a lot in terms of the effects on climate change.

Gerald Joyce:
I want to come back to the point that Dan (Nocera) got to start it on which is this more all-in calculation. It's a question that Ben (Feringa) talked

about. If we talk about something like plant waste as an input to make the feedstocks to then make plastics (basically polypropylene, polyethylene). As chemists, we can imagine what the cost is on the front end for processing those molecules to get to where we want them to go. But how do we even do the calculation of what was the fertilizer cost of managing those fields? What were the transport costs of getting the fertilizer to the fields, getting the plant waste out of the fields? What is the processing cost? How do we learn what those numbers are as chemists? Do we make alliances with economists? Do we believe what they say? How do we get to "an all-in number" you're talking about?

James Liao:

I think when we talk about scales, it depends on the audience. In a global sense, the politicians and the societies are looking at immediate impact on the carbon footprint and that means large scale. No question about it, we have to worry about scale, and the scale is gigaton scale. If you look at whatever product we have, if it's not at gigaton scale, we make less than 0.1% of the carbon footprint. But on the other hand, anything has to start small, which we all recognize that. So when you want to develop a large scale process, you better start on something small. But as Dan (Nocera) said, and many people echoed, you have to worry about the footprint along the way in small scale and that requires many detailed analyses of efficiency. Even though that's not in our formal training as a chemist, chemical engineers have been doing that based on mass balance and energy balance. Because in this space, mass and energy are conserved. You cannot escape that. Unlike in information science, you can create information, but we're dealing with materials, mass is always balanced, whatever material you put in, it has a scale. Whatever waste you produce, it will also multiply in scale. Once you capture those principles, you can do some "back of envelope calculation" to see where you're going. Once you have done that, then you're in a position to justify whatever you do in small scale. Scaling up in the future may not be at your lab but working with someone else and maybe many steps in order to make a real impact.

Ben Feringa:

You are absolutely right. In principle, what we should do is calculate the whole chain from start to finish. In this particular case, this came out a bit about this challenge to make an acrylate replacement from wood remains, it came also a little bit from the challenge and also frustration, because we saw in our country that they were taking all kinds of woods and burning it in coal plants for energy. And I said "Come on, this cannot be true. Nature did its best to make all this beautiful wood, all these materials, we should do more than just burning it." And then we went to see if we could use this wood remains and make acrylates and make a coating. But of course, we'd never calculated the whole circle, all the other aspects. You are absolutely right, at the end, you should do that but we were already happy that we could show the general public and in particularly the politicians that you can do something else than simply burning it. Take advantage of what mother Nature did. I think this is also a duty of us as scientists and as chemists to demonstrate that there are alternatives. But of course, at the end, you have to make the whole picture.

Gerald Joyce:
You had a nice example, because you know that you play a game, because you are just looking at the delta. But then to go to scale, you really do have to do the calculation, and how do chemists find the right people to get that information? We are in our own discipline. I don't think we should depend on the politicians to tell us how to make the calculation. We want to talk to economists, agronomists, atmospheric scientists. I don't know where it all connects.

Ben Feringa:
You can tell me who to consult.

Martyn Poliakoff:
First of all, chemists need to take a much more systems view of their activities. I strongly believe in embedding chemical engineers into chemistry research groups. Our research group at the moment has two chemical engineers, because I think chemical engineers not only can answer these bigger questions, but they can make very sensible suggestions about how we are even doing small scale processes. The other thing, which comes back to my proposal of Moore's law, and reflecting on the discussion of the Haber process, is that a vast amount of nitrate fertilizer is wasted, because it goes on the field but it never gets to the plants because it is

washed off or metabolized by bacteria and released as N_2O, which is a powerful greenhouse gas in its own right. So, we need to think how we're going to use these chemicals. Some of you may be aware of the UNIDO initiative called "Chemical Leasing" which talks about trying to deliver the chemicals to where they're needed, and to make sure they're not wasted in a profligate manner. Because if you use only one third of the ammonia in the fields, then you immediately cut the environmental impact of your manufacturing by a factor of two thirds, because you need to make less. All the way around, we are flushing pharmaceuticals into the sewers, because we're not absorbing the medicines very efficiently. So we do need to think of a much broader systems view.

Daniel Nocera:

Dave (MacMillan): just returning to your question, initially. You have an example; it was in Martyn's talk (Poliakoff). He didn't use the words PMI. But you guys did it — I mean, the pharma organic route. That was a small-scale, but you did PMI and that had a lot of impact, guiding even small-scale science. I guess one thing I wish the community would do is, in addition to PMI, do an EI which is *energy intensity*. Whenever you talk about PMI, you also have EI and that would be at least the first step of small scale that is really important when you go to large scale.

David MacMillan:

It's interesting that they are correlated though because, in those industries, the way you get rid of waste is you actually end up burning it, which ends up being the vast majority of the energy cost. So it turns out that if you drop PMI, it dramatically impacts the energy costs.

Daniel Nocera:

But like we said before, clean energy isn't free. So when you finally go to scale, that's going to be an enormous cost. That's why I think you need to start putting EI in with PMI.

Martyn Poliakoff:

I think that the Carbon Neutral Laboratory that my university has built, has energy monitoring for usage. When you do something in the lab, it's in principle monitored and can be quantified. The whole building uses about 75% less energy than the conventional chemistry lab. I would just like to reply to this idea of wonderful and huge wind turbines. Nobody has solved what to do with these huge turbine blades at the end of life. When they were smaller, you could construct climbing frames for children in playgrounds but there's a limit to the number of playgrounds that need wind turbines. So, we are not yet applying a full life cycle idea on how to use these wind turbines. Essentially, we have to use less of everything. We may make bigger wind turbines, but we have to be conscious what's going to happen to them at the end.

Bert Meijer:

The success of chemistry is that we can produce many molecules. So much that the scale that we do, it goes up every year. If you carefully look to the world, there's a big chance that we continue to grow and that means we make more waste. So we also have to think about getting the scale down, by one way or the other making chemicals more expensive. They're just too cheap. As a result, the Western society is using so enormous number and amounts of chemicals for nothing. I think that's a part we have to take into account. I always think that if you ask, for one hour, a lawyer to help you, you can buy almost all of the chemicals you use in the whole year. Something has to change there and it's also our duty to do this. One of the things could be using waste much more. That also means that you have to think about catalysis that can deal with starting materials that are not pure. I can give you one example on the PET recycling. Every company has his own flame retardant, the oxydant and everything in it. As a result, it's sometimes very difficult to find a catalyst that can deal with all of these different polyethylene terephthalates at the same time because every catalyst is very sensitive to something. So I would prefer to also look at catalysts that really can work with very dirty starting materials instead of a clean molecule in which you do it very correctly.

Martyn Poliakoff:

Could I just respond to your suggestion of making chemicals more expensive as a way of controlling consumption? I think that this is a totally immoral proposal because, at the moment, there are huge proportions of the world's population who are profoundly poor and who deserve to use more chemicals. By raising the price, we are pricing out a large proportion of the world's population from benefiting from these chemicals. What we have to do is to use smaller amounts of these chemicals, so that the price per dose stays the same. But the manufacturer makes more money because they have more units of action from the same amount of chemicals. We really have to look at the poorest people in the world, rather than just fixing our minds on the USA or Europe because we are living to a standard which is shameful compared to many areas of the world.

Peter Palese:
We heard that there will be an increase in the energy footprint. If that is the case, my question to the panel is: if there is really an increase in energy needs, will there be a future with or without nuclear power possible?

Karthish Manthiram:
In many ways, nuclear power is certainly one possibility that remains, although the way in which the public views the technology is only going to make it increasingly hard to use. So part of the question is how do we achieve the baseline power that we need? Of course, solar and wind have struggles with intermittency. If we only power these chemical manufacturing plants, in the future, when the sun shines or the wind blows, we're not going to get the utilization factor that's needed to depreciate the capital equipment. You really need something that can work around the clock to make use of that capital investment. That means that electrical energy storage needs to get really cheap. We need to be able to store electrical energy in lithium ion batteries, or some other competing technology, to be able to then operate these plants around the clock. The issue is that, at present, that's a very expensive way of storing electrical energy. But it turns out that if you look at these plants very carefully, there are other ways of storing that energy. For instance, you can actually do your air separation during the day and store that as liquefied nitrogen. You can produce hydrogen and store it during the day and then operate some of the other non-electrically driven parts at night and actually get capital equipment depreciation. There are creative ways of storing energy that can also be thought about in these contexts.

David MacMillan:

Dan's (Nocera) section is going to be discussing energy in great detail and I don't want you to take away too much of the momentum and the underpinning discussions that are going to show up in that meeting. So I want to keep it slightly more towards the catalysis or the sustainable catalysis side of things. Not to close anyone down on any given sort of topic, but I think just to bring us a little bit more generally on this theme. One point I wanted to make real fast, and then we'll get to Omar (Yaghi), is that, at least in my world, there are many types of catalysis. There are certainly the cases of scale, where you worry about the scale of the actual volume or the amount of material or the energy consumption associated with it. There's another scale associated with catalysis and that's the scale of people. Every day on Earth, there's more than 70,000 human beings who perform a different palladium-catalyzed cross-coupling reaction from each other. They are completely different reactions but there are cross-couplings. That's an enormous number of people but they're doing it on very small scale. The question is: should we be worrying about those people? Should they be thinking about inventing reactions for diversity of synthesis, as much as singularity of synthesis? Is that important? That's obviously going to come from an academic environment as to how we think about that. So the question is: do we think about sustainability as we think about diversity of transformations at the same time?

Omar Yaghi:

I would like to ask the panel whether there is a definition for what sustainable means from a chemistry point of view. What are the intellectual questions that need to be addressed to get to that level of sustainability? For example, you mentioned moving from precious metals to more available metals and things like that. But what bonds should we be avoiding to make? And what bonds should we be focusing on? What strategies can chemists develop to make their chemical reactions more sustainable? I'm trying to get into the "nitty gritty" of what would a graduate student do if they come into our lab and say "I want to work on sustainable chemistry"? First, what is it? And second, what are the real challenging questions on the molecular level that we need to be able to address?

David MacMillan:

First of all, you hit the nail on the head. Sustainable is sort of an amorphous term. It's an umbrella term that incorporates a lot of things that fall beneath it. I think one of the purposes of today is to try and think about what all those different parts are. What to work on is a more challenging question because there's so many different components to this. The key parts are: what are the main intellectual ones? What are the main intellectual questions you can start to think about? I think that boils down to what are the top problems we should be caring about? Defining that is obviously going to be critical, to sort what are the questions and the intellectual problems associated with it.

Martyn Poliakoff:

I teach sustainable chemistry and one of the repeating questions that the students are asked is: what is the difference between green chemistry and sustainable chemistry? Green chemistry is governed by a number of principles (12 principles) while sustainability has three pillars: there is environmental, social and economic. The slick definition is that green chemistry looks at the environmental and social impacts, whereas sustainable chemistry also looks at the economic impact. Whatever green process you make, it cannot be run to financial loss because this is not a basis of a long- term sustainable industry. I got really excited when we got exam answers from students that actually say "this process is quite green but I don't think it's sustainable", which I think is just the way that we have to get people to think. To answer Omar's (Yaghi) question, students have to take whatever the problem is and ask if they can devise a method which is both greener and economic to run. Part of sustainability that one has to think about is the available supply of the elements. We're now spending far more of the petroleum energy getting it out of the ground than we did a century ago. When this conference started, you needed one barrel of oil to get 100 barrels out of the ground. It's now about one barrel of oil to get five barrels out of the ground and that number is going to get smaller and smaller. So we have to think about using the resources that we've got more efficiently. Because I come from Nottingham, the home of Robin Hood, I have encapsulated this in what I call "the Robin Hood question", which is: "how can we give to the poor without robbing the rich"? This is the way that chemists have to go in the future.

Matthew Kanan:

I'm going to try to answer your main questions, Dave (MacMillan), and touch on some of these themes. I think there's merit to worrying about sustainability on the small scale and trying to innovate to address that. That has to be balanced against the pace of knowledge acquisition. As for people working on a small scale and trying to make a molecule to answer a particular question for drug development, I think burdening them with trying to devise a more sustainable way to get to that piece of information may not always make sense and may not always end up saving any energy or resources. But devising ways that can improve the material or energy efficiency of pharma for a scaled up drug is certainly something that advances human progress. I think that's great. I don't see those as stepping stones to going after gigaton scale problems. I don't view that as the early adoption of the future of sustainable production because the natures of the problems are very different. Learning how to electrify a drug doesn't teach me how to electrify a fuel for example. If you want to go after gigaton products, super commodities and the fuels, dig in and look at how those are made, identify the opportunities to intervene and either switch them to green or improve efficiency. There aren't that many targets to make. For ammonia, most of the energy is in the hydrogen. So one could argue that we are in the process of greening hydrogen production. We are at the beginning of what I think is an inflection point for green hydrogen. As was done 100 years ago, you could start to use electrolytic hydrogen to make ammonia again and as the manufacturing capacity and cost of hydrogen comes down, that would maybe be more and more competitive. Same thing with fuel and the Pearl GTL plant in Qatar that Ben (Feringa) alluded to. All of those products (gasoline, middle distillates, waxes) they're all coming from syngas. So if you green syngas, there's infrastructure, there's know how already in place today to turn that syngas into, from a volume perspective, just about everything that we use. That problem is different than C-N or C-C bond formation in pharma. They're both addressing different problems. But I think if you really want to go after the gigaton, it makes sense to think about that now and then figure out the best way to improve that.

David MacMillan:

It's a great point. Can I maybe springboard off of that and put a different question out to this chemistry audience? Picking up Matt's (Kanan) point, I was just talking about diversity in my world. We care about invention of new transformations and new types of bond forming processes. Being able to make things that we couldn't make in a different way previously is inordinately important. To some extent, I agree with you. I'm not sure how much you should care about sustainability with respect to that invention. Are there transformations that have not been envisioned that we should be thinking about on scale that don't exist right now? Or should we be thinking about our invention be around new catalysts to optimize what we already have? I assume everyone's going to say a blend of the two. What are the transformations that we don't have right now and that we should be envisioning? Karen (Goldberg) gave a really nice example during her talk but I was wondering if maybe people had other thoughts around the table of other types of processes that could be valuable in that context.

Daniel Nocera:

That's an important question because this conference started when the chemical industry started and energy was free, basically. You were using oil and nobody cared, one barrel to a hundred. And you didn't care about CO_2. For all your processes, you had no thought about CO_2 or energy. I was actually going to ask Karthish (Manthiram) and the panel, and I agree with Matt (Kanan) and Sossina (Haile), we have a big legacy infrastructure which you want to tap into. But maybe there are other products? I always say this to organic chemists. So, what do we do in oil industry? We take super long energy intensive things; then we break them down to C_1, C_2 or C_3, and then we rebuild everything. There's an energy penalty for that. I always say to my organic colleagues at Harvard: if I wanted to make diesel, instead of breaking it down and rebuilding it, could I just take a very simple freshman organic chemistry reaction, hexane and hexane and in one step make a C_{12}, for instance? The question comes: do you want all these targets of a legacy world, which, again, tapping into for economic reasons is important? But it might not be the targets of a future when you're worried about energy and about CO_2. I have some numbers. You actually have a fair amount of energy in CO_2 for breaking things down and rebuilding, even though most of the carbon content is in the things you're making, that you're digging out of the ground. I acknowledge that but I think there's real value to what you're saying to rethink different targets that aren't defined by a legacy 100-year of infrastructure.

Karthish Manthiram:

Just to add to what Dan (Nocera) mentioned there, I think that there are many opportunities to think about how we make the same molecule while having a very different overall reaction that leads to it. While we're still perhaps looking at making a molecule like ammonia, which is familiar, that molecule can have a very different overall reaction compared to what we're making today. Rather than starting with nitrogen, methane and water, and getting to CO_2 and ammonia, to instead be able to make just nitrogen and water into ammonia and oxygen. You thereby remove this externality (CO_2) that we're of course concerned about. Just going back to a point that Matt (Kanan) made earlier about whether decarbonizing pharmaceuticals and getting sustainability is linked to commodities, I think you're absolutely right Matt that there are definitely differences in the ways that these chemicals need to be approached. That being said, I think there's some shared platform aspects. For instance, being able to use hydrogen, sustainably produced from water electrolysis. We're seeing examples of the pharmaceutical industry to be able to make some of these chemicals on site rather than sourcing them externally. There are increased margins to be able to introduce some of those synthetic steps in those contexts and then, as those technologies become cheaper and cheaper, to implement them at bigger scale. Some technologies are separate and will be made without shared basis. Others can work their way up this pipeline. So it'd be very careful in deciding where those shared elements are, and where things are entirely separate as well.

Robert Knowles:

I was going to follow up on things that Dan (Nocera) said, and also that Bert (Meijer) and Ben (Feringa) had mentioned previously in their comments, and thinking about plastics. We take small molecules and we polymerize them into these large macromolecules. Now we have all this accumulated waste. I think a really interesting prospect for catalysis is thinking about how do we design really good reactions to turn all those things back into feedstocks that we can then repurpose for all sorts of different applications. That's definitely something that has gotten a lot of attention but I think there still are a lot of fundamental chemistry solutions that need to be devised and developed. There are certainly sort of dream reactions that you could imagine being able to take waste plastic and turning it into feedstocks for all sorts of other processes as well.

Martyn Poliakoff:

If I can respond to our colleague from Stanford (Matthew Kanan), I don't think that it's necessarily reasonable to do some small scale reaction sustainably if you just want to make something quickly. However, I feel that the person doing it should have the mindset to know what sustainable chemistry is and to deliberately decide that, on this occasion, they won't do this sustainably. Just in the same way that, when you're crossing the road, you may decide the risks outweigh the need to go to the pedestrian crossing, and you will jaywalk across here, because you need to get somewhere quickly. But I think we need to get people to think carefully. I also take what you say about existing infrastructure of the chemical plants and just point out that, a hundred years ago when this first meeting took place, if people were asked about the future of military shipping, they would say we've got all this investment in battleships and therefore, we should have the future designed entirely around battleships. There may be occasions when we will have to write off these plants because whatever a disruptive technology is coming. We cannot all the time be harnessed by the conservatism of the chemical industry that Chad (Mirkin) mentioned. On the other hand, we need to think sensibly because we don't want to throw away resources if they are valuable.

Kurt Wüthrich:

I am quite intrigued by the suggestion that we should go for catalysts that could work on waste and on impure materials. As an outsider, I would think that heterogeneous catalysis could hardly work because the catalysts would be poisoned in no time. So what is the future? How could this demand be satisfied? Would it rather be homogeneous catalysis or heterogeneous catalysis?

Robert Knowles:

It probably depends on the types of materials. I think one of the big challenges with things like thermoset materials is they're insoluble. Doing heterogeneous catalysis on molecules that don't come in solution is a really big challenge. Whereas other polymers are soluble and you can imagine doing heterogeneous types of approaches. I'm not sure there's a "one size fits all" but I think things that could be heterogenized are probably ideal. But I don't think it's going to be possible for all cases.

Karen Goldberg:

I just wanted to go back and comment on this idea of waste material and what we're producing. When we first started producing plastics, we clearly weren't anticipating the stockpiles of plastic that have accumulated. Same thing with these wind turbines that are now piling up and we have to deal with. And the solar panels, you could say the same thing, what are we going to do when we're going to have these stockpiles of solar panels. So I do think that we should have learned from our plastic experience that we must think about these wind turbines and the solar panels, and how we're going to work with those materials and extract things from them and recycle those materials. So, I think that's part of sustainable future.

Matthew Kanan:

To quickly respond to Martyn's (Poliakoff) comment, I think you're talking about the educational aspect of embedding sustainability at the lab scale. I agree that's valuable, I think the danger that I see in the literature is that sustainability becomes what I write in my intro to justify whatever I'm going to do in the rest of the paper that may or may not be ultimately aimed at anything remotely feasible or sustainable. So I think that has to be balanced in terms of educating people to think critically about sustainability as it applies to one problem and whether or not those principles are at all applicable to another bigger sustainability problem. With respect to plastic, I think this is another great example of a case to think critically about what the landscape of options is today. There's a plastic waste, handling waste management problem. Plastic in the environment is undeniably atrocious and should be addressed. That's not going to be addressed by a new catalyst that would somehow convert a grocery store bag back into monomers. Those are sort of two separate problems. But I think it has to be balanced. What could I do if I accumulate a bunch of waste? What are the options I have today? I could try to figure out how to devise a catalyst to somehow depolymerize C-C bonds. It's an unbelievable and fascinating intellectual challenge. But we have to think about what that could possibly look like from a chemistry or even engineering point of view. I'm trying to process this heterogeneous material back into monomers, to then transform it back into plastic, using alternative ways compared to the more engineering or thermal breakdown cracking point of view, i.e. that I can transform the materials back into monomers with which we already know how to work with. What's the footprint of that, relative to the conceivable footprint, even if I imagine a best-case scenario for a new catalytic process? At the outset, maybe I can't do that perfectly, but I think that those considerations have to be brought to the fore. Just being able to say I have some transformations that turn plastic back into monomer, therefore, that's better than anything else I could do with the plastic. However, I don't think that's true, the devil is in the details with the life cycle of those two pathways.

Nicholas Turner:

To return to Dave's (MacMillan) question, should we invent more catalysts and more reactions or should we improve the catalysts we already have? I tend to see things through the prism of a biochemist. If you look at Nature, it basically uses a much smaller set of organic transformations than organic chemists have successfully invented in the last 100 years but it uses those to make an incredible diversity of products. If it wants to make carbon-carbon bonds, it doesn't have anywhere near the complement of catalysts or reagents that organic chemists have been invented. It tends to use aldol reactions or polyketide synthase reactions, and it uses them over and over again. Of course, Nature produces incredible natural product diversity. Reaction diversity doesn't necessarily equate to product diversity. You can get to beautiful natural products that have inspired people in terms of pharmaceuticals and agrochemicals by judicious and clever use of reactions, often in an iterative way. If you've only got a minimal set of reactions, you need to use them over and over again. I wonder if that approach has been fully explored by the organic chemistry community who loves to create new reactions, obviously, because that usually results in fantastic publications, and it demonstrates your ability to think and invent new chemical reactions. But are there other ways of doing synthesis? Nature obviously perfects catalysts, that clearly happens. It evolves enzymes to be supreme catalysts. So it's gone down the route of taken its minimal reaction toolbox and making those enzymes work incredibly well.

David MacMillan:
That's extraordinary impressive the fact that Nature can do with so little. If Nature had all the reactions that we had invented, just think how fantastic life forms we would be right now. Anyway, that's a whole different thing. But the point is you do see this mimicry of Nature and it's astonishing to me sometimes how long it takes to mimic Nature. I'll give an example. A cascade catalysis, the idea that you can have four or five different catalysts in the same vessel to dramatically improve the efficiency from the simple start inputs to the molecular complexity that you're looking for, has only really started to happen over the last 15 years. That's stunning if you think about how long catalysis has been around. I will also point out that one of the great things with biocatalysis is how it is expanding beyond Nature's toolbox. It really is going in lots of new directions that Nature doesn't do, but clearly enzymes can do.

Ben Feringa:
I fully agree with Dave (MacMillan) and there is one other aspect that has not been mentioned and where we have the ability to design in the future and that's adaptive catalysts, catalysts that respond to their environment. If there are other components in the mixture, they adapt their behavior and they can still handle it. There are now some beautiful examples in the literature where you have responsive or adaptive catalysts, where they can do multiple functions, not only in a cascade but also adapt to a situation where the starting materials slightly differ or where you have a mixture of components. I think there is tremendous open field to discover and to build in these kinds of responsive function. See catalysis or, in general, Chemistry as an information Science. What information does get in there

and how the system responds to it? We will get to these systems this afternoon.

Sossina Haile:
I'm wondering if we've really addressed Omar's (Yaghi) question because if I were asked "what is sustainable chemistry?", I'm not sure that I would have a good answer for that. Does it mean that we're doing the full life cycle analysis? Are we saying that anything, any chemical that's used as an input we see where it's going to end up? Are we defining that in terms of the energy use for the catalysis product? Are we defining it in terms of avoiding precious metal catalysts? I'm curious from this panel what that answer is.

Karthish Manthiram:
I think I hold maybe an overly simplified definition close to heart but I'll state it anyway. For me, sustainability means the ability to practice a chemistry for eternity while minimizing harm. In an oversimplified way, that captures that I want chemistries that we can do essentially forever without having harm, however we define it. From a CO_2 footprint perspective, it could mean from the discharges we have but there's so much that's left to the mind of whoever hears this to determine what types of harm they're concerned about. But I'd like to be as comprehensive as possible on these things and I think this is why life cycle analyses need to be more comprehensive than they are today to draw that box as big as we can to consider all the ramifications. But we have to admit that what we care about in harm is so values based and that's part of why we have to have discussions of the sort that help us set forth how we do define that harm and what eternity means to all of us together.

Martyn Poliakoff:
The problem with life cycle analysis is that you can't really do it until you have built the process and have it running. This is why people need to develop proxies like the E factor or PMI or whatever, which will show you whether you're on the right lines or not. It's like some of the simple things that are used in thermodynamics: Kapustinskii's equation to estimate lattice energy so that you can see whether your Born-Haber cycle is going to

be approximately endothermic or exothermic. If it comes out close to zero, you know that you need to use more precise measurements. But you can't build a full scale plant for every potential process and do a life cycle assessment. I think one has to make some approximations. It is just an awareness of some of these issues which we really need to get as part of the things that chemists think about. I agree with Matt (Kanan) that you don't want people just pretending that their work is sustainable so that they can get it published. But on the other hand, we want more and more chemists to think and genuinely care about the sustainable implications of what they're doing. Sometimes they may find a brilliant reaction that's not very sustainable, but demonstrating that it's possible may then inspire people to think how it could be done more sustainably. It would be disastrous if sustainability became a sort of chemical equivalent of political correctness that would forbid people from doing reactions. But we need to consider it.

Joachim Sauer:

One can ask: is it sustainable chemistry or are we also considering chemistry contributions to a sustainable society? I would say we need both. We need to address the genuine catalysis questions which could then contribute to problem solving, which are needed to cooperating with engineers and economists. Two of these genuine problems have already been mentioned. One by Kurt Wüthrich: how are we dealing with mixtures? There was a lot of activity in the past. Times, we have been very proud to have a selective catalyst for a given feed. But now, we have to ask the question: how we are dealing with mixed feed and impurities? Or as Ben (Feringa) was saying, how we are dealing with adaptation? Another thing is how we are dealing with self-repair which is so great in Nature? So, I would say there are these two aspects if the student is coming and asking, and we can tell him to try to find the catalyst which deals with impurities and poisons and resistant to this one.

Chad Mirkin:

Actually, I really loved your analysis of green chemistry versus sustainable chemistry. The problem with it, and I think you really made the right clarification at the end, is this whole issue of setting a standard. Because, at least as I understood it the way you described it, sustainability is based upon current markets and current limitations on the economic side of things, whether it's energy or actually dollars. And those changes. So going back to this issue of what we should be doing and how we should be training students, to me at the academic level, it's about pursuit of knowledge. What's sustainable today or not, that may or may not be the case tomorrow. I love the clarification that it would be a shame if this became a type of political correctness that got implemented in how you chose to pursue knowledge at university because I think it would lead to us unnecessarily curtailing our growth and understanding as an organization and as a profession.

Clare Grey:

Following on Karthish's (Manthiram) comment, a lot of the motivation of the session was about CO_2 capture. We can have a long debate about what that does. When you're trying to say is it sustainable, I think we need to be very clear on how we motivate it, because if the CO_2 capture is motivating the oil companies to continue... We can discuss this tomorrow, business as usual. So, I think it's important to say what are the fundamental aspects, like Chad (Mirkin) has just articulated. Coming from the battery industry, we had for a very long time, particularly in the US, goals set by the car companies which completely stymied research. It wasn't until some of them were just removed, that the field opened up and brought in different areas that this field (battery industry) was able to grow. So, I think one needs to be very clear on making sure to take into account that there are also political aspects in all of this too. Carbon capture has political aspects in it, too.

Karen Goldberg:

I wanted to comment on this idea of sustainability being forever. I think that's a little bit dangerous, because there are so many unintended consequences of whatever we develop and what we think is going to be sustainable. Then we find out later that we've created other problems. So I think we have to keep in mind that we're not going to know everything and what the unintended consequences are going to be.

Andrew Turberfield:
On the theme of scale and energy use, another industry that's a vast consumer of energy is cement production. Is there conceivably a role for catalysis in cement and why haven't we talked about it?

Karen Goldberg:
I think people are looking into a lot of things like putting plastics into cement and trying to figure out other ways that we can deal with things. But yes, it definitely is on scale of our CO_2 so we have to deal with that.

Karthish Manthiram:
I'll just add that one challenge is that the overall reaction involving making calcium carbonate into what's used in cement intrinsically involves CO_2 production. So it's difficult to conceive how we would alter that by itself if we remain with that reaction.

Andrew Turberfield:
But it's also a huge consumer of energy.

Karthish Manthiram:
Exactly, yes. So there's multiple facets to it. We need to make it more energy efficient. We also need to think of ways of utilizing that CO_2 and where it becomes challenging is that if you can envision the products you'd want to make from that CO_2, you would very easily flood most markets for useful products. So you're almost forced then to make that CO_2 into a fuel if you want to make something that won't simply flood the market that comes out of it. So there's this aspect of scale one needs to consider in terms of what you want to make, in addition to making progress on the energy efficiency, as you mentioned, but there are opportunities across the board for doing that.

David MacMillan:
We're down in the last five minutes and rather than keep going in the vein, I just want to go around the room and ask one last quick question. Are there any aspects of sustainability that we haven't discussed that people feel we should put at the top of our heads before we leave this meeting?

Omar Yaghi:

I'd like us to think more about resiliency rather than just sustainability. We should be designing systems that could work in different contexts. Just like what was mentioned in biology, there are few rules but they're applied in different contexts and they've done very well. So I think about resiliency and adaptiveness, as was mentioned by Ben Feringa. Let's not forget about how things are connected, in the end, to build a cycle. I used to occupy the office of Melvin Calvin at Berkeley who always thought in terms of cycles. So I think it's useful for us to think about a cycle for whatever we're making. Where we start, where we end up and how, what are the byproducts and how can we minimize harmful byproducts. So I would say resiliency, adaptiveness, things that can work in different contexts. And let's not forget about the connectivity through these cycles. Thank you.

Martyn Poliakoff:

I would say that Nature is not terribly resilient and that organisms face extinction when circumstances change so that we don't have the diversity in chemistry to deal with even the modest change of circumstances at the moment. But I would like to say that this has been a really good discussion. I would very much hope that, in the sessions to come, sustainability will still feature, perhaps not as the prime topic. But we should bear it in mind for the rest of the conference because I think it is a really important issue that is becoming higher and higher on the agenda of both academics and industrial chemists.

Daniel Nocera:

Dave (MacMillan), could I mention one thing, because you guys are all leaders, and I heard this a lot today about precious metals? Of course, if you have a really great catalyst, you don't worry about it as much but, as chemists, we forget our human society a lot, that's part of the sustainability. One really important reason for Earth-abundant metals, because that's been around in my field for a long time, it gets a little off track but it's conflict mining. That's really up front, and it gets to Martyn's (Poliakoff) issue, because it's mostly in the poor parts of the world and there's huge geopolitical instability. So having Earth-abundant metals is super important at a social level because of conflict mining. The problem is we don't see that part of the world because we don't live in it. But that's one other thing you guys could add to your slides about Earth abundancy.

David MacMillan:
I think we're getting to a natural end of this discussion here and I think people are probably hungry. I just want to finish off by summarizing. This has been a great discussion. One thing I would point out is that, in Chemistry and Science, we often talk about grand challenges, which are sort of predefined questions for people to think about. I would suggest that the people in this room, no pun intended, are catalysts for the community to decide to work on any given problems. We do have a responsibility because we do tend to work on things that many other groups around the world tend to become interested and excited in. That just is the nature of the beast. I would say that this question we get into, about envisioning what to work on, and Dan (Nocera) had a nice example of hexane plus hexane, that the more of those questions that we put out there, regardless of whether we have solutions for them or not, are actually going to be as important as our own efforts individually within labs, to get the world thinking about questions that would be extraordinarily valuable in this area of sustainable catalysis. Even if we have no idea about how you solve it. I think the setting of the stage is going to be as important as the executors. With that being said, I want to thank all the panelists and I'll turn it over to Kurt (Wüthrich) to finish.

Kurt Wüthrich:
It seems to me that we have not as yet given due consideration to transport of materials to the catalysts. Especially when we talk about capturing molecules from the gaseous state, it is not trivial to get enough material to the catalysts. This is more an engineering problem, probably, than a chemical problem but I suggest that we keep in mind for the following sessions that, on a large scale, efficient material transport to catalysts may be as important as having top catalysts.

David MacMillan:
All right, I think we'll break for lunch there but thanks everyone for participation. Thanks a lot.

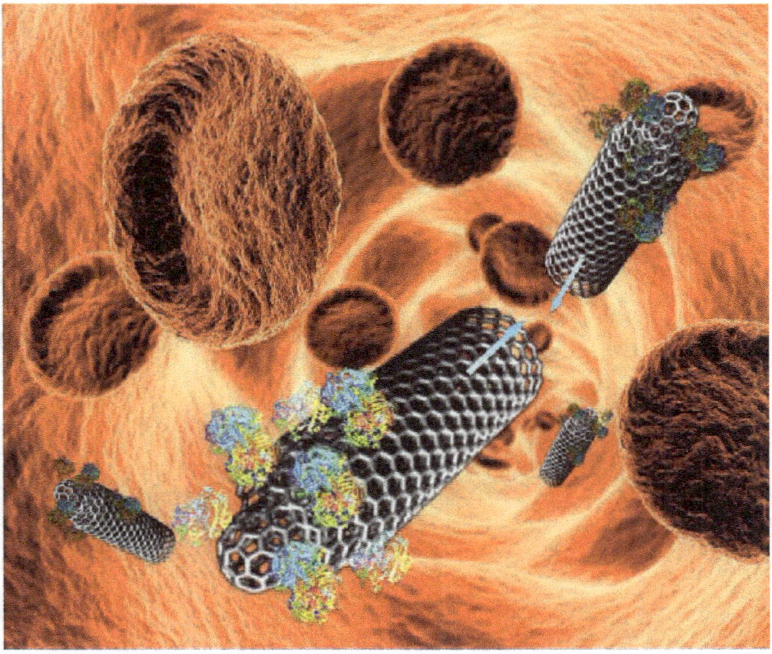

Image courtesy of Ben Feringa

Session 2

From Molecules to Dynamic Supramolecular Systems

https://doi.org/10.1142/9789811282324_0008

From Molecules to Dynamic Supramolecular Systems

Ben L. Feringa

Stratingh Institute for Chemistry & Zernike Institute for Advanced Materials, University of Groningen, Nijenborgh 4, 9747 AG Groningen, The Netherlands

Facing our future, chemistry as the creating science *par-excellence* will play a central role in the sustainable developments of the 21st century. Addressing fundamental challenges associated with the introduction of sustainable processes and products, re-inventing materials and energy carriers, going from trial and error to molecular design in medicinal chemistry, or dealing with molecular complexity, molecular systems, and out-of-equilibrium behaviour provides numerous opportunities for society and industry. Despite the amazing success of modern chemistry, the quest for dynamic self-assembled functional systems, mimicking the intricacy of natural systems, will guide the molecular explorer to the next-generation technologies. In this session, the focus is on dynamic (supramolecular) chemistry ranging from stimuli-responsive materials and molecular machines to biomimetic and adaptive systems as well as autonomous and out-of-equilibrium behaviour.

Going beyond molecules to dynamic supramolecular systems, it should be noted that one has to deal with a multifaceted approach, including structure, reactivity, organization, multi-component assembly, hierarchical levels, amplification, and interface phenomena. Ultimately, specific functions and tasks have to be performed at distinct time and length scales as prominently seen in the "Machinery of Life". Key is also the delicate interplay of covalent, dynamic covalent, and non-covalent bonds in the assembly process and adaptive behaviour. The use of computational methods, artificial intelligence, automation, and robotics will also be essential to address these multi-parameter challenges in the chemistry lab of the future ultimately building life-like systems.

Illustrative examples of important current trends and emerging fields are among others:

(i) Supramolecular assembly and organization; from precision polymers to complex and adaptive materials and multi-component systems. Going from passive matter while increasing the level of complexity and spatial and temporal self-organization, it is envisioned that designing dynamic matter ultimately could result in living or thinking matter.

(ii) Adaptive materials and molecular systems; here the focus is on life-like i.e., bio-inspired and out-of-equilibrium systems.

(iii) Hybrid materials and interfaces; it is evident that they will be central to many advanced applications ranging from sensors and smart coatings to biomedical materials. Control over hierarchical co-assembly and responsive surfaces will enable a range of dynamic functions including self-healing and adaptive behaviour.

(iv) DNA nanotechnology; taking advantage of the power of (supra) molecular assembly provided by nature's "information" building blocks in hybrid systems has proven extremely powerful for high precision construction of complex architectures, sensing, delivery, and motion and molecular computing.

(v) Peptide-based bio-hybrid materials; merging natural and synthetic fragments allow for supramolecular (co-)assembly with a wide range of functions in aqueous environment ranging from hydrogels for drug delivery to artificial muscles. In particular, opportunities arise for biomedical materials and interfacing these with living systems with fascinating opportunities such as tissue repair for neural function or blood vessels.

All these developments have in common the design and exploration of materials and molecular systems. "Learning from Life" features concepts for active, adaptive, and autonomous systems, but one is not limited by the amazingly versatile though rather limited set of building blocks e.g., nucleotides, amino acids, sugars, and lipids that were the basis for chembiogenesis preceding Darwinian evolution. Among the key questions one is addressing, one is as follows: "Can systems chemistry unravel the mysteries of the chemical origins of life?"

A future important prospect is the design of molecular devices formulated by Balzani in 2007 as follows:

> "an assembly of a discrete number of molecular components designed to achieve a specific function. Each molecular component performs a single act, while the entire supramolecular assembly performs a more complex function, which results from the cooperation of various components".

Central to these developments, shaping the future of chemistry is the transition from static to dynamic molecular systems i.e., the control of motion and the development of artificial molecular motors and machines. Pioneering efforts aiming at the control of translation and rotary motion have resulted in the discovery of rotaxanes and catenanes (Sauvage & Stoddart) featuring mechanically interlocked molecules and light-driven unidirectional rotary motors (Feringa). It should be emphasized that there are distinct differences between molecular machines in the macroworld and at the nanoscale i.e., biological machines. For instance, besides the difference in length scale between a robot in a car manufacturing plant (~2 m) and the natural ribosome, the protein-producing robot in our body (24 nm), there is the nature of the materials i.e., hard and soft materials, respectively. Furthermore, in designing molecular motors, it is not so much a matter of getting motion but how to control motion realizing one operates at low Reynolds numbers and in an environment where Brownian motion rules.

Among the wide range of designs and molecular architectures for controlled motion and machine-like function e.g., molecular pumps, assemblers, artificial muscles, responsive surfaces, adaptive catalysts, delivery systems, motorized reticular materials, adaptive polymers, and supramolecular assemblies, nanopores and walkers are just illustrative how widespread new opportunities have been identified in recent years. Controlling motion and mechanical function is also a basis for

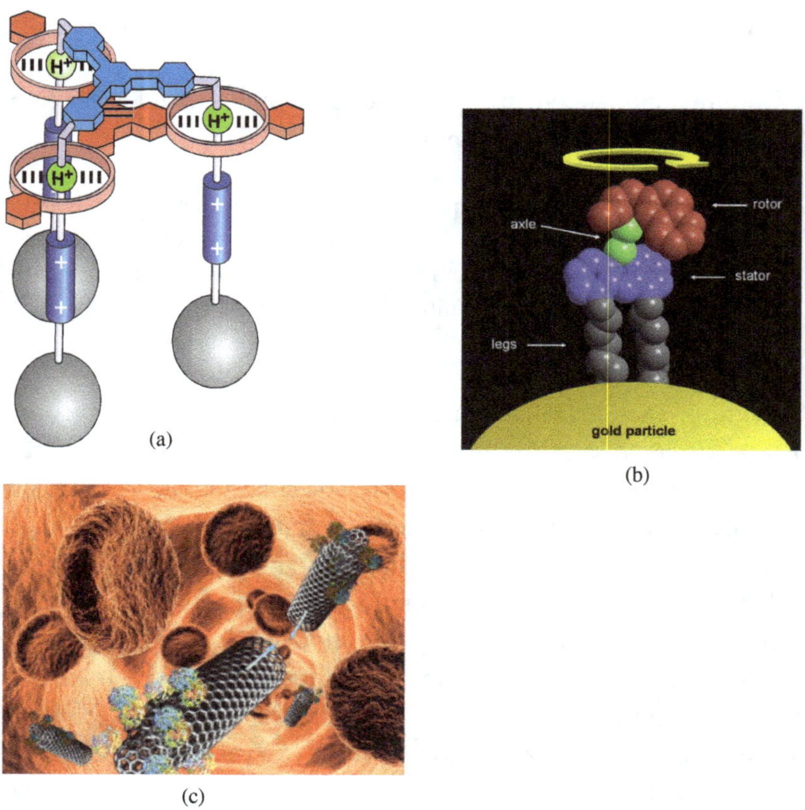

Fig. 1. (a) molecular elevator; (b) light-driven rotary motor on surface; (c) autonomous chemical propulsion system based on carbon nanotubes fueled by sugar.

out-of-equilibrium systems akin to bio-machines representing a major challenge for chemistry in the decades ahead.

Illustrative examples of molecular machines are shown in Fig. 1(a) (molecular elevator), Fig. 1(b) (light-powered rotary motor), and Fig. 1(c) (a catalytic chemical-driven propulsion system for translational motion using sugar as a fuel).

It is evident that in designing molecular motors as a core activity while moving from molecules to dynamic molecular systems, one is faced with multiple parameters ranging from structure to function but also controlling time and length scales. But the prospects are bright and numerous opportunities arise, as summarized in Fig. 2. The amazing possibilities emerging from materials and bio-hybrids to soft robotics and molecular information science will be a source of inspiration and provide reference points to the chemical explorer on the journey towards functional life-like molecular systems.

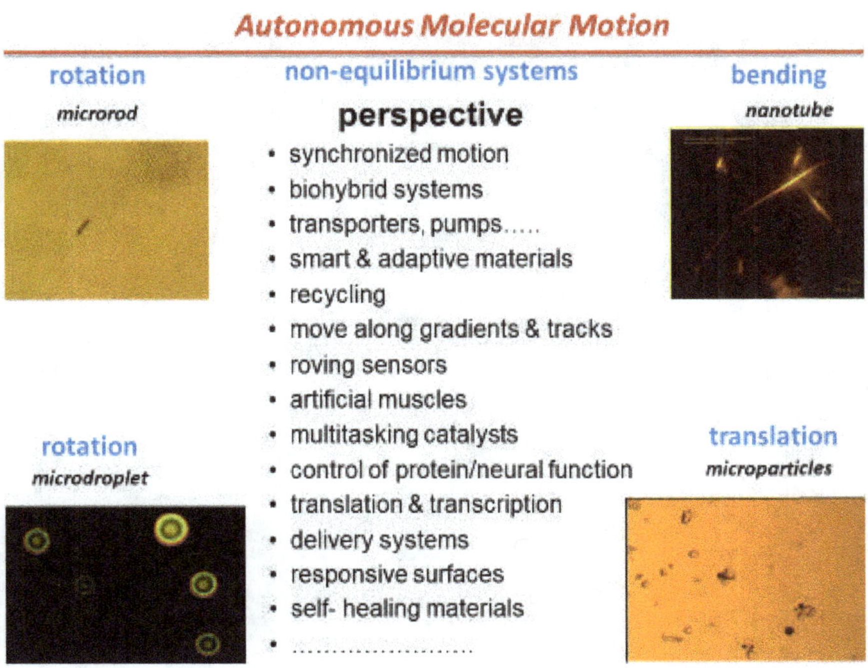

Fig. 2. Control of autonomous molecular motion, opportunities, and future perspective.

https://doi.org/10.1142/9789811282324_0009

Driving Supramolecular Systems Uphill with Artificial Molecular Motors

Nicolas Giuseppone[a,b,c]

[a]*Department of Chemistry, University of Strasbourg*
[b]*Institut Universitaire de France (IUF)*
[c]*Institut Charles Sadron — CNRS, 23 rue du Loess, 67000 Strasbourg, France*

Molecular motors as components of active chemical systems

Although the nature of life and its origin are still a matter of debate, there is a scientific consensus on the peculiar capacity of living systems to self-organize from simple molecules to more complex (supra)molecular systems that support the emergence of key dynamic functions (e.g., self-regulation, adaptation, and self-replication) [1]. In particular, one can mention an important common principle through the words of physicist and Nobel Laureate Erwin Schrödinger in his famous 1944 paper *What Is Life? The Physical Aspect of the Living Cell*, in which he states that "Living matter avoids to decay to equilibrium" [2]. Indeed, it is now well understood that life self-organizes matter out of thermodynamic equilibrium through a number of molecular processes and machineries. In this thermodynamic context, living systems can be qualified as "active" because they produce physical work on their environment. For instance, they can create concentration gradients by powering ion pumps, drive complex endergonic chemical transformations and reaction networks, transport cargo between well-defined locations, contract muscles, generate flagellar motion for propelling cells, and ensure the dynamics of the microtubules during cell division. In all these processes, the active systems are thermodynamically driven uphill through energy consumption and (transient) energy storage, and they subsequently generate the necessary functions when needed. Interestingly, this escapement of thermodynamic equilibrium is often mediated by biomolecular motors which can perform tasks from nano- to macroscale through the controlled mechanical actuation of their elementary parts when fueled by an appropriate source of energy [3].

In the past 40 years, often inspired by living systems, synthetic and supramolecular chemists have achieved a series of breakthroughs leading to the design and first syntheses of fully artificial molecular machines and motors [4]. In particular, the Nobel Prize for Chemistry 2016 was awarded to Profs. J.-P. Sauvage, F. Stoddart, and B.L. Feringa for their seminal works on this topic. For instance, Sauvage and Stoddart managed to introduce a new kind of chemical bond, named today as "mechanical bond", [5] which allows controlled movements of large amplitudes within molecular components, such as catenanes and rotaxanes. Since then, a series of advanced works by the groups of Stoddart, Leigh, Credi, and others have demonstrated the possible ratcheting of these mechanical bonds out of equilibrium upon various sources of energy (including light, redox, and chemical fuels). Following a different approach using photoswitches, the pioneering works of Feringa gave birth to the first example of an artificial molecular motor in the literature, capable of constant and unidirectional rotation under light irradiation with frequencies up to the MHz regime [6].

By looking at the current state of the art in this field of research, one can expect that the next generation of artificial (supra)molecular systems and materials will advantageously implement such "active" molecular machines in order to escape "dead" thermodynamic equilibrium and to generate advanced functions that mimic living processes.

Our recent research contributions on artificial molecular machines that work on all scales

With my collaborators, we recently focused on a series of investigations to amplify the motion and the mechanical work produced by artificial molecular machines and motors up to the macroscopic scale [7].

First, we developed covalent and supramolecular polymers made of bistable [*c*2]daisy chain rotaxanes to access contractile materials sensitive to pH variations (Fig. 1) [8].

In acidic conditions, the macrocycles of the [*c*2]daisy chain preferentially bind the ammonium stations and the molecule stands in an extended state; however, under basic conditions, the secondary ammonium stations are deprotonated, and the macrocycles preferentially bind the triazolium stations, resulting in a contracted conformation. At the molecular scale, this sliding motion occurs over approximately 1 nm, but when the [*c*2] daisy chains are integrated into main chain polymers or into bundles of polymer chains and crosslinked networks, the collective nanometric

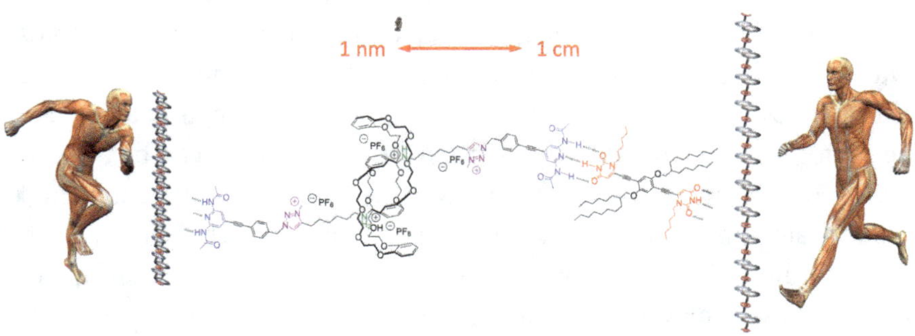

Fig. 1. A supramolecular polymer which integrates the collective actuation of multiple mechanical bonds. In this generic example, the supramolecular polymerization of bistable [*c*2]daisy chain rotaxane monomers leads to contractile polymer chains. The molecular structure is represented in its extended state with the macrocycles bound to the ammonium stations.

displacements are correlated through space and can be amplified up to the macroscopic scale. For instance, we have shown that polymer gels made of [*c*2]daisy chain rotaxanes can contract and expand by approximately 60% of their initial volume. It should be mentioned that in these particular systems, the [*c*2]daisy chain acts back and forth reversibly, and the work produced in one direction is undone during the reverse process. Hence, this switching situation where the machine influences its surrounding only as a function of its state is not sufficient to progressively increase the mechanical work performed on its environment, as opposed to what can be achieved in an out-of-equilibrium manner by living systems.

To go one step further in this direction, we have demonstrated that light-driven rotary motors can be introduced as active reticulating units in polymer networks, and they can induce — upon autonomous cycling rotation — the twisting of the polymer chains up to the macroscopic contraction of the material (Fig. 2(a)) [9].

Because the Feringa's motors we used rotate unidirectionally (Fig. 2(c)(i)), the coiling of the polymer chains leads to the formation of a variable number of chiral twists which preserve the information on the motors' trajectories. A system of second generation was subsequently designed to make it fully reversible [10]. It uses elementary modules made of a combination of two different mechanically active molecular units: one rotary motor and one rotary modulator linked by polymer chains (Fig. 2(b)). The rotary modulator is based on a tetrasubstituted dithienylethene photoswitch (Fig. 2(c)(ii)). Its role is to function as an on-demand elastic releaser that operates at a wavelength different from the wavelength that actuates the motor. As the individual cycling units of the module can be actuated out-of-equilibrium at two different wavelengths, the work output at a steady state can be finely tuned by modulation of frequencies with the motor turning clockwise and the modulator anticlockwise. The importance of these achievements compared to the state of the art was twofold: (i) the demonstration that these directional out-of-equilibrium cycling motions can be coupled to polymer chains and amplified up to the macroscale and (ii) the proof that molecular motors can influence a whole system as a function of their trajectories and not only as a function of their state as usually observed with switches. This unique out-of-equilibrium control of motor–polymer conjugates was also implemented in the

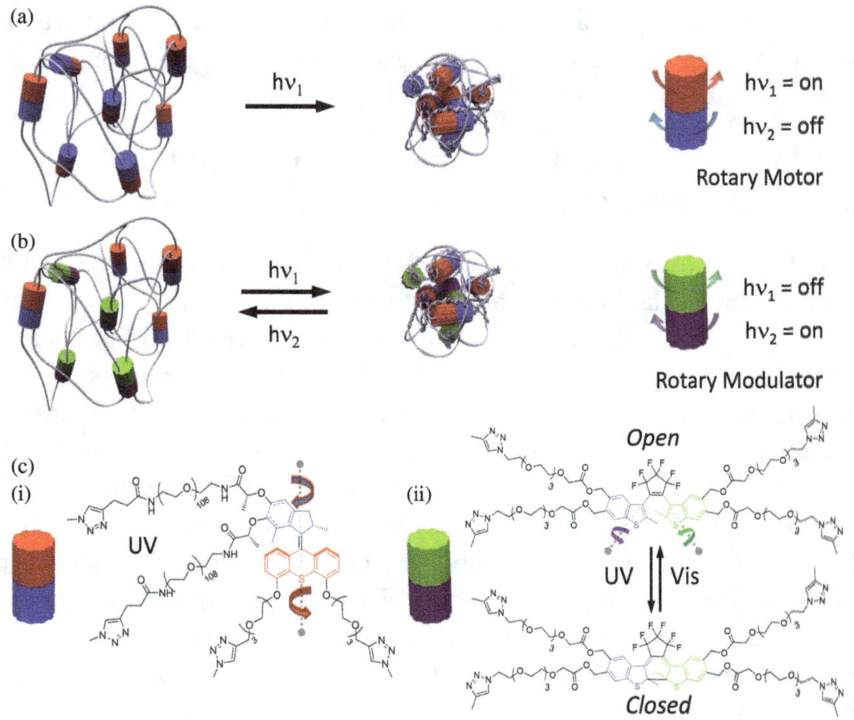

Fig. 2. Mechanically active motor–polymer networks. (a) Schematic representation of a polymer–motor network that, under UV ($h\nu_1$), produces a continuous, unidirectional rotation of the motor part (red and blue cylinders) with subsequent winding of the polymer chains leading to a macroscopic contraction of the entire network. (b) Schematic representation of a polymer–motor–modulator gel that proceeds as in (a) when exposed to UV light but unwinds the polymer chains when exposed to visible light ($h\nu_2$) by activating the modulator part (green and purple cylinders). (c) Chemical structures that compose the polymer–motor–modulator involving (i) enantiopure crowded alkenes as motors and (ii) photoswitchable dithienylethenes as modulators.

mechanotransduction of living cells [11], in the transport of metal ions through phospholipid bilayers [12], in the determination of the work, torque, and force that can be generated by a light-driven rotary motor [13], and in the engineering of macroscopic bending actuators [14].

These series of works illustrate the possibility to couple molecular machines, including switches and motors with their environment through (supramolecular) polymers chains, and to possibly push these systems progressively energetically uphill, far from thermodynamic equilibrium, in order to store and make use of their mechanical work therefrom.

Outlook to future developments of research on mechanically active supramolecular systems

Over the past forty years, a series of fantastic scientific achievements by synthetic molecular chemists has led to the first syntheses of molecular machines and motors. However, up to now, these fascinating objects have mainly been designed and synthesized as individual units and essentially studied on their own. This was understandably requested by the imperative need to understand and control their mechanism of actuation, but a remaining question is still recurrent in the scientific community and beyond: "What could molecular machines be useful for?" I believe that this question will be answered only if we understand how to interconnect these machines with their environment. By building further on our original works, we not only want to apply already existing knowledge on molecular motors, but we also want to answer the remaining fundamental questions that have not yet been answered even theoretically: (i) Will the integration of molecular motors in (supra)molecular systems and materials be limited to a few and specific examples, or can it be implemented by the design of more general and rationalized approaches? In particular, how to preserve — or even amplify — the out-of-equilibrium ratcheting actuation of a molecular motor when integrated into a material, possibly up to the macroscopic scale? (ii) Which efficiency can be reached by these systems, in terms of energy conversion, motion's trajectory, workload, speed of actuation, and power? Would these materials offer decisive advantages compared to simpler switching materials in terms of functioning autonomy and work production? (iii) How to couple such molecularly motorized materials with appropriate devices in order to transduce their mechanical actuation in other types of energy? How to integrate them with other molecular elements (or macroscopic segments) to access multitasking materials? How to engineer them and to interface them with their environment in order to control them and/or to generate autonomous feedback loops leading to adaptation? Can we push such systems sufficiently far away from equilibrium to generate self-organizing emergent behaviours (reaching, for instance, bifurcation points in their trajectories)? These questions should support new endeavours in the course to reach advanced dynamic systems capable of mimicking complex functions such as those encountered in living systems.

Acknowledgements

I am deeply grateful to all my past and present collaborators, who have vastly contributed to build part of the knowledge presented in this proceeding. I also want to thank the various institutions and funding agencies which supported our research works, including the University of Strasbourg and the CNRS, the Institut Universitaire de France (IUF), the European Commission (in the framework of the ERC, ITN-MSCA, and FET-open programs), the ANR, the LabEx CSC of Strasbourg, and the Jean-Marie Lehn Foundation.

References

[1] P. L. Luisi, *The Emergence of Life*, Cambridge University Press, Cambridge (2006).

[2] E. Schrödinger, *What is Life? The Physical Aspect of the Living Cell*, Cambridge University Press, Cambridge (1944).

[3] M. Schliwa and G. Woehlke, *Nature* **422**, 759 (2003).

[4] S. Erbas-Cakmak, D. A. Leigh, C. T. McTernan, and A. L. Nussbaumer, *Chem. Rev.* **115**, 1081 (2015).

[5] C. J. Bruns, and J. F. Stoddart, *The Nature of the Mechanical Bond*, John Wiley & Sons, Inc., Hoboken, NJ, USA (2016).

[6] N. Koumura, R. W. J. Zijlstra, R. A. van Delden, N. Harada, and B. L. Feringa, *Nature* **401**, 152 (1999).

[7] D. Dattler, G. Fuks, J. Heiser, E. Moulin, A. Perrot, X. Yao, and N. Giuseppone, *Chem. Rev.* **120**, 310 (2020).

[8] E. Moulin, L. Faour, C. C. Carmona-Vargas, and N. Giuseppone, *Adv. Mater.* **32**, 1906036 (2020).

[9] Q. Li, G. Fuks, E. Moulin, M. Maaloum, M. Rawiso, I. Kulic, J. T. Foy, and N. Giuseppone, *Nat. Nanotechnol.* **10**, 161 (2015).

[10] J. T. Foy, Q. Li, A. Goujon, J.-R. Colard-Itté, G. Fuks, E. Moulin, O. Schiffmann, D. Dattler, D. P. Funeriu, and N. Giuseppone, *Nat. Nanotechnol.* **12**, 540 (2017).

[11] Y. Zheng, M. K. L. Han, R. Zhao, J. Blass, J. Zhang, Zhou, *et al.*, *Nat. Commun.* **12**, 3580 (2021).

[12] W.-Z. Wang, L.-B. Huang, S.-P. Zheng, E. Moulin, O. Gavat, M. Barboiu, and N. Giuseppone, *J. Am. Chem. Soc.* **143**, 15653 (2021).

[13] C. Gao, A. Vargas Jentzsch, E. Moulin, and N. Giuseppone, *J. Am. Chem. Soc.* **144**, 9845 (2022).

[14] A. Perrot, W.-Z. Wang, E. Buhler, E. Moulin, N. Giuseppone, *Angew. Chem. Int. Ed.* **62**, 202300263 (2023).

https://doi.org/10.1142/9789811282324_0010

From Molecules to Motile Chemical Systems: Setting Chemistry in Motion

Nathalie Katsonis

Stratingh Institute for Chemistry, University of Groningen, Nijenborgh 7, 9747 AG, Groningen, The Netherlands
n.h.katsonis@rug.nl

From chemical reactions to microscopic motility

In the man-made world, motion is caused by mechanical forces that operate at the macroscopic length scale. In contrast, it is the combination of physical and chemical mechanisms, involving molecules and their interactions, that sets living matter in motion. Whether biogenic movement is fuelled by chemical transformations, or whether it is directed by the complex operation of molecular machines, (supra)molecular interactions are ultimately responsible for any and all biogenic movement and associated purposeful motility [1].

On the way to unravelling some of the rules that govern motion at all length scales [2], we investigate the fundamental rules that impart purposeful motion to active supramolecular compartments, that is, to microscopic objects that are also chemically active. One main motivation for this is that the cell is the unit structure of life, thus its motile behaviour is also essential to its function, e.g., the motility of white cells for immunity, the motility of red blood cells for respiration, and the concerted movement of many cells contracting muscles. Further, motivation comes with the possibility of a plausible link between evolutionary advantages and the motility of chemically active prebiotic cells. Finally, achieving control over the motile behaviour of compartments would give us greater control over mass transport in complex and dynamic molecular environments, at length scales between those of molecules and materials.

Hereafter, I will discuss how chemistry can set supramolecular compartments in motion and, conversely, how motion can alter the chemistry of these compartments. The reciprocal (mutualistic) interaction between chemistry and motion has implications for a variety of chemistry fields, including bio-industrial applications with the development of membrane-free artificial organelles, smart carriers, neuromorphic materials, and any other processes that involve controlling mass transport in complex supramolecular systems and at microscopic length scales.

Physical chemistry of droplets out of equilibrium

It is remarkable how little systemic complexity is required for purposeful movement to emerge, though only within a limited range of parameter space. In a heterogeneous oil-in-water system, the presence of a minimal concentration of lipids in water is sufficient to confer motility to oil droplets (Fig. 1). Hereafter, I refer to this mechanism of motility as Marangoni propulsion, and one of its distinguishing characteristics is its relevance to a wide range of systems and environments.

Once oil droplets are immersed in a lipid-containing aqueous solution, the lipids organize at the droplet interface and stabilize the droplet by decreasing the liquid–liquid interfacial tension. Above a minimal concentration, lipids also self-assemble into micelles. At even larger concentrations, these micelles can set a droplet in motion: the micelles take up oil molecules into their hydrophobic core, and this pinching sets

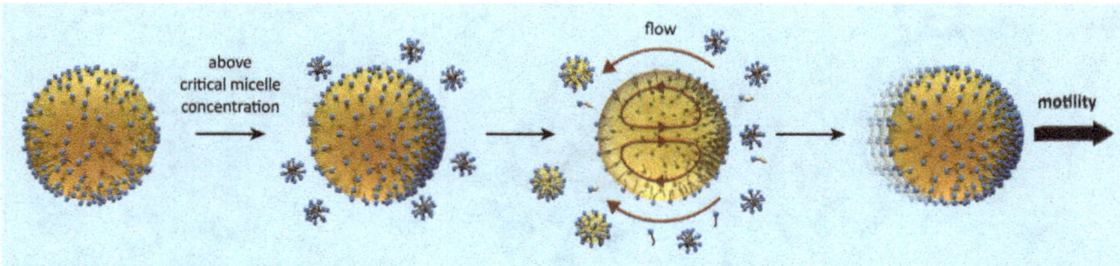

Fig. 1. Motile behaviour of droplets in lipid-rich aqueous solutions. Above critical micellar concentration, and in the presence of micelles, the uptake of oil from the droplet causes local instabilities in the interfacial tension. Gradients of interfacial tension lead to mass transport and eventually to the establishment of Marangoni flows that propel the droplet forward.

the droplets out of their equilibrated organization. In particular, the disruption of the lipid coverage at the interface induces the formation of a flow both inside and outside the droplet, and the combination of these flows propels the droplet forward. The droplets move towards higher concentrations of micelles that sustain the interfacial tension gradient and hence motion. Therefore, this movement is chemotactic towards sources of micelles [3].

Flows created as a response to gradients in interfacial tension are known as Marangoni flows [4], and therefore, when applied to droplets, the mechanism is best described as Marangoni propulsion. In this motile system, small gradients in interfacial tension drive substantial displacements. The movement stops once the lipid distribution on the droplet is homogenous and the droplet reaches equilibrium, which happens either when there are no empty micelles in the system anymore or once the droplet diameter becomes too small.

Emergence of complex motility in the course of chemical reactions

In a heterogeneous environment, the rate of chemical reactions can be enhanced by active supramolecular motion (Fig. 2). The thio-Michael reaction between 1-hexanethiol and water-soluble 2-methacryloyloxyethyl

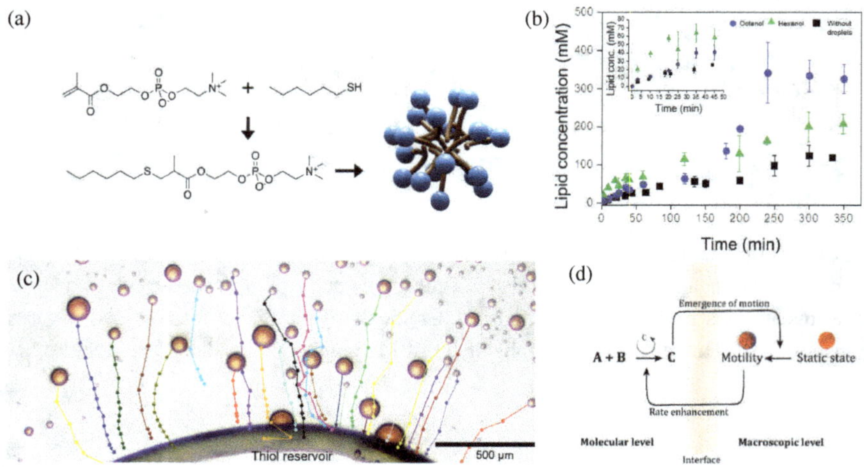

Fig. 2. Mutualism between lipid reproduction and microscopic motion [7]. (a) The reaction between 1-hexanethiol and 2-methacryloyloxyethyl phosphorylcholine (MPC) yields a surface-active lipid, which self-assembles to form micelles in water. (b) The rate of the lipid-forming reaction is enhanced in the presence of motile droplets. The kinetics are reported in the presence of motile droplets (150 μm, circle), in the presence of stationary droplets of hexanol (~132 μm, triangle), and in the absence of droplets (square). (c) The rate enhancement is attributed to the fact that octanol droplets move chemotactically towards regions of high micelle concentration, which also corresponds to regions of high thiol concentration. (d) Mutualism between chemistry and motion.

phosphorylcholine (MPC) produces a surface-active lipid that self-assembles into micelles. This system installs micellar auto-catalysis, where the products of the reaction form micelles, and the presence of the micelles accelerates the reaction further [5, 6]. Octanol droplets added to the reaction medium are initially stationary, but later, once the reaction produces enough lipids to create an asymmetry in the interfacial tension, they start moving chemotactically. We observed that the rate of the bond-forming reaction was enhanced (Fig. 2(b)). We hypothesise that this acceleration results from two complementary effects: (i) the droplets actively and chemotactically attract the reactive interfacial areas to the reagents and (ii) after solubilization, the resulting filled micelles are larger than the empty micelles. As the surface area of a sphere scales quadratically with radius, filled micelles may be able to provide a greater interface for micellar catalysis [7].

Overall, a lipid-producing chemical reaction initiates the chemotactic movement of droplets and, reciprocally, the chemotactic motion of these droplets enhances the production of lipids. In this symbiotic relationship between chemistry and motion, droplet motility can catalyze bond-forming chemical reactions — we call this phenomenon kinematocatalysis.

Although chemotactic, the movement of chemically propelled droplets lacks persistence — typically, it results in erratic motion, with a notable lack of control over rotational degrees of freedom. In contrast, swimming cells and bacteria exhibit persistent movement in fluid, turbulent, and chaotic environments by utilizing deterministic mechanisms to alter their propagation direction. Helical pathways are an evolutionary trait shared by these cells that actively move through water — in other words, they follow the threads of a screw. To explore their environment effectively, these cells have also evolved chirality-based reorientation processes, such as the run and tumble of *Escherichia coli* [9] or even more intriguing mechanisms, such as the helical klinotaxis of sperm cells and chlamydobacteria, which is deterministic [10]. Overall, it appears that helical trajectories are at the origin of the purposeful movement of swimming cells.

Chiral liquid crystal droplets also exhibit helical motility (Fig. 3). We have shown that cholesteric liquid crystal droplets exhibit a distinctive spiral pattern [11]. We created such droplets by adding a small amount of molecular motors as dopants to a nematic liquid crystal. Using an artificial molecular motor as a dopant is obviously not without benefit, as we know that such liquid crystals undergo helix inversion [12]. These chiral droplets follow helical trajectories, with a handedness that is determined entirely by the handedness of the helix-shaped molecular motor dopants (Fig. 3). As soon as the chiral droplet is irradiated, the droplet and its trajectory undergo chirality inversion and, also, the direction of propagation is modified dramatically. This deterministic reorientation is entirely encoded by a molecular process and can be controlled by adjusting the concentration of dopant or the intensity of light.

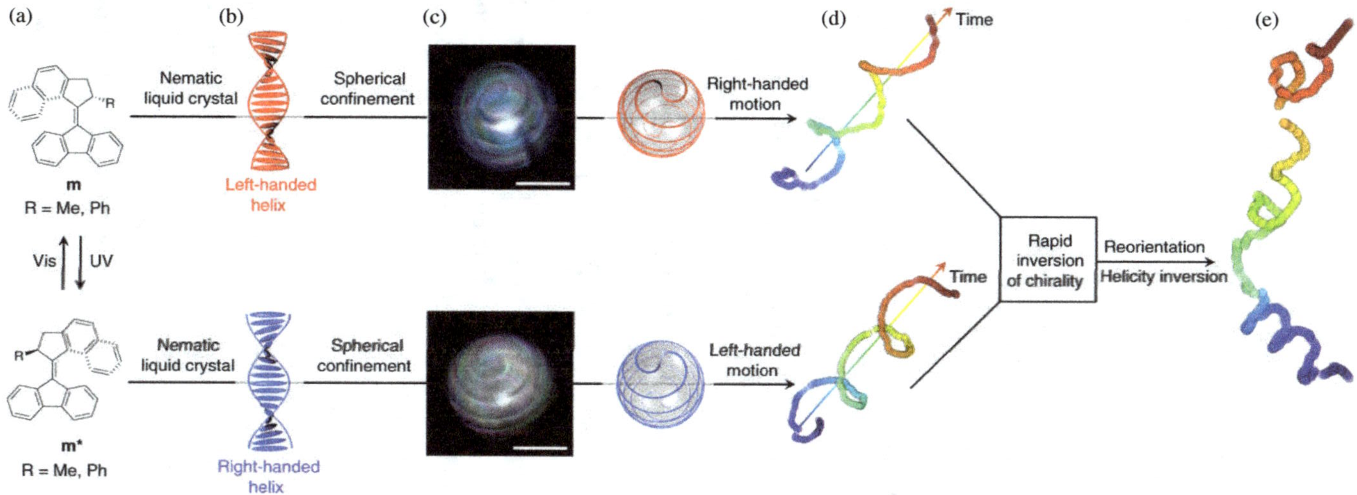

Fig. 3. Reorientation of light-responsive spiral droplets [8]. (a) Under illumination with UV light, artificial molecular motor **m** converts into **m*** with opposite helical shape. (b) In a nematic liquid crystal, both forms of the motor twist the liquid crystal and thus a supramolecular helix forms with a mesoscopic handedness that is defined by the helical shape of the motor. (c) In spherical confinement and with the liquid crystal molecules aligning perpendicularly to the interface, a double spiral disclination line forms at the surface of the droplet. (d) These chiral droplets move along helical trajectories with a handedness that is opposite to that of the droplets. (e) Once the molecular motor is illuminated, the photo-inversion of handedness reorients the helical trajectory and the new propagation direction correlates with the number of spiral turns on the droplet.

Outlook

Active mass transport can be controlled by chemical reactions in supramolecular systems. This mass transport at the microscopic length scale can be harnessed to guide interactions between large ensembles of molecules with specific functions — what we refer to as complex molecular systems. In general, systems chemistry is concerned with approaching the chemistry of life. This step forward into complexity requires chemists to shift their focus from optimization and the tackling of independent problems, towards complex systems that integrate orthogonal, parallel-operating functions. Beyond diffusion, I argue that the future of systems chemistry will involve active mass transport, as remaining out of equilibrium necessitates transporting matter.

Controlling supramolecular motility may also prompt the development of new materials for brain-inspired computing. Motion at intermediate length scales serves as means of reorganizing molecules within supramolecular architectures, a crucial step in the evolution of artificial cognitive systems. The active transport of mass will be essential to the design and synthesis of complex and functional molecular systems because if there is to be communication between two distinct compartments, it will be necessary to manipulate matter, alter its form, deliver fuel for chemical processes, remove waste, etc. In such complex systems, spatially separated molecular modules — compartmentalization — will be essential.

The emergence of movement by interfacial instability requires only an emulsion and surface-active molecules, both of which must have existed at the onset of life on Earth [13]. Consequently, the interplay between movement and chemical reactivity as we describe it is also relevant to prebiotic chemical evolution. Primitive compartments, whether droplets, coacervates, vesicles, or a combination of these would have required motility to effectively exchange molecules and react with their environments. Current micro-organisms are propelled by entirely different mechanisms than early life compartments would have been, as molecular machines and protein motors have emerged much later. Nevertheless, there must have been a time when primitive life forms and movement mechanisms coexisted.

Overall, I argue that interfacial mechanisms set chemistry in motion. The generation of flows at interfaces is a primitive mechanism that can endow supramolecular systems with autonomous and directional movement. Such purposeful movement is fundamental to the mechanisms of life and will undoubtedly prove to be equally significant in designing increasingly complex and functional chemical systems.

References

[1] F. Lancia, A. Ryabchun, and N. Katsonis, *Nat. Rev. Chem.* **3**, 536 (2019).
[2] S. Iamsaard, S. Aßhoff, B. Matt, T. Kudernac, J. J. L. M. Cornelissen, S. P. Fletcher, and N. Katsonis, *Nat. Chem.* **6**, 229 (2014).
[3] D. Babu, N. Katsonis, F. Lancia, R. Plamont, and A. Ryabchun, *Nat. Rev. Chem.* **6**, 377 (2002).
[4] C. Marangoni, Sull'espansione delle goccie d'un liquido galleggianti sulla superficie di altro liquido [On the Expansion of a Droplet of a Liquid Floating on the Surface of Another Liquid], Fratelli Fusi, Pavia, Italy (1869).
[5] A. Bissette and S. P. Fletcher, *Angew. Chem. Inter. Ed.* **52**, 12800 (2013).
[6] A. Bissette, B. Odell, and S. P. Fletcher, *Nat. Commun.* **5**, 4607 (2014).
[7] D. Babu, R. J. H. Scanes, R. Plamont, A. Ryabchun, F. Lancia, T. Kudernac, S. P. Fletcher, and N. Katsonis, *Nat. Commun.* **12**, 2959 (2021).
[8] F. Lancia, T. Yamamoto, A. Ryabchun, T. Yamaguchi, M. Sano, and N. Katsonis, *Nat. Commun.* **10**, 5238 (2019).
[9] S. Chattopadhyay, R. Moldovan, C. Yeung, and X. L. Wu, *Proc. Nat. Acad. Sci. USA* **103**, 13712 (2006).
[10] K. F. Jarrell and M. J. McBride, *Nat. Rev. Microbiol.* **6**, 466 (2008).
[11] T. Orlova, S. Aßhoff, T. Yamaguchi, N. Katsonis, and E. Brasselet, *Nat. Commun.* **6**, 7603 (2015).
[12] R. Eelkema, M. M. Pollard, J. Vicario, N. Katsonis, B. Serrano Ramon, C. W. M. Bastiaansen, D. J. Broer, and B. L. Feringa, *Nature* **440**, 163 (2006).
[13] A. Kahana and D. Lancet, *Nat. Rev. Chem.* **5**, 870 (2021).

https://doi.org/10.1142/9789811282324_0011

The Challenges of the Non-Covalent Synthesis of Functional Life-Like Supramolecular Systems

Ghislaine Vantomme and E. W. Meijer

Institute for Complex Molecular Systems
Eindhoven University of Technology
5600 MB Eindhoven, The Netherlands

Life-like supramolecular systems

The field of supramolecular chemistry is based on the intermolecular non-covalent interactions driven by the self-recognition programmed into the covalent framework of molecules [1]. Following the seminal contribution in the area "from molecules to dynamic supramolecular systems", we now aim to synthesize complex life-like supramolecular systems and materials. Inspired by Richard Feynman's statement "What I cannot create, I do not understand" [2], our goals are to gain a molecular understanding of life using synthetic systems and to make synthetic systems and materials that can function in a biological environment because they are indistinguishable from their natural counterparts. Based on the knowledge accumulated so far, the field of life-like supramolecular systems and materials is progressing in a stunning way with more and more scientists from different backgrounds being active [3–7]. Many biological structures and functions have been created with non-biological systems, such as replication [3], unidirectional motion [4], multi-scale architectures [5], soft robots [6], oscillating films [7], sensory nerves [8], and artificial tissues and cells [9, 10] to name a few. At the same time, the field of supramolecular chemistry is aware that several important issues are emerging that require a next step in the approach. First, the words self-assembly and self-organization imply too much that the fabrication of supramolecular systems goes by itself, which masks their inherent complexity. Recent work on pathway complexity and kinetic traps in cooperative assembly processes [11, 12] shows that it is more appropriate to use the term non-covalent synthesis [13–15], which also emphasizes reproducibility. Second, the various forms of dynamics in supramolecular systems and materials, both at their formation and during their use, constitute the hidden language in the field [16–18]. In this short contribution, we would like to highlight these two important points for future research in the field "from molecules to dynamic supramolecular systems".

Dynamic supramolecular systems and materials: Challenges and pitfalls

From the moment one-dimensional supramolecular assembly of monomers into supramolecular polymers yielded polymeric materials that possess material properties like macromolecules, the field attracted worldwide attention [19]. While macromolecules are based on the strong covalent bonding of monomers, their supramolecular analogues possess a highly dynamic nature due to the non-covalent bonding between the monomers. This unique property leads to self-healing and easy processing, making supramolecular polymers attractive for many applications, for instance, as biomaterials in regenerative medicine [20]. Over the years, many supramolecular polymers have been made, which can be divided into random-coil polymers assembled by an isodesmic mechanism and highly ordered and often helical polymers assembled by a cooperative or nucleation-elongation mechanism [21]. In the 25 years of studying supramolecular polymers, it is also becoming crystal clear that we arrived at an important point in the further development and understanding of these materials. In recent years, we have shown that the cooperative mechanism of the formation of these supramolecular polymers gives rise to the difficulty of pathway complexity. Similar to polymorphism of crystals, different nuclei can be formed resulting in the formation of different structures of the ordered polymers in the elongation, often with opposite optical activity. However, the dynamics of the system determines whether and how the kinetically formed structures are transformed into the thermodynamically stable structure. In highly dynamic systems, the kinetic form is often not observed in sharp contrast to highly static systems that transform only after annealing. An example of pathway complexity is given in Fig. 1 [11]. It is often unclear how many different structures can be formed and how many can be present simultaneously. In the systems studied so far, these problems are obviously not an issue, but in the future when the complexity will increase, this challenge must be considered and then the words self-assembly and self-organization may be too simple.

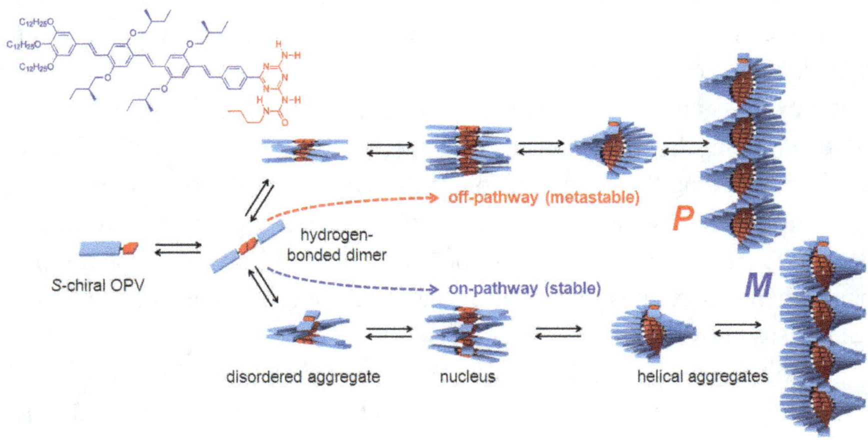

Fig. 1. Pathway complexity in the formation of helical supramolecular polymers. Adapted from Ref. [11].

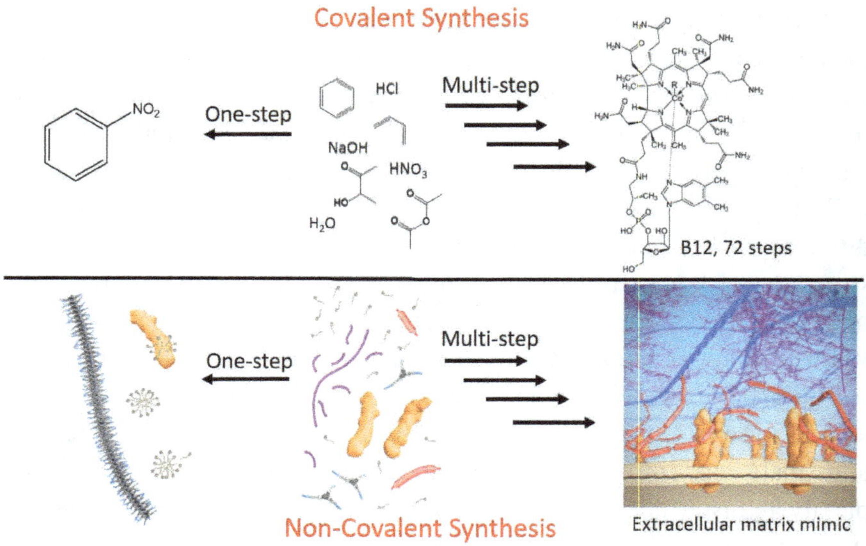

Fig. 2. The proposed paradigm shift in synthetic chemistry; non-covalent synthesis instead of self-assembly and self-organization. Adapted from Ref. [15].

Pathway complexity as presented above becomes even more of a challenge when more than one cooperative association must be controlled simultaneously. In recent work, we showed that the competition between micelles of surfactants and one-dimensional supramolecular polymers in water exhibits an unusual dilution-induced assembly process [22]. At high concentrations, the surfactant can disassemble the polymer by co-assembly. Then by reducing the concentration of the system, while the ratio between the components remains the same, this interference of the surfactant decreases with the polymer being formed. With the right choice of components, their ratios, and concentrations, we were able to obtain a gel-sol-gel-sol transition by dilution.

Another unexpected influence we have recently noted is the role of impurities and the effects of solvents. When studying a rather simple monomer that assembles in hydrocarbons at lower temperatures, its behaviour was found to be highly seasonable and thus poorly reproducible [23]. It took more than a year to discover that minute amounts (20–40 ppm) of water have a major impact on the cooperative supramolecular polymerization and when and if the helicity of the chiral polymer switched by temperature. When the experiment was conducted in a glovebox with water concentrations below 8 ppm, the assembly process that formed the supramolecular polymer showed a logical and often observed mechanism. As soon as more water entered the medium, the helical polymer switched from left-handed to right-handed structures at certain temperatures. The temperatures at which this highly cooperative transition occurred were strongly dependent on the water concentration. Detailed investigations revealed that water, as a comonomer, became part of the supramolecular polymer below room temperature. Hence, the system changed from a rigid polymer chain at 30°C into a more flexible chain at 10°C, another counterintuitive finding. For some other examples from our group, we refer to a recent article "How subtle changes can make a difference: reproducibility in complex supramolecular systems" [24].

The examples given above prompt us to rethink the future of dynamic supramolecular systems and materials and how we can fabricate life-like supramolecular systems. Although the term non-covalent synthesis has been used before by Whitesides [13] and Reinhoudt [14], it has never been embraced by our community. Therefore, we reintroduced the term to illustrate today's challenges in a perspective article in Science [15].

Outlook on the future of dynamic supramolecular systems

More than ever, supramolecular chemistry faces the same dilemmas that synthetic covalent synthesis faced in the past: which structural elements are robust and will always work, what mechanistical insights are needed to predict the outcome of the assembly process, how sensitive is the system to impurities, is the process window wide enough or is it very sensitive to concentration and temperature, how can the structure of the formed product be proven and assigned, how do we investigate the dynamics of the system, similar to conformational aspects of molecules, to name just a few of the challenges. Compared to the total synthesis of complex organic molecules such as vitamin B12, we can also design concepts similar to retro-synthetic and protecting-group approaches, multi-step synthesis, capture unstable intermediates, and the design of catalysts for the non-covalent synthesis of our systems. With these ideas, we will not only be able to increase the complexity of the dynamic structures created but also arrive at systems and materials that resemble life, allowing us to use them as look-a-likes of natural biomaterials, and integrate them with living systems to remove the dogma that natural and synthetic molecules are different. Equally important, we will gain a full understanding of the principles of life. The latter can eventually lead us to an understanding on the origin of life or create artificial life.

Acknowledgements

We like to thank the many PhD students and postdocs in our group for their seminal contributions to arrive at our views expressed here.

References

[1] J.-M. Lehn, *Angew. Chem. Int. Ed.* **52**, 2836 (2013).
[2] R. Feynman, *The Feynman Lectures on Physics*, Addison-Wesley Pub. (1963).
[3] J. M. A. Carnall, C. A. Waudby, A. M. Belenguer, M. C. A. Stuart, J. J. P. Peyralans, *et al.*, *Science* **327**, 1502 (2010).
[4] N. Koumura, E. M. Geertsema, A. Meetsma, and B. L. Feringa, *J. Am. Chem. Soc.* **122**, 12005 (2000).
[5] W. L. Noorduin, A. Grinthal, L. Mahadevan, and J. Aizenberg, *Science* **340**, 832 (2013).
[6] M. Wehner, R. L. Truby, D. J. Fitzgerald, B. Mosadegh, G. M. Whitesides, *et al.*, *Nature* **536**, 451 (2016).
[7] A. H. Gelebart, D. J. Mulder, M. Varga, A. Konya, G. Vantomme, *et al.*, *Nature* **546**, 632 (2017).
[8] Y. Kim, A. Chortos, W. Xu, Y. Luo, J. Y. OH, *et al.*, *Science* **360**, 998 (2018).
[9] Z. Yu, F. Tantakitti, T. Yu, L. C. Palmer, G. C. Schatz, and S. I. Stupp, *Science* **351**, 497 (2016).
[10] N. Reznikov, J. A. M. Steele, P. Fratzl, and M. M. Stevens, *Nat. Rev. Mater.* **1**, 16041 (2016).

[11] P. A. Korevaar, S. J. George, A. J. Markvoort, M. M. J. Smulders, P. Hilbers, A. P. H. J. Schenning, *et al.*, *Nature* **481**, 492 (2012).

[12] P. Jonkheijm, P. van der Schoot, A. P. H. J. Schenning, and E. W. Meijer, *Science* **313**, 80 (2006).

[13] G. M. Whitesides, E. E. Simanek, J. P. Mathias, C. T. Seto, D. Chin, M. Mammen, *et al.*, *Acc. Chem. Res.* **28**, 37 (1995).

[14] L. J. Prins, J. Huskens, F. de Jong, P. Timmerman, and D. N. Reinhoudt, *Nature* **398**, 498 (1999).

[15] G. Vantomme and E. W. Meijer, *Science* **361**, 1396 (2019).

[16] S. M. C. Schoenmakers, A. J. H. Spiering, S. Herziger, C. Böttcher, R. Haag, *et al.*, *ACS Macro Lett.* **11**, 711 (2022).

[17] Z. Álvarez, A. N. Kolberg-Edelbrock, I. R. Sasselli, J. A. Ortega, R. Qiu, *et al.*, *Science* **374**, 848 (2021).

[18] Y. Cho, T. Christoff-Tempesta, S. J. Kaser, and J. H. Ortony, *Soft Matter* **17**, 5850 (2021).

[19] R. P. Sijbesma, F. H. Beijer, L. Brunsveld, B. J. B. Folmer, K. J. H. K. Hirschberg, *et al.*, *Science* **278**, 1601 (1997).

[20] T. Mes, A. Serrero, H. S. Bauer, M. A. J. Cox, A.W. Bosman *et al.*, *Mater. Today* **52**, 175 (2022).

[21] T. Aida and E. W. Meijer, *Israel J. Chem.* **60**, 33 (2020).

[22] L. Su, J. Mosquera, M. F. J. Mabesoone, S. M. C. Schoenmakers, C. Muller, *et al.*, *Science* **337**, 213 (2022).

[23] N. J. Van Zee, B. Adelizzi, M. F. J. Mabesoone, X. Meng, A. Aloi, *et al.*, *Nature* **558**, 100 (2018).

[24] T. Schnitzer, M. D. Preuss, J. van Basten, S. M. C. Schoenmakers, A. J. H. Spiering, *et al.*, *Angew. Chem. Int. Ed.* **61**, e202206738 (2022).

https://doi.org/10.1142/9789811282324_0012

Dynamic Supramolecular Systems Built by Nucleic Acid Self-Assembly: Quick, Flexible, and Programmable

Andrew J. Turberfield

Department of Physics, University of Oxford,
Clarendon Laboratory, Parks Road, Oxford OX1 3PU, UK
Kavli Institute for Nanoscience Discovery, University of Oxford,
Dorothy Crowfoot Hodgkin Building, South Parks Rd., Oxford OX1 3QU, UK

Introduction

The capacity of nucleic acids to store and transmit information plays a central role in the dynamic chemistry of life and must therefore have great potential in the creation and control of synthetic supramolecular systems. Oligonucleotides store information digitally using an alphabet of four bases: what distinguishes them from other information-rich molecules, such as proteins, is the facile control of interactions afforded by sequence-specific base-pairing. Control of intra- and intermolecular bonding through rational design of base sequences underpins research into the reprogramming of natural systems (synthetic biology), devices and systems built by the assembly of oligonucleotide components (DNA and

RNA nanotechnology), archival information storage, and molecular computation. There is natural precedent for each of these applications, even molecular construction: nucleic acids are not only the material of genes, they play key roles in the regulation of gene expression and constitute the molecular machinery for ribosomal protein production.

This chapter provides a broad-brush, personal perspective of a powerful but idiosyncratic technology that is still discovering its place in the physical and life sciences.

Building dynamic systems with nucleic acids

Nucleic acid nanofabrication [1], using information encoded in base sequence to program assembly, is rapid and versatile. Nanostructures made from DNA make use of a small set of structural motifs: the double helix provides rigidity and the opportunity to link complementary domains of two molecules; interhelical strand crossovers serve as vertices in mesh-like structures [2, 3] or bind together parallel helices to fill space and increase rigidity [4, 5]. Assembly, typically by annealing multiple oligonucleotide components, is programmed through the control of patterns of base pairing through sequence design. It is usual to assign a unique sequence to each double-helical segment, even in otherwise symmetrical objects, such that the target assembly is unambiguously the most stable product. The simplicity and modularity of this construction principle mean that computational design tools can be used to automate sequence design [6, 7]. It also means that DNA nanofabrication is extremely versatile — the same set of structural motifs can be tweaked to improve a design or reconfigured to make something completely different. Construction with RNA adds a richer set of tertiary structural interactions [8] and, for intracellular applications, poses the challenge of design for assembly by out-of-equilibrium co-transcriptional folding of a single strand of RNA [9]. The field relies on the quality, speed, and rapidly decreasing price of commercial DNA synthesis: a DNA origami nanostructure (typically 4.5MDa) can be designed, ordered, and assembled within a few weeks with a material cost of the order of €1000.

Most synthetic nucleic acid systems rely on a very limited repertoire of chemical reactions: base-pairing and other non-covalent interactions that determine DNA and RNA secondary and tertiary structure, sometimes supplemented by enzyme-catalyzed covalent reactions (restriction, ligation, and templated polymerization). RNA [10, 11] and even DNA [12] have catalytic potential which can be used in dynamic devices [13, 14]. Structural, physical, and chemical diversity can be added through the use of covalently modified oligonucleotides to add lipids, fluorophores, backbone inserts, multivalent vertices, reactive linker groups and thereby peptides, proteins, nanoparticles, etc. [15]. Because each part of a nucleic acid nanostructure can usually be uniquely identified by a local sequence of base pairs, the stoichiometry and position of each modification can be precisely controlled. An interesting application is DNA-templated synthesis, in which dynamic systems of chemically modified oligonucleotides are used to control a sequence of covalent reactions by bringing reactive building blocks into proximity in a programmed sequence [16].

A significant advantage of nucleic acids in the creation of supramolecular systems is that the sequence-specific base pairing that controls nanostructure assembly can be used to mediate dynamic interactions. A simple example is a strand-displacement reaction in which one oligonucleotide is displaced from a duplex by another with a domain of similar sequence [17, 18]. Strand displacement can be used to change nanostructure conformation, for example, to open and close tweezer- or cage-like structures [17, 19], to move a "walker" along a track [20], or to initiate a templated coupling reaction [21]. The rates and equilibria of strand-displacement reactions can be widely tuned by breaking the symmetry between the initial and final duplexes by adding or subtracting terminal base pairs [22], introducing or healing base-pairing defects [23, 24], or using transient interactions to co-localize reactants [25, 26]. Kinetic control of strand-displacement reactions can be used to create molecular machinery that operates autonomously [27, 28] and systems for molecular computation [29, 30]. Molecular computation based on strand-displacement reactions can be surprisingly powerful in practice because it lends itself to composable design: circuit components replicated with different sequences can operate more or less orthogonally and oligonucleotides embodying information-carrying signals can act as both inputs and outputs of elementary operations.

An important class of dynamic systems makes use of natural DNA- and RNA-modifying enzymes [31]. Operations include ligation (which can be conditional on hybridization to a complementary "splint"), extension by a polymerase of a primer hybridized to a template, restriction (nicking or complete cutting) of a duplex incorporating a precisely defined sequence motif, and degradation (hydrolysis of the backbone). More closely biomimetic genetic circuits make use of, for example, riboswitches or ribosomal production of transcription factors to control gene expression in cell-free systems, including geometrically controlled artificial environments [32], or in cells [33].

Dynamic systems of nucleic acids can also be designed to respond to chemical stimuli, typically through competition between a ligand-binding motif, such as an aptamer, G-quadruplex or i-motif, and an alternative base-paired secondary structure, and to light. Such systems can form complex adaptive reaction networks [34].

There is a formidable strand of application-focused research into the use of DNA for archival information storage [35]. This is motivated by its chemical stability, information storage density, and certain protection from the obsolescence of technologies for manipulating and reading DNA. At its most straightforward, this technology involves only base-by-base solid-support synthesis and sequencing technologies, but there is scope for more complex synthesis schemes and context-dependent information retrieval that build on techniques of DNA computation [36].

Some advantages and frustrations of building with nucleic acids

Compared to conventional top-down nanofabrication by lithography, DNA and RNA self-assembly offers considerably greater resolution, enabling construction with isotropic sub-nanometre precision [37] of micrometre-scale objects [38]. However, assembly defects are ubiquitous and increase dramatically when patterning at larger length scales.

Compared to covalent synthesis and synthetic protein production, construction by DNA and RNA self-assembly is quicker and more flexible, capable of producing larger and more complex systems (in dimension and number of distinct components), and has an embedded capacity for embedded information and reprogrammable interaction that is difficult to emulate. It is not well adapted to provide the fine spatial control (on sub-nanometre length scales) required to create a synthetic enzyme, offers much less chemical functionality, and currently operates at very low synthesis scales (typically in the picomole to nanomole range).

Perspective

Applications of nucleic acids in nanostructure assembly, synthetic molecular machinery, dynamic reaction networks, and computation are all programmed — and can be reprogrammed — through information encoded in base sequence. For this reason, nucleic acids provide a uniquely flexible system for the exploration of the fundamental science of supramolecular assembly and dynamic molecular systems. For the purpose of creating and testing model systems, the limitations of dynamic systems of nucleic acids are outweighed by their flexibility and speed.

In developing more practical applications of synthetic systems of nucleic acids, it is important to recognize their limitations. DNA synthesis costs are falling dramatically and it is already reasonable to contemplate nanostructure fabrication on gramme scales [39]. However, it is likely that complex DNA and RNA nanostructures will be limited to high-value applications. One promising (but static) example is the use of DNA templates to lay out the components of three-dimensional molecular electronic circuits. This would be a technological revolution, — but the drive to discover a radically new way to continue to miniaturize electronics in the spirit of Moore's law will soon become irresistible.

A second class of application where economic drive has the potential to overcome economic drawback is in medicine. Synthetic nucleic acid systems have the potential to combine sensing and computation and actuation to create theranostic devices, that is, autonomous systems that combine local diagnosis and therapy. Nucleic acids provide a natural interface to natural genetic control systems and to RNA-directed gene editing: potential applications include tissue- and cell-specific treatment for genetic disease or senescence. Pursuit of applications within living systems suggests the use of RNA nanostructures folded within the cell from RNA transcribed *in situ*. This creates new challenges and opportunities to develop dynamic assembly techniques: co-transcriptional folding from a single RNA strand is a non-equilibrium process in which kinetic traps must be overcome or exploited [9].

DNA-templated chemistry can be applied to enable sequence-controlled oligomer synthesis and could form the basis of a discovery technology to explore new chemical spaces [16, 40]. Synthetic molecular machines, made from DNA and programmed by synthetic "genes", would generate libraries of oligomers from natural and non-natural building blocks in parallel, autonomously, in a one-pot reaction. Retention of the connection between programming gene and product would enable amplification and readout of the genes attached to the few products selected (for example, for receptor binding or catalytic activity), resynthesis, and even mutation and evolution. Selected products would then be synthesized conventionally for characterization and application — programmed synthesis by molecular machinery would only be used for discovery, for which current synthesis scales are already appropriate.

A related potential technology builds on Drexler's vision for additive manufacture with molecular or atomic control [41], using DNA-programmed machinery built from DNA to position a "write head" with nanometre precision. A two-dimensional molecular printing system has been demonstrated [42] — extension to three dimensions and to more useful chemistry is work in progress. The question of whether such a manufacturing technology could be useful given limitations related to DNA synthesis scales is unresolved: it is straightforward now to operate 10^{12} devices in parallel, ten or more orders of magnitude more than a typical number of conventional production lines operating in parallel but fewer by at least the same factor than the number of parallel reactions in bulk chemical synthesis. An appealing possibility is that DNA machinery could be used to bootstrap a second generation of molecular printers using materials with less intrinsic programmability but more desirable physical characteristics.

Elements of DNA computation are embedded in all of the applications discussed above. Design for effective nanostructure assembly involves control of assembly pathways that are closely connected to computational strand-displacement cascades, dynamic reaction networks map onto electronic circuits, a theranostic device must compute its sensor inputs to determine its mode and level of output, and context-specific information retrieval from a DNA archive is a form of computation. DNA strand-displacement networks are capable of executing much higher-level programs using quantities of material so small that they could be distributed throughout a structure or material: future uses of this extraordinary capability are difficult to predict.

References

[1] N. C. Seeman and H. F. Sleiman, *Nat. Rev. Mater.* **3**, 17068 (2018).
[2] J. H. Chen and N. C. Seeman, *Nature* **350**, 631 (1991).
[3] R. P. Goodman, I. A. T. Schaap, C. F. Tardin, C. M. Erben, *et al.*, *Science* **310**, 1661 (2005).
[4] P. W. K. Rothemund, *Nature* **440**, 297 (2006).
[5] S. M. Douglas, H. Dietz, T. Liedl, B. Hogberg, *et al.*, *Nature* **459**, 414 (2009).
[6] S. M. Douglas, A. H. Marblestone, S. Teerapittayanon, A. Vazquez, *et al.*, *Nucleic Acids Res.* **37**, 5001 (2009).
[7] E. Benson, A. Mohammed, J. Gardell, S. Masich, *et al.*, *Nature* **523**, 441 (2015).
[8] A. Chworos, I. Severcan, A. Y. Koyfman, P. Weinkam, *et al.*, *Science* **306**, 2068 (2004).
[9] C. Geary, G. Grossi, E. K. S. McRae, P. W. K. Rothemund, *et al.*, *Nat. Chem.* **13**, 549 (2021).
[10] P. Nissen, J. Hansen, N. Ban, P. B. Moore, *et al.*, *Science* **289**, 920 (2000).
[11] D. S. Wilson and J. W. Szostak, *Annu. Rev. Biochem.* **68**, 611 (1999).
[12] S. W. Santoro and G. F. Joyce, *Proc Natl. Acad. Sci. USA* **94**, 4262 (1997).
[13] M. N. Stojanovic and D. Stefanovic, *Nat. Biotechnol.* **21**, 1069 (2003).
[14] K. Lund, A. J. Manzo, N. Dabby, N. Michelotti, *et al.*, *Nature* **465**, 206 (2010).
[15] C. K. McLaughlin, G. D. Hamblin, and H. F. Sleiman, *Chem. Soc. Rev.* **40**, 5647 (2011).
[16] X. Y. Li and D. R. Liu, *Angew. Chem. Int. Ed.* **43**, 4848 (2004).
[17] B. Yurke, A. J. Turberfield, A. P. Mills, F. C. Simmel, *et al.*, *Nature* **406**, 605 (2000).
[18] D. Y. Zhang and G. Seelig, *Nat. Chem.* **3**, 103 (2011).
[19] R. P. Goodman, M. Heilemann, S. Doose, C. M. Erben, *et al.*, *Nat. Nanotechnol.* **3**, 93 (2008).
[20] J. S. Shin and N. A. Pierce, *J. Am. Chem. Soc.* **126**, 10834 (2004).

[21] W. Meng, R. A. Muscat, M. L. McKee, P. J. Milnes, *et al.*, *Nat. Chem.* **8**, 542 (2016).

[22] B. Yurke and A. P. Mills, *Genet. Program. Evol. M* **4**, 111 (2003).

[23] R. R. F. Machinek, T. E. Ouldridge, N. E. C. Haley, J. Bath, *et al.*, *Nat. Commun.* **5**, 5324 (2014).

[24] N. E. C. Haley, T. E. Ouldridge, I. M. Ruiz, A. Geraldini, *et al.*, *Nat. Commun.* **11**, 2562 (2020).

[25] A. J. Genot, D. Y. Zhang, J. Bath, and A. J. Turberfield, *J. Am. Chem. Soc.* **133**, 2177 (2011).

[26] J. Cabello-Garcia, W. Bae, G. B. V. Stan, and T. E. Ouldridge, *ACS Nano* **15**, 3272 (2021).

[27] A. J. Turberfield, J. C. Mitchell, B. Yurke, A. P. Mills, *et al.*, *Phys. Rev. Lett.* **90**, 118102 (2003).

[28] S. J. Green, J. Bath, and A. J. Turberfield, *Phys. Rev. Lett.* **101**, 238101 (2008).

[29] D. Soloveichik, G. Seelig, and E. Winfree, *Proc. Natl. Acad. Sci. USA* **107**, 5393 (2010).

[30] L. Qian and E. Winfree, *Science* **332**, 1196 (2011).

[31] K. Montagne, R. Plasson, Y. Sakai, T. Fujii, *et al.*, *Mol. Syst. Biol.* **7**, 466 (2011).

[32] E. Karzbrun, A. M. Tayar, V. Noireaux, and R. H. Bar-Ziv, *Science* **345**, 829 (2014).

[33] M. N. Win and C. D. Smolke, *Science* **322**, 456 (2008).

[34] Z. Z. Li, J. B. Wang, and I. Willner, *Adv Funct. Mater.* 2200799.

[35] L. Ceze, J. Nivala, and K. Strauss, *Nat. Rev. Genet.* **20**, 456 (2019).

[36] C. Bee, Y. J. Chen, M. Queen, D. Ward, *et al.*, *Nat. Commun.* **12**, 4764 (2021).

[37] M. Kube, F. Kohler, E. Feigl, B. Nagel-Yuksel, *et al.*, *Nat. Commun.* **11**, 6229 (2020).

[38] K. F. Wagenbauer, C. Sigl, and H. Dietz, *Nature* **552**, 78 (2017).

[39] F. Praetorius, B. Kick, K. L. Behler, M. N. Honemann, *et al.*, *Nature* **552**, 84 (2017).

[40] R. K. O'Reilly, A. J. Turberfield, and T. R. Wilks, *Acc. Chem. Res.* **50**, 2496 (2017).

[41] K. E. Drexler, *Proc. Natl. Acad. Sci. USA* **78**, 5275 (1981).

[42] E. Benson, R. C. Marzo, J. Bath, and A. J. Turberfield, *Sci Robot.* **7**, eabn5459 (2022).

https://doi.org/10.1142/9789811282324_0013

Non-equilibrium Structure and Dynamics: Giving Chemistry Direction

David A. Leigh

Department of Chemistry, University of Manchester, Manchester M13 9PL, UK

Chemistry out-of-equilibrium

Although physical laws are universal, the relative influence of different forces changes according to scale. Principles that focus on Newtonian mechanics (momentum, inertia, etc.) for controlling dynamics are not useful at the nanoscale, where statistical mechanisms dominate [1]. The idea of using random thermal fluctuations to drive directed motion has its origins in the visionary works of von Smoluchowski [2] and Feynman [3], but breakthroughs in the 1990s led to the invention of ratchet mechanisms [4] for rectifying the motion of Brownian particles. In the 2000s, it was demonstrated [5–7] that ratchet mechanisms could be used to control molecular structure and dynamics, imparting kinetic asymmetry [8] that drives chemical systems away from equilibrium. Increasingly, it is being recognized that these concepts can be applied widely in chemistry, not only to the conformational changes of molecular machinery but also to other stochastic chemical exchange processes.

From molecules to dynamic supramolecular systems

Although ratchet mechanisms were originally invented to rectify the random motion of Brownian particles, their principles can be applied to other kinds of stochastic exchange processes. In short, these are the mechanisms that give chemistry direction, from molecular machines to reaction cycles to dissipative materials. Molecular ratchets are effectively chemical engines [9] in which the catalysis of "fuel" to "waste" is used to drive another chemical process, causing directional impetus in what otherwise would be chemical systems at equilibrium. For example, all biological pumps and motors are autonomous catalysts; they maintain the out-of-equilibrium conditions of the cell by harnessing the energy released from their catalytic decomposition of a chemical fuel [10] (typically ATP). Recently, an artificial autonomous chemically fuelled information ratchet was described that in the presence of a chemical fuel (**2**) continuously pumps molecular rings (**3**) from bulk solution onto a molecular axle (**1**; Fig. 1) [11].

There is nothing inherently special about the chemistry of mechanically interlocked molecules. They are bound by the same chemical principles regarding conformational changes and transformations as any other molecular structure. It follows that Brownian ratchet mechanisms are also the means by which other molecular systems are driven away from equilibrium. We recently demonstrated [12] that a simple 26-atom biaryl compound (1-phenylpyrrole 2,2′-dicarboxylic acid) is a catalysis-driven motor [9] that continuously transduces energy from a chemical fuel [10] to induce repetitive 360° directional rotation of the two aromatic rings around the covalent single bond that connects them (Fig. 2). Upon treatment of the motor molecule with a carbodiimide fuel, intramolecular anhydride formation between the rings, and the anhydride's hydrolysis, both occur incessantly. Both the anhydride formation and hydrolysis reactions are directionally biased (they occur faster through particular rotamers of the biaryl system). In this way, catalysis of the hydration of the carbodiimide fuel by the motor continually drives net directional rotation of the rotor around the stator. The motor's remarkable simplicity may

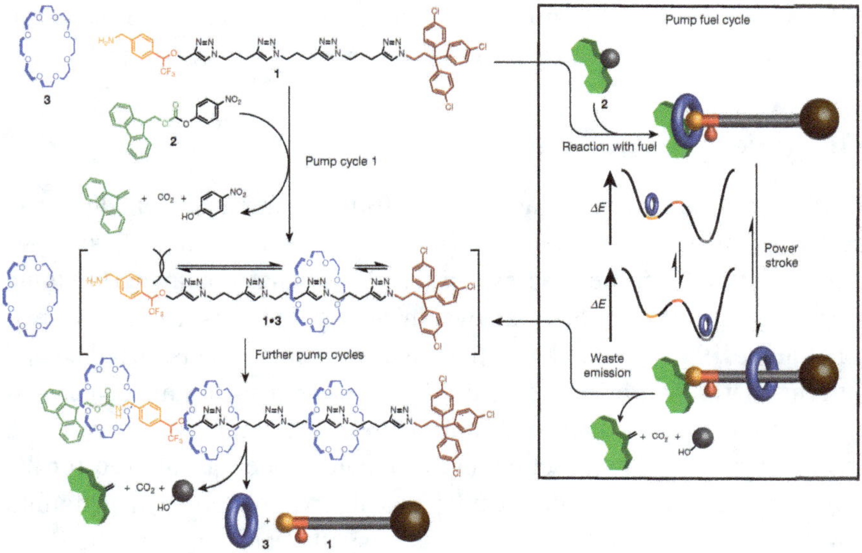

Fig. 1. An artificial catalysis-driven molecular pump [11]. Continuous pumping of 24-crown-8 rings (**3**) onto a molecular axle (**1**) in the presence of fuel (**2**) causes the rings to be dissipatively captured in the form of high energy [*n*]rotaxanes. Once all the fuel has been consumed, the rings dethread and the system returns to equilibrium.

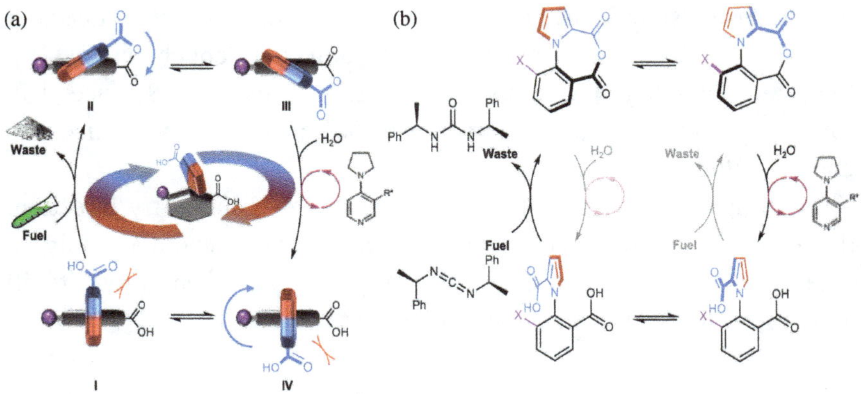

Fig. 2. An artificial catalysis-driven molecular rotary motor [12]. Chemically powered rotation of the rotor (pyrrole-2-carbonyl) carbonyl (blue) past the 6-position (purple) of the stator (phenyl-2-carbonyl) is favoured in the acid form (IV⇌I), whereas passing of the rotor carbonyl past the stator carbonyl is only possible in the tethered anhydride form (II⇌III). The chiral (*R*,*R*) fuel kinetically favours anhydride formation I→II over IV→III, whereas the chiral hydrolysis catalyst is chosen to kinetically favour III→IV over II→I. Once all the fuel has been consumed, the aryl groups can no longer rotate past each other.

facilitate its interfacing with other components for the performance of work and tasks at the molecular level.

Outlook

Ratchet mechanisms are rapidly transforming the understanding and design of molecular systems — biological, chemical, and physical — moving away from the macroscopic analogies that dominated thinking regarding aspects of molecular dynamics (e.g., power strokes, and pistons) in the 1990s and 2000s to the scale-relevant concepts, insights, and designs that underpin out-of-equilibrium research in the molecular sciences today. It has established molecular nanotechnology as a research frontier for energy transduction and metabolism and has allowed for the reverse engineering of biomolecular machinery, delivering insights into how molecules "walk" and track-based synthesizers operate, how the acceleration of chemical reactions enables energy to be transduced by catalysts (such as motor proteins, and in the future by synthetic catalysts), and how dynamic (supra)molecular systems are driven away from equilibrium through catalysis. The recognition of ratchet mechanisms in biological systems, and their invention in synthetic systems, is proving of ever-increasing significance in numerous areas, including supramolecular chemistry [13], systems chemistry [14], dynamic covalent chemistry [15], DNA nanotechnology [16], polymer science [17], materials science [18], molecular biology [19], and numerous other areas of nanoscience and nanotechnology.

Put simply, Brownian ratchet mechanisms give chemistry direction. Kinetic asymmetry, the underlying principle behind ratchet mechanisms, is the dynamic counterpart of structural asymmetry (i.e., chirality). It will surely prove just as fundamentally important.

Acknowledgements

DAL is a Royal Society Research Professor.

References

[1] E. R. Kay, D. A. Leigh, and F. Zerbetto, *Angew. Chem. Int. Ed.* **46**, 72 (2007).
[2] M. von Smoluchowski, *Phys. Z* **13**, 1069 (1912).
[3] R. P. Feynman, R. B. Leighton, and M. L. Sands, *The Feynman Lectures on Physics*, Vol. 1, Addison-Wesley, Reading, MA, Chapter 46 (1963).
[4] P. Reimann, *Phys. Rep.* **361**, 57 (2002).
[5] J. V. Hernández, E. R. Kay, and D. A. Leigh, *Science* **306**, 1532 (2004).
[6] M. N. Chatterjee, E. R. Kay, and D. A. Leigh, *J. Am. Chem. Soc.* **128**, 4058 (2006).
[7] V. Serreli, C.-F. Lee, E. R. Kay, and D. A. Leigh, *Nature* **445**, 523 (2007).
[8] R. D. Astumian, *Nat. Commun.* **10**, 3837 (2019).
[9] S. Amano, S. Borsley, D. A. Leigh, and Z. Sun, *Nat. Nanotechnol.* **16**, 1057 (2021).
[10] S. Borsley, D. A. Leigh, and B. M. W. Roberts, *Nat. Chem.* **14**, 728 (2022).
[11] S. Amano, S. D. P. Fielden, and D. A. Leigh, *Nature* **594**, 529 (2021).
[12] S. Borsley, E. Kreidt, D. A. Leigh, and B. M. W. Roberts, *Nature* **604**, 80 (2022).
[13] K. Das, L. Gabrielli, and L. J. Prins, *Angew. Chem. Int. Ed.* **60**, 20120 (2021).
[14] J. H. van Esch, R. Klajn, and S. Otto, *Chem. Soc. Rev.* **46**, 5474 (2017).
[15] L. S. Kariyawasam, M. M. Hossain, and C. S. Hartley, *Angew. Chem. Int. Ed.* **60**, 12648 (2021).
[16] S. J. Green, J. Bath, and A. J. Turberfield, *Phys. Rev. Lett.* **101**, 238101 (2008).
[17] A. Sorrenti, J. Leira-Iglesias, A. J. Markvoort, T. F. A. de Greef, and T. M. Hermans, *Chem. Soc. Rev.* **46**, 5476 (2017).
[18] B. Rieß, R. Grötsch, and J. Boekhoven, *Chem.* **6**, 552 (2020).
[19] B. Lau, O. Kedem, J. Schwabacher, D. Kwasnieski, and E. A. Weiss, *Mater. Horiz.* **4**, 310 (2017).

https://doi.org/10.1142/9789811282324_0014

Discussions of Session 2 — From Molecules to Dynamic Supramolecular Systems

Chair: Ben Feringa
Auditors: M. Surin*, H. Valkenier†

*Laboratory for Chemistry of Novel Materials, Université de Mons — UMONS, Belgium
†Engineering of Molecular NanoSystems, Université libre de Bruxelles, Belgium

Ben Feringa:
Welcome back everyone, my name is Ben Feringa as you know now, University of Groningen. So, we learned this afternoon a lot about dynamical systems and complex systems in a more general context. I would like to start with that. We heard that this is a problem, a challenge for us for the coming decades. At least, we think it is a challenge to go to more complex dynamic systems, that we have all these parameters that will influence how such a system will operate and going all the way from structure design to dynamics and to organization, length scales, time scales. Bert Meijer said 'non-covalent synthesis'; we have to discover a whole new field. Do we have the tools available to do this, the methodology, the way we are going to handle that? That is my question to the panel

and maybe they want to react on it. Because we are all trained a bit in traditional chemistry with reactions and the kind of things we do currently. Do we also need to develop new ways on how we think about it, how we measure things, how we do things, what are we missing? And so, to me sometimes (except maybe for the DNA that Andrew mentioned where the designing rules are probably more clear because they are already established for us by mother Nature), we explore them. But how to go from "trial and error" to a bit of design in this field? I would like to ask that and maybe then everybody can comment on it. We start the discussion this way. Maybe I can start on the far left with Bert Meijer?

Bert Meijer:

Thank you very much. For design, I think it's very important what you want to design at the end and not just the molecular structure. And it depends also strongly on the application you have in mind. If you want to use materials that are used in your body or in tissue generation, it's a different thing as if you want to go into energy or another material for mechanical properties. So, it's important to realize what the design should be. The point I wanted to make today in the talk is that, if you go to multicomponent systems and bringing them together, and especially as in some of the aggregation phenomena there's a cooperativity involved, then there are tipping points in which suddenly things change completely. And before we understand this correctly: it is very difficult to design. Now, from the 20–30 years I am active in this field, I have come to the conclusion that: to find out the systems that work always in a good way and to understand how it goes, is difficult. There are many examples, also from our laboratory, that work well but that is a unique part. You can do it one or two times if you do it exactly according to that way. But then, there is another student coming and wants to reproduce it, and it doesn't go. And we have no clue where it comes from. Some of the assembly studies we do, are done perfectly, and then we buy a new bottle of a solvent, and it doesn't go anymore. And it is similar to catalysis, and it is similar to crystallization. But it is not, in the field of supramolecular chemistry, so much acknowledged, that that is a point as well. I think it is very important, before you come to design, to say what are actually the structures that you can use on a regular basis, where it always works. That's also the reason, probably, that DNA origami works nicely, because we know now how that assembly process goes, at least that's what I think. In the artificial world, it is more difficult to find which ones are working very well. I can tell you that with supramolecular polymers, we have made a very simple one, with a quadruple hydrogen bonding unit, very simple to make, works always and I think, I don't know 20 – 50 laboratories around the world are making and using that molecule as well. So that's a robust system that always forms a bond with a certain strength and a certain dynamics. I think we need more of those and then you have to also have them available by Aldrich or whatever other compagnies, so you can make that progress. I am always a little bit jealous of those who are doing catalysis. Someone finds a new catalyst that works and about, I would say half a year later, you can buy it somewhere. And then people are using it, it is used by everyone in the world. In our field it is too much individual. Molecules are used by certain laboratories, you can't get them easily other than that you have to synthetize it completely. So, your motor should be available by, I don't know, *Aldrich* or by the Ben Feringa company or whatever company so that everyone can use it.

Ben Feringa:

I am happy to hear this because, not about the motors, but the fact that you will have the ligands for instance, so you can buy 150 different ligands from Strem or from Aldrich or whatever. That you would have a collection of building blocks that people in supramolecular chemistry and systems chemistry would use to explore more. So that there are more rules sets, based on these building blocks like I think in catalysis or in synthesis etc.

Omar Yaghi:

I'm sort of put in this group, even though I don't consider myself a supramolecular chemist. I have spent all my life linking molecules by covalent bonds, a strong bond, and there is no major reproducibility problem. Some of these things are made by *BASF* in multi-ton quantities, and also you can buy them from *Aldrich*. But what I really want to talk about is more a productive discussion about what we heard from our colleagues in supramolecular chemistry, where they are designing systems that self-propel or are highly dynamic. What is needed is directionality and purpose. And what I was saying to my colleagues like Bert is that, to do that, you need reticular chemistry. And I don't mean that they should turn and come back tomorrow and occupy my session. But what I mean is that I think you can get the best of both worlds by coupling the supramolecular onto that reticular grid. Imagine a system that they are working on or a molecule that is dynamic or whatsoever. By attaching that to a grid that, let's say, I make, you're attaching let's call it "heterogeneity onto order". And so now you have a complex order that you can potentially design to have molecules moving in a circuitry or in a direction in a purposeful way. You now can imagine a catalyst or a substrate, moving in a circuitry and visiting different catalysts along the way. Well, you need a grid for that. This thing needs to be moving on an otherwise ordered system, but itself could be highly disordered. I think that's where a lot of the systems that we heard could work a lot better. Otherwise, jokingly I said to Bert, otherwise you're making goo, and in noncovalent synthesis of goo and this stuff, using multi-step non covalency will produce even more goo. So, I think that, by superimposing this on a grid, these two areas are ideally suited to be coupled to produce beautiful systems.

Ben Feringa:

You are making life not easier for us because I heard from Bert, balance robustness versus responsive behavior and from you I hear now, balance order and disorder, but Nathalie probably has the answer.

Nathalie Katsonis:

One thing I can say about purpose and directionality in molecular systems is that if we look at concepts from the biological world, that's typically something that chirality brings, so that's probably something to look into: 'chirality at all scales'. And probably another thing is ratcheting mechanisms, that give directionality and purpose to systems of functioning molecules, in addition, of course, to other options. Now, to come back to Ben's question. So typically, movement is an emerging function that emerges in a very narrow parameter space, a narrow range of parameters. And in that sense, it's very difficult to design it. It's not actually a property that comes by design, but essentially a systemic property. And that's I think how the concept of systems chemistry emerged and probably systems chemistry requires different tools than classical chemistry. One of the tools is automatization, so that we can vary a broad range of parameters in parallel. And maybe also something that we need to consider is how to connect the molecular world to the macroscopic world. So we have techniques that characterize materials and techniques that characterize molecular structures but to characterize both worlds in a dynamic fashion is, I think, quite challenging. There are, I think, very exciting developments, not from me, but in the microscopy world at the moment, which, I think, will allow us to really move forward. And one last thing, I think also it is not a technique that we need but maybe a different mindset. It is that traditionally we look at the chemical reactivity disjoined from the physical environment and if something we can learn from chemical evolution is that, actually, chemical reactivity cannot be disjoined from the physical environment. It's something that we need to learn, really, to look at the chemical reactions and chemical reactivity in heterogeneous systems. It's a good idea to start doing chemistry, not necessarily in solution, not in homogeneous systems, but we need to start thinking about heterogeneous conditions and mixtures.

Ben Feringa:
Thanks very much. That wakes me up again to a point that was made this morning about the interface of heterogeneous and homogeneous catalysis, where for many years we were facing the same kind of issues and questions, about heterogeneity versus homogeneity. Anyway, I go to my neighbour here, Nicolas. Do you want to make a comment on this design versus trial and errors and the faith of our field?

Nicolas Giuseppone:
Thank you, Ben. So first of all, I think that regarding the synthetic tools that we have at the moment in supramolecular chemistry: they are highly diverse, so we can do many things using non covalent bonds from ion pairing, from hydrogen bonds, from halogen bonds, or from Vanderwaals interactions. In addition to that, we have the toolbox using reversible covalent bonds which gave birth to dynamic covalent chemistry and dynamic combinatorial chemistry. So, we really have the power to create a lot of diversity under thermodynamic equilibrium and we are able to synthesize dynamic systems which are robust enough to be analysed for a given period of time, at least in certain conditions. So, for these we have the

toolbox for making dynamic supramolecular systems. We have also many analytic tools from the molecular scale up to the macroscopic one, as Nathalie said, I think we have a lot to learn from physicists, from soft matter, to analyse these systems. Particularly using scattering techniques or rheology aspects, etc. Then the directionality is important. I think we are only at the beginning of the control of the directionality at nanoscale with all these works on ratcheting mechanisms, which are quite complex and still difficult to harness. You have to create order from Brownian motion. This is not something that easy at that scale. But then we are going to enter in complex systems, so, for having complex systems we need multiplicity of interactions, we need multiplicity of compounds, and we need integration of these interactions. And there we come to very complex systems, mixtures, exchanging components, which become more difficult to analyse. So there are two things, maybe, we should concentrate also on function: which function can emerge from a complex system, so we can directly measure the function? The second point, I think, is that we have to be helped by theoreticians: how to design our complex system? It is not only a question of mixing components to extract an important function. We need to know what theoretically could emerge from feedback loops, from a network of reactions and there's a lot to learn from theoreticians on that aspect. And finally, if we want to push them out of equilibrium, you need to inject energy and we need to find relays to transform this energy and fuel the system in an interesting way that can lead to nonlinear behaviours.

Ben Feringa:
Thank you very much. For these functions that have to emerge, focus on there. That's a very important message from your side. And 'cooperation', that's also what I think of.

Andrew Turberfield:
So first a disclaimer: I'm a low temperature solid-state physicist by background, so I'm not hampered by too much traditional chemical knowledge. In my field, designing very big and very complex stable structures is almost trivial. What we don't do very well at all is design the pathways by which they assemble, so designing for fully efficient assembly is

something we're just beginning to explore. That's a big challenge. So when we, the community, show you in a cryo-electron micrograph, gorgeous, huge — I mean hundred nanometer dimension — assembled structures, they have been selected. Those are the successes, the failures are stuck at the top of a gel in the well, and we cut them out and throw them away. So efficient assembly is a big challenge. Static structures are relatively straightforward, notwithstanding, dynamic systems are much harder. Because typically when we design complex interacting systems with many components, if we tweak one component we disrupt everything else. So we are developing a toolkit of methods, introducing base pairing defects, doing not entirely obvious things. But it is very hard to design rationally for dynamics and in fact the best tools we have, which are becoming extremely good, are computer simulations. So coarse-grained simulation is the way to go and really the only limitation on that is computing power. We can practically simulate assembled huge systems, but the dynamic behaviour of only very small ones. And, as you know, inevitably, these programs get more efficient and computing power grows, we will become more and more capable of designing our systems through the medium of simulation. So, just on directionality. You've got directionality in time, directionality in space. For directionality in time, you've got to dissipate energy. In our systems, generally that's done by catalysis and control of hybridization reactions. So, in fact, you're using the creation of base pairs as your energy source. That's very programmable, very controllable, very flexible. It's a wonderful system. It's a bit frustrating that we can't interface our system readily to small molecule reactions. We'd like them to go much, much faster, and to use small molecule fuels that you can have present in millimolar quantities rather than the micromolar that we typically use for oligos, so that's a frustration. Directionality in space, that's something actually self-assembled systems made from DNA and RNA come with for free, so we can make a planar, reticular if you like, surface on which to build our systems. We can address every pixel on that system individually because they're all identified by different DNA sequences. So that's one of the aspects of the structures that we make, that's actually very rewarding. Every point on a 100 nanometer self-assembled DNA structure is individually addressable and we can make something walk around it.

Ben Feringa:

Great, as you focused on the dynamics, I have an additional question for you. From a physics point of view, do we have enough tools to study in detail these dynamics? Nowadays, we have time resolved AFM, for instance, which helps us a lot, but there are other tools. Would you ever dream, that you say that or that would help us tremendously in this field?

Andrew Turberfield:

We have done time-resolved AFM. On the right system, it works wonderfully well. AFM works beautifully on very planar systems, but as soon as you have got anything lumpy or three dimensional, then it works very badly. So really, what we would like is the equivalent of time resolved AFM that would work. There's electron microscopy but, obviously, every time you use that you're looking at a frozen, dead system. So yes, something, some tool for three-dimensional imaging, that will give you dynamics. We look at snapshots and that's all we do. And then, dynamics we get with very high time resolution by strapping fluorophores to our structures and using time-resolved FRET. But that gives you very limited information: it will tell you whether two particular points on your structure are approaching each other or are far apart, and it can give you that with very high time resolution. But really detailed three-dimensional structural information is very hard.

Ben Feringa:

Thank you very much. Okay, I want to go to the general discussion, but first I want to ask one more question. But as Kurt Wüthrich is our president, he is allowed to ask his question.

Kurt Wüthrich:

Are you suggesting that kinetic energy of stochastic Brownian motions is funnelled into directed kinetic energy, which is possibly even periodic? This is a fantastic idea. Are we going to solve the energy problem that way?

Nicolas Giuseppone:

So, what I claimed is that the kinetic asymmetry in these ratcheting systems can create directionality, but you need to furnish energy to the system to make it directional, you cannot have it for free. So because you go against entropy doing that, you cannot solve the energetic problem. You will always need to furnish more energy to your system than what you will recover as an output of your system. There is dissipation, things like that, that enters into play. So I don't say that we are creating energy, we are creating directionality from energy. And then, this directionality can be transduced to other functions in a cascade of processes where you lose energy all the way.

Kurt Wüthrich:
Yes, but the energy is available. At 300 K we have high thermal energy that leads to the stochastic Brownian motions.

Nicolas Giuseppone:
You cannot extract directionality from one base of energy. You need the thermal energy plus another source of energy to bias the random thermal motion in one particular direction. Maybe Ben will correct me.

Ben Feringa:
Of course, in the systems that we studied, that we showed here and also you showed and Natalie and so, of course you put in either chemical energy or you put in photochemical energy. Energy from the light or from a chemical conversion. To make that possible, you have this extra energy to make it propulsing in a certain direction.

Andrew Turberfield:
Well as a physicist I'm forbidden from violating the second law of thermodynamics. The only comment is that obviously, yes, we have external energy input. By playing with low energy chemical inputs, I think you have an opportunity to be very efficient. Blue photons are squandering energy, but low energy noncovalent reactions allow you to rectify Brownian motion in a very efficient manner. So that's some sort of response to the energy crisis.

Kurt Wüthrich:
So, would you only need to lower the temperature of the solvent in order to feed a lot of energy to your macromolecular system? There, I think, would be no problem with thermodynamics.

Andrew Turberfield:
Only if your molecular system has one leg in your cold bath and one leg in the hot one.

Nicolas Giuseppone:
That's right. So if you can have a gradient of temperature at the scale of a molecular motor, yes, but it's difficult!

Ben Feringa:
I would like, before we go to the general discussion, ask one more question to each panel member here. That is a question I often get when I discuss, either with chemists or also with more general scientists and the public: Why do we need molecular systems?

Bert Meijer:
All around us is molecular systems, everything that we use is a molecular system. Many molecules at the same time, if we go to the shower, if we eat, everything is a combination of molecules. And so far, the understanding of making these molecules together into their function and their property is based on trial and error. And I think that's the reason why we have to do this, especially that's the part of the dynamics I'm interested in: to see how you can tune the dynamics, from robust and stable for as long as you want to have it stable, and having it dynamic in order to remove it when you want to remove it. And I think there's an enormous need but it's a material-oriented need, and therefore, it's everywhere. Everything that you use and buy is a molecular system, a combination of molecules and if you see how the messenger RNA vaccine is made by surfactants and messenger RNA, it's a beautiful way of doing, but it's also a bit of lucky that it works so well and we have to realize that only 8% of the messenger RNA in the cell is active and it can go to 100% if we know not only how to bring it there, but also how to get it released just at the right moment. And that means that molecular system that goes into the cell is responsive to the cell to deliver all the messenger RNA at the right moment. It's a very hot topic at the moment, but it's just an example. It's everywhere to get that control.

Omar Yaghi:
I wasn't sure if you're asking about molecular or extended systems?

Ben Feringa:
No, I'm more asking about the system. And on the one side we have the most complex system, which is probably the cell, with all these functions which are more complex than the whole city of Brussels, I think. On the other side, we have simply hair shampoo that you use for washing your hair in the morning. Still, that's a complex system and you look at all the components that work together there to do something that you get this nice hair, beautiful hair, etc. I mean it is amazing and when you talk with industry, people say to me, oh, yeah, we mix things together and we look at the properties and that's about it. No, that's not entirely true, but it's less complex than the cell, let's admit that. So, this is why I asked when you look at systems, there are different graduations in what we call the system, but it's certainly not one single molecule. They are molecules that do things together.

Omar Yaghi:
I mean, I'm always surprised by the popularity of molecular chemistry. Because for the last 100 years, we've been working on molecules, but yet the world around us, including your own body, requires extended structures like bone structures, right? So, you can't build an airplane from just discrete molecules, from soft matter. I think there's a lot of examples around us that require us to think more about building structures that, well, first protect living systems, but also are used in many applications, not the least of which are transportation, computing, and things like that. I'm not saying that molecules are not important. I'm just saying that the other side of chemistry has been terribly neglected. And we are focusing a lot on molecules and soft systems, when in fact, as I mentioned earlier, if we think about extended systems, be it heterogeneous catalysts or alloys, they are absolutely necessary for the molecular world to operate in a way that would be productive.

Ben Feringa:
I'm happy to hear that. In my opinion it doesn't exclude each other but Bert wants to comment on this.

Bert Meijer:
If you would bring all the molecules out of this room, what do you think will be left?

Omar Yaghi:
Wait, I want to be very clear. Everything is made of molecules, but they're not discrete molecules. Some of them are interacting through weak interactions some of them are interacting through covalent interactions, right? And the latter ones have been used to make extended systems, have been longer neglected, that's what I'm saying.

Ben Feringa:
Wait, guys, you'll get your chance in a moment, you see this is already very interesting and debatable. Nathalie, can you comment on this from your perspective? You work on systems. How would you convince the people that they should work on this?

Nathalie Katsonis:
I don't know how to convince them, but I think we need complex answers to complex questions. So of course, we need to look into complex systems and I think the key here is complexity. Let's say one concept, I think the idea is that instead of taking molecules separately, we want to make them work together in space and time so they can become more than the sum of their parts. And when we find a way to make molecules more than the sum of their parts, I think then we can do a lot of interesting things. But that line of thought didn't bring me that much money so far, so I don't recommend it particularly.

Ben Feringa:
Okay, to getting grants, that's another problem, we can discuss later. But okay, maybe Nicolas, you also want to say something?

Nicolas Giuseppone:
So why I believe it's interesting to go to systems is because we want chemical systems to be responsive, capable of sensing their environment, adapt and regulate. And I think for the regulation, you need to have many

interactions in well-organized networks of reactions or networks of interactions. This is what you see in living system, for allostery, what you see with negative or positive feedback loops, this is what regulates the cells for adapting to their environment and finally to evolve. Obviously in living systems you have also to add self-replication processes, etc. Hence if we are able to do that in artificial systems, it will be important for the stability of the network as a whole to be in a complex system. And I think that if we are able to build that from scratch with simpler molecules, we will understand aspects of the origin of life and also we will use these tools to make smarter and more active materials and in that sense, we will create new technologies.

Ben Feringa:
Thank you very much. Yes, Andrew.

Andrew Turberfield:
I'm a physicist, and my colleagues often want me to study simple, boring, periodic systems. I come from the opposite extreme, I ask myself, what could we do? Given that we can make things that move, can walk and move along the track directionally, can pick things up, can make them go, can you construct something by making a molecular machine tool, an assembler? Can you program it? Can you build something that you couldn't build otherwise? Can you make an autonomous agent that would actually be useful medically? Something that would sense within a particular cell, decide whether to do something or not, and if it found a condition, activate something that was therapeutic in nature. All of these things are challenges that you could only conceivably solve, address, not with one molecule but with a complex interacting system. So, complex dynamic systems are, obviously to me, where the interest lies. What are the limitations? I'm sure there are no limitations. It's just a question of time before we gain the capabilities to do all of these things. What can we do, how do we get there?

Ben Feringa:
Thank you very much. Nobody asked me but I fully agree with what was mentioned here. And I want to add where my excitement came, maybe

from a fundamental point of view and that is: we make all these materials, we know all these beautiful molecules and we enjoy that. And then suddenly I realized: how to make things move? Because that is what life does, as Natalie mentioned already, how to repair itself, how to adapt itself, how to respond to input from outside and sense and adapt, etc? And when we made the first molecule that showed some controlled movement by energy input, and we made the first piece of plastic that we cut and then it repaired within 10 minutes, I got so excited that I thought there must be something in chemistry for the future there, you know. So, with that, I would like to open the general discussion.

Peter Palese:
A question for Doctor Yaghi. Is there any commercial product available already, which uses your molecular weaving technology at this point?

Omar Yaghi:
Not yet, but there is a couple of startups that are commercializing that technology. It's only five years old, since our discovery.

Peter Palese:
Can you give us a feeling? What can you say? What kind of direct applications do you see coming first?

Omar Yaghi:
I think for the molecular weaving: it doesn't interfere with the way you make the material, the original material, let's say the advanced polymer or the structural material. It is just an additive and you're adding it at 1%, 2% or 3% maximum by weight. So, it basically does not interfere in the actual process. But, at the nano level, it is distributed throughout the material, and therefore it could then operate in terms of managing the stress on a material in the way I described. So structural materials, anti bulletproof vests, this kind of thing, anything that you can make: where usually with these materials, when you increase strength, you also increase the fracture ability of the material. But, with the weaving, you're able to combine strength with less fracture. And so, it just has,

I think, tremendous implication on materials. Needless to say: also tissues, bone.

Ben Feringa:

Okay, thank you. I think it's an important point.

Chad Mirkin:

So very interesting discussion. I was kind of surprised. To me, this field needs a couple of things. One is a good set of drivers. So, most of what I've heard, is of course, a part of any field, you know: what is possible? Can I make a motor? Is it possible to make a molecular machine, which is an interesting question? But then what do I do with it that ultimately people care about? How do I do it? And how do I use the tools that we currently have? And one of the things that is of course very limiting in this area is soft matter in interface with the tools that were used to process hard matter. So, the semiconductor industry is probably that greatest example of one of the greatest engineering feats in all of time. And it is remarkable but it is not set up to work with soft matter. Well, in fact, most people will run if you'd move towards their fabs with soft matter. So how do you scale it down even further, right? Because you have to, if you really want to make use of many of these types of things. How do you address all of the individual structures? It's very different from seeing things walk on a sheet of DNA, having that sheet of DNA interfaced in a particular way, that I know right over here, one molecular process is occurring and over here, a different one is occurring, and I want those two at a specific position. We don't have tools that allow us to do that. So, it sure seems to me that if you're going to move this field forward, you're going to need to think about: what can I make that would show people this has a lot of legs? Because there's been a lot of kind of gee whiz type of things that have occurred, which again, that's not criticism, is a part of any field. But to take the next step, you really have to think, to do what I call 'a Bob Letsinger experiment'. I used to work with my 84-year-old collaborator Bob Letsinger. He said: "Chad, let's assume everything works the way we think it's going to work. You know, was it worth doing? Can we actually get to something that the world is going to care about?" I think it's worth doing that experiment now.

Andrew Turberfield:

I haven't mentioned it, because it's not dynamic as an application, but one of the things that is driving what we are trying to do at the moment is the idea of templating molecular electronics, so it is precisely the semiconductor industry beyond Moore's law. And one of the things that has held back the whole field of molecular electronics is that there's no good way of assembling molecular electronic components. And actually, the DNA self-assembly technology is just begging to be used for this. It's a way of positioning things in three dimensions in a programmed way. It is a breadboard, you lay things out in a three-dimensional fashion, and in fact, having got them in the right place, you then wire them up by forming covalent bonds between them, so templated chemistry. So that is a vision. It falls short of your requirement for a gee whiz driver, in an interesting way: if we could do it, it would replace the entire semiconductor industry. And if you look at the magnitude of that statement, you see the problem which is that the activation barrier to getting there, it's just far too high. And I would love it if someone told me what intermediate product we could aim to make, that would demonstrate that technology, without requiring us to replace, you know, a billion-dollar semiconductor fab.

Chad Mirkin:

It seems like you have to do the analysis he did, which was he made a similar issue with respect to CO_2 reduction or the Haber process, coming up with an electrocatalytic version of this. Pick something like the pharmaceutical industry where, again, you don't have such a big barrier. I agree with you, if you try to go head-to-head with the semiconductor industry, you're going to be knocking on a lot of doors with very few opening, right?

Andrew Turberfield:

Yes, I mean, that's the problem I'm posing. I'm just not sure what that industry is.

Bert Meijer:

Obviously, I'm not in that area of moving small objects. But I asked myself the question: if you compare non-covalent bonds and covalent bonds, when is the covalent bond good and when is the non-covalent bond good? In many applications of materials, we have too many covalent bonds that are so robust that they will stay there forever. And, with the knowledge that we now have about supramolecular chemistry, we can make molecules that are smaller, have more specific interactions that you can tune, in which the balance between covalent bonds and non-covalent is changing to more non-covalent, in which the dynamicity of their temporary use is much higher. So, for instance, our materials are used in tissue engineering, in human beings, because we can tune the mechanical properties and the degradation independently of each other and, therefore, the body can take care of removing that material just at the moment that its role is taken over by the cells. I think that is also the same for what I had mentioned just before this. If you want to have debonding on demand, which is bond continuously, but if you want to get rid of that strong bonding, you have to tune the number of covalent bonds and non-covalent bonds. And with that, obviously, dynamic covalent bonding is also an option. The whole idea is that you build up three-dimensional material with better control over covalent and non-covalent bonds, this gets you that dynamic. And the same is, as was just mentioned, in the assembly process. My group is working for almost 35 years already on organic semiconductors. And one way of doing this, is to have full control over how the π system is organized in space and for that you need dynamics in order to get the structure fully formed, and then it has to be there as a material. That is a self-assembly organizing process that will be used and if you go to smaller, you have to do it with structures that are better defined than the ones that are used today. As an example, if you have a hole of let's say 20 nanometers and you want by directed self-assembly to make that a hole of one nanometer, you need an exactly discrete molecule to fill up that hole, because if that is diverse in length, every hole has a different composition. And assembly in confined space is a very complicated thing, you get all kinds of different structures. So, you need to have at the right moment, the right chemical structure for a specific application and I fully agree with you, you need a specific application in order to tune the molecule and then you have to take care that the covalent bond/non-covalent bond ratio is just that what you need for that specific application. And that is new, because polymer chemists actually never did it other than nylons, in a way, and organic chemists were not very interested in materials. So that field where organic chemistry and materials chemists are coming together, is a field that has to do this trick. But it has nothing to do with moving objects.

Yamuna Krishnan:

I have a comment and a question to the panel. My comment is: I think it's very easy to say: what have your jivas done good to you? There have been a lot of ways of looking at DNA nanotechnology as like molecular gymnastics, but I just want to say that it has given us strand displacement, which was the basis of hybridization chain reaction, which very recently -we've all just come through a pandemic- was used for making many COVID tests. So, let's not forget that. There's also DNA sequencing, which is also DNA nanotechnology, and that was used to pinpoint variants of concern that were popping up all across the world and saved many of our lives by informing travel policies. So, before we ask DNA nanotechnology, what is your jivas trick? I think there have been a couple already. And so my comment to the panel is, actually, thank you so much for putting up those five questions. Because those five questions have remained for the past 20 years, when I started out in this field and moved on: how do we control, achieve extrinsic control of nucleation, aggregation, growth and movement in 1D, 2D, 3D and time? To some extent, we kind of try to maybe convince ourselves we have some control, but that's intrinsic control by controlling the shape of the molecule or the composition of the molecule. If we look at how Nature is already achieving nuclear controlled nucleation, aggregation, growth, and all these things in an extrinsically imposed way, we don't fully understand this yet, even by looking at biology. It's usually through multicomponent systems, and many of you have alluded to that, but I just wanted to understand how you see that interfacing happening between systems which give you some level of control but there is not much processivity and robustness, combined with molecules that give you very nice reversibility but we don't have the level of control in 2D and 3D. I just want to see what your vision is.

Bert Meijer:

Thank you. A short answer actually with a back question. That is, there are many people here in catalysis: So why is there only one example of the Soai reaction? That's a complex molecular system, with EEs close to 0.001 for the catalyst and it becomes 100%. Why is it not general? Something apparently we don't understand of a complex molecular system, or maybe someone knows why there are not more?

David MacMillan:

I think, maybe I'm answering the obvious here, but it's because obviously the product molecular structure influences the subsequent catalyst and so the chances of that happening, the probabilities of that happening are remote and that's why it happens at the level that it happens. It has happened more than once, there's more than one example of it.

Ben Feringa:

I am happy that you bring this up, this multicomponent aspect, because in my experience, but the others can comment on it, if you bring several components together and you have to do this kind of hierarchical organization and then you have also to do the dynamics you know, in adaptive etc., it is still extremely difficult to predict. Look at all our examples that were shown today. Still, a lot of it is based on well-defined components and I challenge everybody to tell me the design principles when you have more than three components for instance, or even more than two, you know, it's tough.

Henry Snaith:

It's fascinating, I'm an outsider to this field. It's fascinating seeing the complexity of these systems and what you're trying to control. And my question, related to some of the previous questions asked, but specifically with respect to the molecular motions or these active moving molecular machines: Is there an understanding of what you have to achieve? Sort of like a roadmap, let's say, towards being able to do something useful with them or having a key application? And are there applications that we think this would be useful for and has anyone got any idea on the timeline to get there? Maybe someone will say "yes, they are already present".

Ben Feringa:

Thank you very much. This is a very important question that was asked to me many times and I'm sure Fraser Stoddart can give a much better answer to it, because he has very clear ideas there. But there is, I think, the roadmap, there are so many aspects to it. You know, it was mentioned already, directionality, translating for instance rotation into a translation. I showed examples of catalytic propulsion, using a chemical conversion but then using it to transport something, for instance, in and out of a cell, like biology does, or from point A to point B or using it for assembly in a direction manner. This is still a tough, tough one. Omar was mentioning the beautiful extended molecules he makes, which are these framework reticular materials. This could be a fantastic template for showing how you could transport things at nano or sub-microscales and I think this offers tremendous opportunity. So I'm really happy that we are here together with people from different angles coming together to see how we could do this. Because I cannot do this on my own. I need this kind of framework materials or surfaces, modified surfaces, etc... to do that. And, just to give you an example, you could build a trajectory, like in your muscle or in your body where the motors walk over the filaments, the actin filaments; build a trajectory where you balance the non-covalent interactions so that your molecules don't fly bananas everywhere but walk from A to B over a trajectory, keeping to the trajectory and then also transporting something and doing something. I mean, in the next 10 years, I would be extremely happy if my students would be able to accomplish that, because this is supramolecular chemistry *par excellence*, combined with either photochemistry or catalysis and balancing the interactions and then doing mechanical function. We have been working on it for a while, so we are still at a very early stage. On the other hand, looking at applications, we have now the first examples where we have self-cleaning glasses and surfaces, because they are responsive, and they clean themselves. And also, as I mentioned, they're moving and you can move material on the surface, and so there will probably be applications in the next decades where people will use maybe not our motors — they might be too expensive, but when you know the principles and you have seen Joanna Aizenberg (I showed that on purpose), she has fantastic surfaces where you can have this amplification of motion because she can structure surfaces and then get synchronized motion etc. I think these kinds of materials, when I was in industry, I would keep an eye on that, because I think that will be the first application in my opinion. And many people will be happy if you don't have to wash your car anymore or your window.

Joachim Sauer:

I would like to make a general comment. So, if we would have to prove as a chemist that we understand something, a system, we have to be able to synthesize it. Otherwise, we don't understand it. So, I would see the value there.

Ben Feringa:

I'm happy you make this remark. I think it was already Richard Feynman many years ago that made similar remarks. I really appreciate that. Because, indeed, if we can design systems that have this kind of complexity and have functions, I think it will guide us in many aspects in the field of chemistry. That's my firm opinion. I'm not sure about the application, all the applications, etc. And certainly, yes, we have to think about that, but thanks for this remark, I appreciate it.

Nathalie Katsonis:

I just wanted to mention that I'm not convinced that the field needs a set of drivers to move forward. I think maybe a field needs conceptual innovation and strong concepts and sometimes the drivers are these conceptual innovations. At least that is my feeling, so I wanted to mention that. And maybe something that has not been said and is maybe not super concrete, but molecular machines, biomolecular machines are usually used for anything our body does. And although that's not one specific application, I think that sets the stage for the potential of this approach. And on midterm and applicability basis, I would think about really neuromorphic materials, because we need these kinds of materials if we want to move forward with our technologies. And for learning and neuromorphic materials, we need active repositioning strategies that operate at length scales that are intermediate between the molecule and the material.

Makoto Fujita:

So before going to a complex system I will simply discuss about the terminology. So to describe the dynamic motion or metastable kinetically driven structures, we often use 'non-equilibrium' or 'out of equilibrium'. But I feel that this is not appropriate. Because, even if molecules are pumped up at their high energy states or if we kinetically trap the metastable structures, they always miraculously equilibrate up to the metastable state. So, just regarding the dissipative structures, we can say that this is 'non-equilibrium' or 'far from equilibrium', but in a dynamic system, if molecules are trapped at a local minimum, molecules are equilibrated there. You don't agree? So, in my feeling, it's not appropriate to use 'non-equilibrium' or 'out of equilibrium' to describe this.

Andrew Turberfield:

It's a question of loose use of the word equilibrium. So, in true equilibrium, yes, you would have molecules in your high energy state and they will be arriving at and departing from that state at exactly equal rates and nothing useful can be done with them. So they may be locally equilibrated, if you only look at the system very locally in space or in energy, or in a configuration space. But clearly, globally, if you step back and look at the system, the system is out of equilibrium, which is how it's driven to do something interesting. So I mean, strictly speaking, I think we're using the word 'out of equilibrium' correctly. I think you're using it correctly in a very limited sense, which is that if you restrict your view just in the vicinity of the molecule that you're looking at, maybe locally, yes, it is an equilibrium but globally not.

Makoto Fujita:

Also, are there any differences between 'non-equilibrium', 'out of equilibrium' and 'far from equilibrium'? We need some clear definition for these terms.

Ben Feringa:

Yes, I would agree with that, because there is a lot of discussion and there is a lot of use of these terminologies, maybe not at the correct manner. I agree with Andrew on this point when you look at the thermodynamics and kinetics of the system. But, definitely, we need maybe to clarify this, and to the community, maybe we should write a perspective on that to say: what are the correct uses of this, and where are we heading for? That's maybe an important message, here, for all of us.

Andrew Turberfield:

It's a bit like the word symmetry. I mean, strictly speaking, something is either symmetric or it isn't. There is no close symmetry, I mean symmetry is absolute, is mathematical. And I think the same is true of equilibrium. You can always retreat to absolute rigor. Equilibrium goes with the principle of detailed balance and if you haven't got that then you haven't got equilibrium.

Jack Szostak:

So a question and maybe a challenge, primarily for Andrew, but maybe for other members of the panel. You have shown us very impressive, extended, well defined but static structures, and I would be interested in the question whether you could make something like a cytoskeleton that is extremely dynamic and controlled out of nucleic acids instead of out of proteins. What do you think is needed to get to that goal?

Andrew Turberfield:

We can. When I say 'we', I mean the field. It depends on what length scale you're talking. We're okay on a length scale of 100 nanometers, and maybe up to a micron; bigger than that we don't do very well at all. We can actuate single hinges on that length scale, I mean, if we try. But, slowly, we could sort of diffuse in signal molecules which could make things bend and unbend. So we can do something, but it's crude. We could make the cytoskeleton for something of the size of E. coli and we could actuate it, but not with a very great degree of local control. Certainly not with rapid control. And we could do that now. The question is precisely what challenge you're setting for us.

Jack Szostak:
Well, thinking about early stages in the evolution of life, you would want something that would help to control cell division for example. And I think that probably evolved with RNA machinery, but it would be nice to see a demonstration, something dynamic in the lab.

Andrew Turberfield:
I don't think that's totally impossible. I think knowing what we know now, we could give that a good go. But the signalling, I mean the energy input, would be crude and slow. We couldn't power that with a small molecule reaction. We could power it with strand displacement reactions.

Jack Szostak:
So the challenge is bringing in chemical energy into the system.

Ben Feringa:
Now I have a question to you, and maybe I should ask this question at your session. But why should we be then limited to these few molecules that Mother Nature uses, because we have unlimited possibilities. You have seen already examples in this field of supramolecular. So why should we design it based on an artificial molecule?

Jack Szostak:
I totally agree. There are several people in the field trying to develop genetic polymers out of completely different molecules and if you could build other parts of a synthetic cell that way, that would be great. But at least we know that somehow life got started using things like RNA. Maybe it's an easier challenge.

Ben Feringa:
But there could have been other try-outs, no?

Jack Szostak:
I doubt it.

Ben Feringa:
You doubt it? Okay, we will discuss this later.

Bert Meijer:

It's a very good question, but I like to see it from the extra-cellular matrix. There's a journey ongoing to make a substitute for matrigel in order to go to stem cells out of organoids. That is a journey that started something like 20 years ago and the progress is slow, but they are coming closer and closer to materials that really grow the stem cells into small organoids by totally artificial ways. But I have to be careful because it still has oligopeptides that are synthesized and put into it, but it is coming closer and closer to that and that would be a huge advantage because that matrigel is always inhomogeneous, every time is different and all of the cell biologists would love to have this on a large scale. The progress has been enormous the last couple of years. It is not inside the cell but outside the cell, but it may be equally important.

John Sutherland:

So, a question to the panel is: How would you build in evolvability, in particular evolvability using phenotype/genotype linkage? If you can do that, you can recapitulate biologists' greatest trick, which is systematic variation and finding the best solution.

Ben Feringa:

I am not a cell biologist, but I have colleagues here that are more knowledgeable in that field than I.

Andrew Turberfield:

I am not sure I'm here to give that sort of answer. One application that I really do think is worthwhile pursuing is the synthetic ribosome angle. The idea there is to make some sort of crude molecular machinery out of DNA or whatever comes to hand, I don't care, that is genetically programmed to again pick monomers from a pool labelled with a bit of DNA, which is the analogue of a tRNA. So the vision is that you have a soup of monomers for real organic chemistry. Each one is covalently attached to an oligo, which identifies it. Our crude molecular machinery will be programmed by a synthetic gene, with codons that aren't triplets, they're probably 10-mers or whatever. It has our own structure, our own format, but the instruction tape, the gene, tells the machinery in which sequence to concatenate these monomers to make an oligomer. And if we can make a 10-mer, that's probably good enough, out of 10 components. So, if you can do that, you can make huge libraries of totally unbiological oligomers with synthetic backbone linkages and totally non-natural side chains. And in that vast chemical space, which has never been looked at, there must be drugs and catalysts and other useful things. If you can do that, you can do evolution, because you can cut, recombine, mutate our synthetic genes. Having discovered something halfway useful, you can recapitulate the synthesis and do it better.

Ben Feringa:

And maybe I add one aspect to this discussion. Now, maybe I'm not in the genotype or phenotype business, but what we did, for instance, is we put our rotary motors as a mono layer on surfaces and then we grow stem cells on it. Stem cells are very sensitive to mechanical force and using these stem cells, we could control, using these rotary motors autonomously, powered by light, we could use the outgrowth of stem cells and the differentiation, to some extent. And at the level of the DNA, at the level of the protein expression and the anchoring to the surface, it affected the stem cells and depending on the rotary motion, you could differentiate stem cells. This is the stage where we are now. So, there are different options, I think, in this in this field.

Kurt Wüthrich:

I'm curious to hear from Andrew, whether he sees relations between his work on nucleic acids and DNA libraries, which have a big impact already in many areas of chemistry.

Andrew Turberfield:

I'm not sure what you mean by "DNA libraries". Are they these combinatorial libraries of small molecules, DNA encoded?

Kurt Wüthrich:

Yes.

Andrew Turberfield:

Right, so the answer I just gave to John Sutherland was related to that. So yes, our projects, our aim to create a system which works as a synthetic ribosome is related to the idea of a DNA encoded library. It's a special case, because it's based on molecular machinery reading a program or reading a gene and synthesizing the corresponding molecule. It has the additional property that you can do exactly what John asked us to do, which is to evolve. So rather than just have a static library of interesting things, which are identifiable through the attached DNA, if you have a genotype-phenotype relationship, if the label actually codes for the synthesis of the product that is labelled, then by mutating or playing around with the gene, with a label, you can change the product. And what that means is that you can do exactly evolution, you can first create a library of 10^{12} products and select two things from that, that might be interesting. Take their genomes, recombine them with themselves and with all sorts of other rubbish you happen to have selected at the same time and you can begin to do exactly what biology does, which is to evolve. So yes, I mean that the synthetic ribosome project that I described is a special case, but a very powerful case of a DNA encoded library. Does that make sense?

Kurt Wüthrich:

Well, I mean, synthetic DNA libraries have already a big impact, and I'm sort of surprised that you're not more directly involved with your work.

Andrew Turberfield:

Well, I came at the subject from a different angle. I'm not a chemist by background. I'm more interested in mechanisms than educated in chemistry, which is why I'm doing what I'm doing.

Daniel Nocera:

I want to return to one of your questions about tools. So when I went to graduate school, I had to read out Hertzberg and I used to do Franck-Condon analysis just to figure out what my excited state looked like with the photon. Now my students just write a proposal to the advanced light source, they bring the molecule, they use the LASER and they actually just get the structure of the excited state. And right now, science is on a nexus with these light sources. So Argonne National Lab and Simon Billinge at Columbia, with the power of computing, they can accelerate Monte Carlo and then with the light source and time resolution, they're almost on the verge of getting structures of molecules in solution, dynamically. Have any of you tried yet, to PDF with an advanced light source to actually get the structures of the dynamic system?

Ben Feringa:
From my own perspective, from our library of motors, yes, we have done it. So far, we do femtosecond LASER spectroscopy to look at this. But indeed, the combination now of the modern, theoretical methods to work on the excited state energy profiles and the landscape, and at the different pathways, etc. together with these laser techniques, you can do amazing things. So recently we submitted a paper, or it is accepted now, where we have boosted the quantum efficiency to 70% by simply doing exactly what you say, using these theoretical methods to tune our excited state profile and to get away from dissipative pathways, etc. So, I think there is a lot to be gained there if we revisit the energy landscapes, and probably maybe David wants to comment on this, also for catalysis, you know, for the redox catalysis, etc. There must be still tremendous opportunities there.

Daniel Nocera:
I guess the thing I was also referring to was with PDF, pair distributional functional analysis, they are literally able to put molecules in solution, they are basically getting the X-ray structure of the molecule in solution. It seems like this field could really benefit from basically getting structures of molecules in solution, and this is literally due to this computing power and the intensity of the light sources for scattering.

Ben Feringa:
I'm happy that you mentioned it. I think there are tremendous opportunities indeed.

Clare Grey:

Just a comment, talking about the PDF, that gets so complicated so quickly. I mean, I am aware of Simon Billinges work, but I agree with you that if you have a light as a trigger to change a conformation, then if you look back at to say the work of Phil Coppens or others, then you start to be able to look at differences and then it becomes interesting. So I think, it's an area where there's more to do. I'm not sure in its current state, but I agree that in general, yes.

Daniel Nocera:

That's on the verge, clearly. And I think with some of these very well-defined molecular motions and motors.

Clare Grey:

Well, some of Omar's systems where there's more periodicity.

Daniel Nocera:

I don't want too much periodicity. I mean, that's where the power is coming out. It's literally without the need for periodicity. Some of this hasn't been published, some has, the beginnings of it, but they are literally getting structures of molecules in solution.

Ben Feringa:

I would love to see it, yes. This is great.

David McMillan:

I was just interested in a slightly different tangent and maybe this is a boring question, but I was interested from the panel to sort of talk about, aspirationally: What do you imagine or think about? What will be the things that you could achieve, and you think is realistic to achieve within five years? And to pursue this is, number one: I always think it's difficult when I see a supramolecular talk because there's a lot of aspirational components, which I love, I can't tell the timeframe. So I'd love to sort of hear what you think about the next five years. And the second part is a quick Bob Grubbs story. I once saw Bob Grubbs being interviewed and the interviewer asked him: "What will you be doing five years from now?" and he said: "well, if I knew that, I'd be doing it now".

Ben Feringa:
This is a great question, Dave, so maybe I ask every panel member what their aspiration is, what their dream is. Let's start with Bert.

Bert Meijer:
So, our materials are now in the last clinical phase, and I hope they will really be used in human beings in the coming five years, based on supramolecular materials. And the other one that I really hope is if there is a way to use spin-controlled chemistry by using chiral electrodes. Not saying that it will be useful, but I don't know whether that's true or not true, or it will not work at all.

Ben Feringa:
Thanks, Bert.

Omar Yaghi:
I would say that, again, I encourage the supramolecular to combine with the periodicity, so if you do that, I would say that the opportunities are endless in terms of systems. First, they can sort molecules very well, because they can design complexity and when that complexity is superimposed on the periodicity, it gives it structure and control in the metrics of that structure. And also control of the ratio of those supramolecular elements and their distribution on the grid. So once that system is able, let's say, to fold, let's say turn from a sheet into a pipe or any other shape, I would say that's possible within five years to have a system that can sort molecules. I don't want to say a system that has a circuitry as I described before, but I think that's more complicated. I think directionality and where the substrate is going is quite doable. I mean, ideally, I would love to have a crystal that is made from, let's say, a MOF or a COF onto which you have in the pores superimposed a supramolecular ensemble that then can separate air into different components and so oxygen goes along that pore, nitrogen goes along this pore and water or CO_2

comes out of the third pore. I think that this is not out of the realm of possibility.

Nathalie Katsonis:
A PhD in the Netherlands is four years. In five years, I hope I have educated good chemistry students. If I do that, I'm already happy. Sciencewise, I hope I understand better the origin of purposeful movement in chemical systems.

Nicolas Giuseppone:
I agree, maybe the field is a bit curiosity-driven at the moment with no clear applications and it is important to make something that we cannot make by other means, to prove it is technologically relevant. So, for applications, I think there is room for making artificial muscles for soft robotics so it is clearly an advantage from the forces that we can generate from these machines and the weight of the muscles we can create, so we can reach probably high force/weight ratio, this is predicted by physicists. I think we are not that far from there. We need also to avoid the production of waste so we need also some sustainability in the production of motion. So maybe this is related to the first session of this morning. The second example I would give is the creation of gradients of ions or molecules between compartments by pumping with a molecular machine. I think we are not that far from doing that and such a pumping/storage of energy is interesting for many potential applications.

Ben Feringa:
Thank you very much.

Andrew Turberfield:
So, a PhD or a DPhil in the UK is three years, and I am, if you ask my students, remarkably poor at predicting what they can achieve in three years. But on five years, I reckon we will have wired up and measured, probably at low temperature, a molecular electronic device, not a very sophisticated one. I think we will have printed a pattern on a surface using directed motion of a write-head and the pattern will have nothing to do

with DNA by the time we have finished. It will be proper chemistry on the surface. And I think we will have made a genetically programmed 10-mer from a soup of possible building blocks.

Ben Feringa:
Thank you very much. And I should challenge the whole committee, I suppose, I can also give my dreams. I gave already one, that is to move autonomously over a trajectory, not covalently but supramolecularly, from A to B, hopefully 20 nanometers or so on a filament, or whatever, maybe the kind of structures that Omar makes. And we made already a motor MOF, by the way, where we have the motors as the pillars in the MOF and so control transport, that is one of the other. We will team up to control transport, that you can select which molecules will pass through such porous materials, as active membranes. The other dream is to see how we can make responsive drugs and I put a lot of effort in that now. We build in these tiny machines into drugs, and we were able, lately, to influence the communication between bacteria. Because bacteria communicate with each other to make biofilms and this is really important for the medical field and for the implants and whatever, etc. And recently, we were also able, together with Japanese colleagues, to change the circadian clock in cells, reversibly by four hours. I don't know about your jetlag, it will not be applied to human beings yet, but I think in about 10–20 years from now. So these are my kind of dreams. And my third dream, that I would like to realize: we made already one type of catalyst, an adaptive catalyst that would change and could do sequentially a number of steps. So my dream is a bit to program such a molecule to do a sequence of steps because the catalyst itself adapts and can do step A, step B, step C. That is a bit of a dream, don't ask me exactly how to do it, but this is what I'm dreaming.

And this is actually the last question I wanted to ask, because I think we have to finish sometime. There are people from the biology field and we heard already some challenges from the biology field, but I challenge you to ask, to give us a message: what should we learn from the most complex system which is a cell, or from biology? What is something that we have not discussed and that you would say, please community, take this in your luggage back home and think about it?

Sabine Flitsch:

I was wondering about compartmentalization. If you look at life, you have cells, you have membranes, you can compartmentalize things, you can use membranes as surfaces and so on and I think you can maybe incorporate that into your systems. I am thinking of Hagen Baileys, cells and so on. If you combine your motors with that sort of system, I wonder whether that would give you some better control as well. It looks more complicated, but it might give you better control.

Ben Feringa:

Fantastic that you bring this up, because I think it was mentioned by a couple of people here before in their talks: compartmentalization, and then also dynamic compartmentalization, adaptive, that's what you see in cells all the time. I don't know maybe there are other suggestions?

Donald Hilvert:

My comment goes in the same line as Sabine. In biology, these partitioning events, controlled partitioning events, are really, really important. And it seems to me it would be directly relevant to understanding how these biological condensates form and what their role might be, and maybe one can even learn how to exploit them in new ways.

Ben Feringa:

Thank you so much.

Andrew Fire:

As a biologist who's lived through the last three years, one thing that would be wonderful would be to have a 'semi-solid' media that would in the presence of a defined and highly organized structure, a.k.a. a Coronavirus, form a macroscopic visible change, a singularity that would be visible. So we would be able to quickly visualize the presence in our environment of microscopic particles, with a high degree of organization, so it should allow all of these structures to form. That would be a huge game changer in both real-time medical diagnostics and in keeping the world going during pandemics. So, I'll thank you in advance for doing that.

Ben Feringa:
We work on it. Maybe it will take a bit of time before we do that.

James Liao:
If you want to take this in the biological area, I think the most important thing is what we just talked about, the relationship between the genotype and phenotype that allows it to become evolvable. And I think there are a few people who talk about it. Particularly Andrew, to address this issue a little bit I was wondering: What is your idea to encode this information into whatever you call the artificial ribosome or whatever genetic code of this synthetic steps? Is there any idea what they're up to now, any imagination?

Andrew Turberfield:
Sure! So, most synthetic machinery made out of DNA, which by biological standards is very crude, is driven by DNA strand displacement reactions. So, what initiates such a reaction is simply an invading strand of DNA, which is slightly different from one that it displaces, and these strands carry information. So, I showed you in my 10 minutes talk an example of a DNA computation, or a system for computation, which is driven by a cascade of such strand displacement reactions. You can use a cascade of strand displacement reactions to read the information carried by a synthetic gene, which is just a concatenation of codons which specifies the sequence in which you wish to do your synthesis, to reveal them in sequence, for example, and allow them to recruit from solution complementary sequences which are attached to the building blocks or the monomers that you wish to or to join to your growing polymer chain. So, you know, it's all based on the idea of DNA templated synthesis, which is the promotion of reaction by forcing proximity between reactants, by bringing them close together using attached DNA handles, which are the analogues of tRNA. So yes, it's a strand displacement cascade, and sequential programmed interaction with the bits of DNA attached and labelling the reactants in solution.

Ben Feringa:
Thank you so much. I look around the table if there are not any urgent questions. Of course, we can continue the discussion, but then I think we

do that in the presence of the fruits of biology. Oh, we leave at 6:10pm, then it's time to stop. I would like to conclude here. We had already presented, all our presenters here, a lot of challenges and questions, but I get the feeling that we have a lot more challenges and questions, and you challenged us a lot. Thank you so much, everybody for the contributions to the discussions, all the speakers here, etc. I hope you enjoyed the afternoon session as much as I did. And we look very much forward to the rest of the week. We will continue on some of these issues, I'm sure, and we look forward to a nice reception tonight. Now I give the floor to our president.

Kurt Wüthrich:
I have nothing more to say. Thank you for doing a great job.

MOF-303

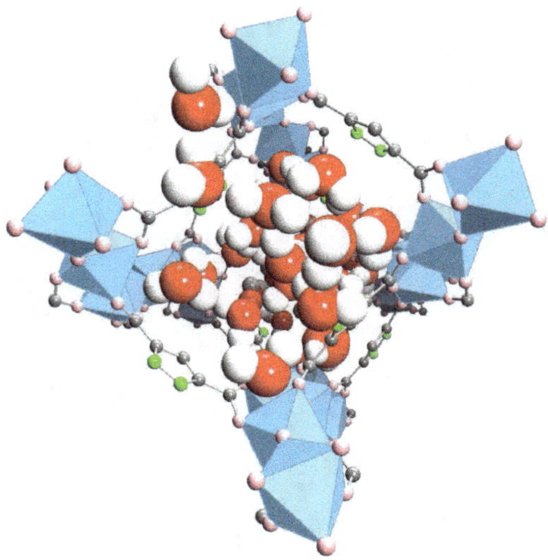

Image courtesy of Omar Yaghi

Session 3

Reticular Chemistry and New Materials

https://doi.org/10.1142/9789811282324_0015

Reticular Chemistry and New Materials

Omar M. Yaghi*,† and Zhiling Zheng*,†

**Department of Chemistry, University of California, Berkeley CA 94720, USA*
†Bakar Institute of Digital Materials for the Planet, College of Computing, Data Science, and Society, University of California, Berkeley, CA 94720, USA

The societal problems we face today in climate change, clean energy, clean air, clean water, food, and health impact everyone on our planet. Solutions to these will require innovations in materials to bring about a more sustainable and resilient future. Some progress in understanding the basic science underlying these problems is being made and it is beginning to drive the invention and discovery of new materials. However, currently, there is a gap between the materials we require and our ability to realize them. This is largely attributed to the fact that traditionally making extended chemical structures has been a trial-and-error activity with very little design expressed on the outcome of solid-state synthesis. This changed with the emergence of reticular chemistry, defined as the chemistry of linking molecular building blocks by strong bonds to make extended structures. As elaborated in the following, this new chemistry laid the foundations for making atomically precise structures, which also could be chemically modified by design. The extension of reticular chemistry to linking nanoparticles and biological molecules further widened the complexity and scope of the building blocks amenable to reticulation, thereby ushering in a new era in making new materials where the chemists' dream around the idea that 'if you think it, you can make it' becomes a reality (see Fig. 1).

Challenges also exist in the characterization of such extended structures when single crystalline samples are not accessible. Diffraction techniques have evolved to provide means of determining the chemical connectivity of crystals in the nano-size regime. I wish to remark that even after addressing the design, synthesis, and characterization challenges, the study of material properties is only the first step in determining the suitability of the material for a target application. It is becoming increasingly important that we as chemists must integrate into our research what might be termed the innovation cycle. This is the fundamental research that goes into not just molecules linked into materials but also materials from factors and their integration into systems (i.e., the innovation cycle of molecule to material to engineering to system to society). These require expanding our knowledge beyond molecules and materials to encompass basic engineering principles. Since the materials we make can be designed with precision, the study of the various parts of the innovation cycle provides a feedback loop between the molecular properties and system performance. It is worth noting that intensive research into each component of the innovation cycle has been carried out, but the science and engineering required for integrating these *a priori* remain underdeveloped.

Attaining the knowledge to execute the innovation cycle in an orchestrated and correlated fashion and the speed in implementing it for new materials and applications will significantly speed up discovery and thereby scale impact. We suggest that computation including artificial intelligence and machine learning will be critical to achieving these objectives, and without a meaningful coupling of the innovation cycle to modern computational tools, we will not capitalize on the potential the building block approach brings to the discovery of materials and their properties at the peril of society. Reticular chemistry's logical approach involving the use of molecular or nano-sized building blocks is ideally suited for extending chemistry to engineering and computer science to

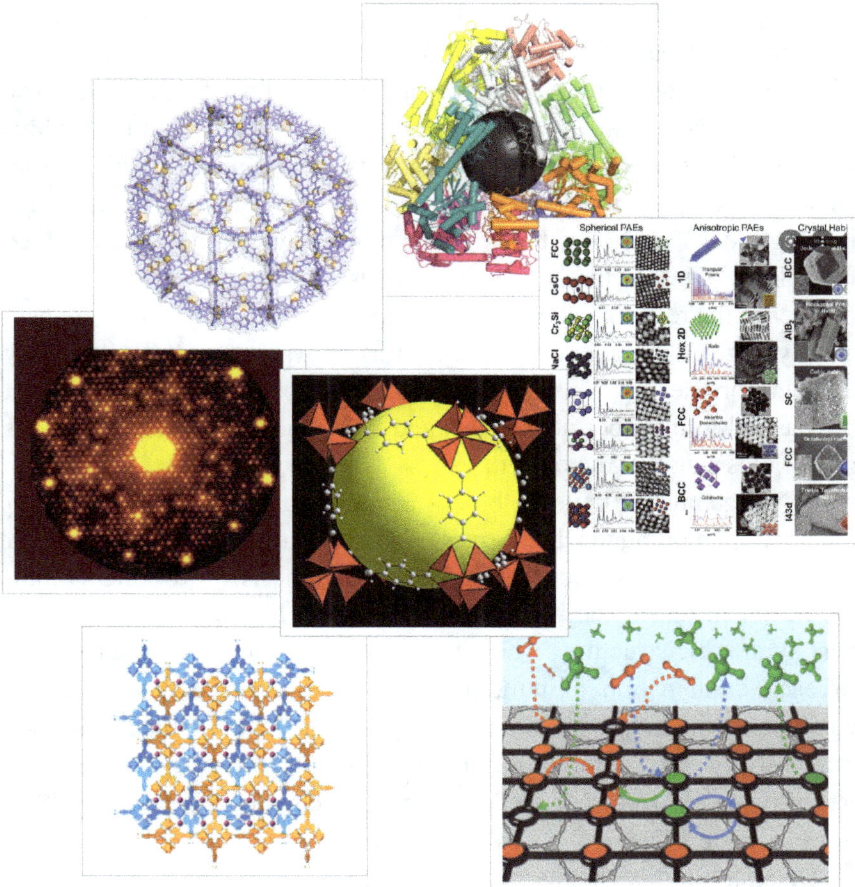

Fig. 1. Clockwise starting at 11 o'clock: Makoto Fuijita's large cage, Todd Yeate's protein cage, Chad Mirkin's nanocrystal assemblages, Joachim Sauer's accurate computations and prediction, Arne Thomas' silicon-based organic framework, Xiaodong Zou's advanced electron diffraction techniques for nanocrystals, and center is Omar Yaghi's metal-organic framework. This collage illustrates the establishment of the building block approach to the synthesis of new materials and the evolving important role of computation and diffraction techniques in illuminating this chemistry.

effectively solve societal problems. This background provides the context for the presentations to be given at our symposium titled "Reticular Chemistry and New Materials" at the 100th Anniversary of the Solvay Conference. The focus here will be on the basic ideas of reticulating crystalline extended chemical structures incorporating molecules, nanoparticles, and proteins as building blocks.

Every extended structure can be divided into linkers and nodes. The linkers join two chemical entities together (2-connected) and the node is 3 or more connected. The compositions of these can vary widely from simple to complex molecules, nanocrystals, and proteins. When such building blocks are linked together, the point at which they join is called the linkage. This is the 'glue' that allows the building blocks to stay together in the resulting extended structure. Such building blocks are necessarily large chemical entities to allow for the expression of directionality and therefore design. This inevitably leads to open structures, an aspect that has led to ultra-porous crystals and consequently many applications in the fields stated at the outset.

When geometric building blocks are reticulated, an important question arises pertaining to which structure among the almost infinite number of possible structures would result from the synthesis. A working hypothesis has been that if the building blocks are highly symmetric, it is the most symmetric structures which would result from the synthesis. Thus, for each combination of building block geometries, a few structures would be considered as reasonable targets. As the building blocks become progressively less symmetric, the possibilities are vastly increased. Presently, most structures made in reticular chemistry belong to those listed in Table 1.

Diversity in this chemistry is achieved by varying the building block geometries, composition, size, and connectivity. Further diversification comes from the linkage variation and addition of functional groups to decorate the interior of the resulting structures. When multiple metal ions or functional organic groups (variants) are incorporated into a specific structure, the diversity space increases exponentially. Although the nature of the functional groups or metal ions and their ratios in the structure are known, their spatial arrangements are unknown. However, the heterogeneous arrangements of these variants give a chemical-rich environment capable of operating better than the sum of the parts. This scenario is conceptually related to DNA, where the nucleotides are superimposed and covalently linked to a repeating backbone: the variants in the reticulated structure are also bound to the ordered backbone to give sequences of

Table 1. Tabulation of the resulting structures from the combination of geometric building units. Unfilled squares indicate no chemical structures have been made for those geometries. This applies to networks of molecules, nanocrystals and proteins.

Building unit 1 / Building unit 2	2-c Linear	3-c Triangle	4-c Square	4-c **tet**	6-c Hexagon	6-c **oct**
3-c Triangle	**srs**	bwt, **pyo**, srs-b, ths-b	**fjh**, fmj, gee, lab, yac, yao	asn, **ept**, ofp	**cys**, dnf*	anh, **ant**, apo, brk, cep*, cml, czz, eea, qom, rtl, tsx, zzz
4-c Square	**nbo**, lvt, rhr	**pto**, tbo	cev, cdl, cdm, cdn, cds, cdz, **mot**, muo, qdl, qzd, ssd, sse, ssf, sst	**pts**	**nts**	myd, **ybh**
4-c **tet**	**dia**, lcs, qtz, sod	**bor**, ctn	fgl, mog, pds, **pth**, pti, ptr, ptt	bnl, byl, cag, cbt, coe, crb, fel, icm, kea, **lon**, pcl, qtz-b, sca, tpd, ucn	–	alw, bix, **cor**, ing, spl, toc
6-c Hexagon	**hxg**	**cys**, dnf	**she**	–	**hxg-b**	–
6-c **oct**	**pcu**, bcs, crs, reo	**pyr**, spn	**soc**	**gar**, iac, ibd, toc	–	**pcu-b**, bcs-b
6-c **trp**	**lcy**, acs	ceq, dag, fmz, hwx, moo, sab, **sit**, ydq	**stp**	fsi, hea, **tpt**	**htp**	**nia**
8-c **cub**	**bcu**	**the**	**scu**, csq, sqc	**flu**	–	**ocu**
12-c **cuo**	**fcu**	**sky**	**ftw**	**edc**	–	–
12-c **ico**	–	–	–	**ith**	–	–
12-c **hpr**	–	**aea**	**shp**	–	–	–
12-c **tte**	–	**ttt**	–	–	**mgc**	–
24-c **tro**	–	–	–	**twf**	–	–

variants running along the entire crystal structure. This 'complex order' is the basis of a whole new and extensive chemical space, whereby the reticulated structure properties are no longer discrete states, but in fact, they are represented on a continuum of states. The implications of complex order of this type are that substrates can sample many different microchemical environments to be bound and/or transformed.

In all these systems (synthetic and biological), the space encompassed by the reticulated structure is useful in selective binding and storage of guest molecules, such as hydrogen, carbon dioxide, water, organics, and drug molecules. It has also been used for catalysis and installment of molecular machines. Given the precision with which structures can be reticulated, the extensive diversity of this chemical space and the new powerful computational tools, it is not inconceivable that we will be able to design new materials representing concept transfer from biology, specifically, materials capable of multiple functions and those that can operate in parallel and in sequence, materials constructed from compartments that are open to each other, yet operate independently or synergistically, and materials programmed to sort and count molecules.

References

[1] M. Fujita, J. Yazaki, and K. Ogura, *J. Am. Chem. Soc.* **112**, 5645 (1990).

[2] J. E. Padilla, C. Colovos, and T. O. Yeates, *Proc. Nat. Acad. of Sci.* **98**, 2217 (2001).

[3] C. A. Mirkin, R. L. Letsinger, R. C. Mucic, and J. J. Storhoff, *Nature* **382**, 607 (1996).

[4] J. Sauer, *Chem. Rev.* **89**, 199 (1989).

[5] P. Kuhn, M. Antonietti, and A. Thomas, *Angew. Chem. Int. Ed.* **47**, 3450 (2008).

[6] X. Zou, S. Hovmöller, and P. Oleynikov, *Electron Crystallography*. Oxford University Press, 2011.

[7] H. Li, M. Eddaoudi, M. O'Keeffe, and O. M. Yaghi, *Nature* **402**, 276 (1999).

https://doi.org/10.1142/9789811282324_0016

Multivariate Catalytic Materials

Arne Thomas

Functional Materials, Department of Chemistry, Technische Universität Berlin, 10623 Berlin, Germany

Learning from nature to create catalytic materials and the present state of research on reticular chemistry

Nature has created fascinating materials whose structures are precisely controlled over several scales of length. Furthermore, nature has developed catalysts that dwarf all human-made catalysts, not only in terms of activity and selectivity but also function under ambient conditions and without harmful or toxic substances, solvents, or reagents. For chemists that want to develop materials for catalysis, nature has thus created a blueprint that can be used in terms of design, structural control, precision, and sustainability.

However, despite the great advances that chemistry has made in recent decades, our synthetic capabilities are far from sufficient to even come close to the perfection of nature's synthetic toolbox — in fact, nature's advantage of having several billion years of time to optimize materials and catalysts can discourage anyone who tries to approach such systems synthetically.

However, that doesn't mean we can't draw inspiration from natural materials and catalysts, especially when it comes to the chemistry of different components within them. Be it the interplay of nano- and micro-structured inorganic crystallites with a soft polymer matrix in biomaterials or of metal cofactors and protein shells in enzymes — the optimized mutual interaction of different functional units in such materials and molecular assemblies ultimately leads to their outstanding properties.

Reticular chemistry is a promise to create novel materials with unprecedented structural control [1]. By linking molecular building blocks via strong covalent bonds to crystalline open scaffold materials, our synthetic toolbox has expanded significantly to create materials with an unprecedented level of control over chemistry, structure, and porosity at the molecular level, thus bringing us one step closer to the functionality of natural materials.

While simple molecular building blocks were initially used, primarily to build new structures and topologies without decisive functions, recent years of progress in reticular chemistry has increasingly focused on incorporating molecular building blocks with a wide variety of functions into open framework structures, thus creating novel materials that could be used in a number of applications [2]. This includes materials for energy storage or conversion, for the purification of drinking water, for the separation of gas mixtures, or for catalysis, i.e., for all areas that are of the highest relevance for the sustainable development of our society.

Materials produced via reticular chemistry will therefore play an important role in solving the Chemistry Challenges of the 21st century.

Our research contributions to reticular chemistry and new materials

Our group deals with the synthesis and application of nanostructured materials, with a special focus on the production of porous functional materials for catalytic applications. We try to push the boundaries of the traditional disciplines of chemistry and materials science by working on inorganic and organic materials, as well as hybrid materials and developing unconventional approaches for their production, e.g., by synthesizing organic materials in inorganic molten salts [3] or incorporating main group elements as structure-building units in organic framework materials [4]. A turning point in our research was the development of metal-free semiconductors that can be used, among other things, as photocatalysts to produce hydrogen from water [5]. The generation of storable chemical energy from the energy of sunlight is one of the great chemistry challenges of the 21st century. Photocatalytic water splitting for the synthesis of hydrogen or the reduction of CO_2 into useful basic chemicals are two important reactions, which would yield independence from fossil raw materials and the reduction of greenhouse gases. The development of photocatalysts based exclusively on abundant elements would enable a new, decentralized production of chemical energy sources and products. After initially investigating polymeric carbon nitrides as photocatalysts, we also succeeded in the first synthesis of a porous organic network at room temperature as a photocatalyst [6]. In recent years, we have increasingly focused on the use of the principles of reticular chemistry to build covalent organic framework materials (COFs) as new organic photocatalysts [7].

As described above, reticular chemistry, inspired by natural catalytic systems, allows various functional units to be incorporated into an open scaffold material with high precision. With such multifunctional or multivariate COFs, structure and properties of a photocatalyst can be tailored, e.g., by controlling its band gap and band positions, improving charge carrier separation and mobility in the polymer backbone, and varying the polarity of the material or introducing molecular co-catalysts (Fig. 1).

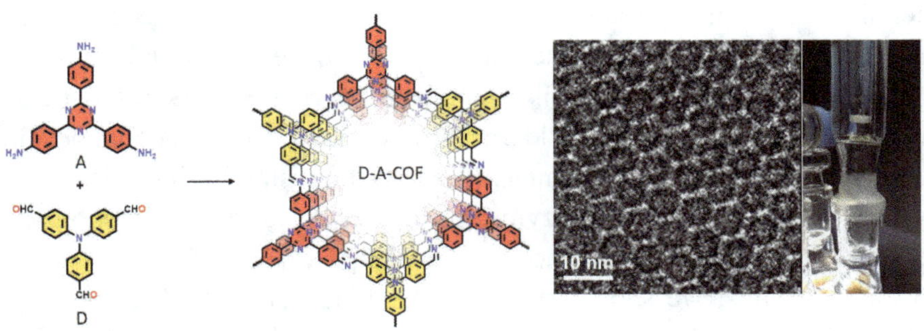

Fig. 1. Synthesis of a COF including electron donor (D) and acceptor (A) moieties, HRTEM micrograph of the D-A-COF, and experiment for photocatalytic hydrogen generation.

We can also learn from nature that the structures of a functional material must be controlled not only at the molecular level but also in the best case over several length scales. Therefore, we have tried to create COFs that have defined structures even on micrometre and even larger length scales by adding additional templates or building composites with other materials, such as graphene [8, 9].

Outlook to future developments of research on reticular chemistry and new materials

The field of materials chemistry has expanded significantly with the development of reticular chemistry. For the first time, we have the opportunity to introduce a large number of different functional units into a single material with high precision in terms of distance, orientation, and electronic coupling to each other. How can this help create advanced materials of the future that are urgently needed for the sustainable development of our society? As an example, let's consider again materials for catalysis. As described in the opening statement, nature has created perfect catalysts, namely enzymes consisting of a complex protein shell and several cofactors whose functions have so far only been partially revealed. However, it can be safely assumed that nature would not have developed such complex assemblies if a much simpler solution would have offered similar catalytic performances and functions. The idea that just because certain metal ions are found in the active centre of an enzyme, the same metals could exert similar reactivity on the surface of a simple support material or an organic ligand is tempting but certainly too short-sighted. Perhaps we should rather consider a systemic approach to material development, i.e., dealing with complexity rather than simplifying things too much. The protein shell of enzymes fulfils a wide variety of functions in the catalytic process and is not only a complex ligand for metal ions. Inspired by this, we should strive for catalytic materials that not only immobilize a catalytic centre but also whose multivariate structure can control the entire catalytic process, including the control of chemical and electrical potentials and the transport of mass and energy to and from the actual active centre. Furthermore, why should we limit ourselves to materials that are only suitable for one single chemical reaction as a catalyst, when we can also consider a cascade of successive or simultaneous chemical or catalytic processes coupled by the transfer of reactants, products, and energy in the material, as we can eventually observe in our cells? Such an approach would require the construction of one or more multivariate materials, which are nevertheless spatially coupled with each other so that different catalytic reactions can take place in separate compartments, which suggests a structural complexity of the materials, which we can potentially achieve through the promises of reticular chemistry. Such approaches will not pre-emptively create a synthetic enzyme or an artificial cell but at least materials that come a significant step closer to this goal.

In Berlin, we are currently investigating such concepts in an interdisciplinary research network entitled "Unifying Systems in Catalysis — UniSysCat", in which more than 60 research groups from 9 institutions with expertise from bio- to inorganic solid-state chemistry and from theory to engineering sciences are involved [10].

Acknowledgements

AT is grateful to the Deutsche Forschungsgemeinschaft (DFG, German Research Foundation) under Germany's Excellence Strategy-EXC 2008/1 (UniSysCat)-390540038.

References

[1] O. M. Yaghi, *ACS Cent. Sci.* **5**, 1295 (2019).

[2] J. Jiang, Y. Zhao, and O. M. Yaghi, *J. Am. Chem. Soc.* **138**, 3255 (2016).

[3] P. Kuhn, M. Antonietti, and A. Thomas, *Angew. Chem. Int. Ed.* **47**, 3450 (2008).

[4] J. Roeser, D. Prill, M. J. Bojdys, P. Fayon, A. Trewin, *et al.*, *Nat. Chem.* **9**, 977 (2017).

[5] X. Wang, K. Maeda, A. Thomas, K. Takanabe, G. Xin, *et al.*, *Nat. Mater.* **8**, 76 (2009).

[6] K. Kailasam, J. Schmidt, H. Bildirir, G. G. Zhang, S. Blechert, *et al.*, *Macromol. Rapid Commun.* **34**, 1008 (2013).

[7] J. Yang, A. Acharjya, M.-Y. Ye, J. Rabeah, S. Li, *et al.*, *Angew. Chem. Int. Ed.* **60**, 19797 (2021).
[8] X. Zhao, P. Pachfule, S. Li, T. Langenhahn, M. Ye, *et al.*, *J. Am. Chem. Soc.* **141**, 6623 (2019).
[9] C. Li, J. Yang, P. Pachfule, S. Li, M. Y. Ye, *et al.*, *Nat. Commun.* **11**, 4712 (2020).
[10] https://www.unisyscat.tu-berlin.de/.

https://doi.org/10.1142/9789811282324_0017

Discovery of Novel Nanoporous Materials Advanced by Electron Crystallography

Xiaodong Zou

Department of Materials and Environmental Chemistry, Stockholm University, 106 91 Stockholm, Sweden

My view of the present state of research on development of electron crystallography for reticular chemistry and new materials discovery

Reticular chemistry and computational materials design are accelerating the discovery of new materials. During the past two decades, nanoporous materials such as zeolites, metal–organic frameworks (MOFs), and covalent organic frameworks (COFs) have undergone tremendous developments. Based on the principles of reticular chemistry, numerous new nanoporous materials with desired topology, pore size, and functionalities have been synthesized. While powder X-ray diffraction (PXRD) and single crystal X-ray diffraction (SC-XRD) are the most commonly used techniques for phase analysis and structure determination, it is challenging to analyse samples containing multiple phases by PXRD and SC-XRD requires large single crystals. During the discovery of new materials, major efforts and time are often spent to optimize synthesis conditions to obtain large crystals and pure samples. On the other hand, electron crystallography has unique advantages in studying nano- and micrometre-sized crystals that are too small for SC-XRD and multiphasic samples too complex for PXRD. All polycrystalline powders are seen as single crystals on a standard transmission electron microscope (TEM).

Since the 1970s, *ab initio* structure solution has already been demonstrated on inorganic and organic crystals using high-resolution transmission electron microscopy (HRTEM) imaging and electron diffraction (ED) taken along crystallographic zone axes [1]. However, the methods have not been widely used until the development of the new 3D electron diffraction (3D ED) techniques [2]. 3D ED data are recorded from an arbitrarily oriented crystal, while the crystal is rotated during data collection. Before 3D ED, it was generally believed that the strong interaction of electrons with matter would make ED intensities dynamical and unusable for structural analysis. The 3D ED data from arbitrary crystal orientations are found to be less dynamical than zonal-axis ED data and can be treated as kinematical data during structure determination in the same way as SC-XRD data. Today, a complete 3D ED dataset can be obtained in a few minutes on a standard TEM, and numerous new structures, ranging from inorganic, organic, and protein crystals, have been determined by 3D ED. During the past decade, 3D ED has played a key role in the discovery of novel zeolites; 80% of the newly discovered zeolite structures since 2015 were solved by 3D ED [3]. Many beam-sensitive materials such as MOFs and COFs are also determined by 3D ED [4]. The 3D ED techniques have revolutionized crystallography and provided new opportunities for discovering novel structures and materials and exploring their properties and applications.

My current research contributions to reticular chemistry and electron crystallographic methods for structural analysis of new materials

For more than 30 years, our group has been developing electron crystallographic methods and applying them for the structure characterization of novel materials that cannot be studied by X-ray diffraction. We developed

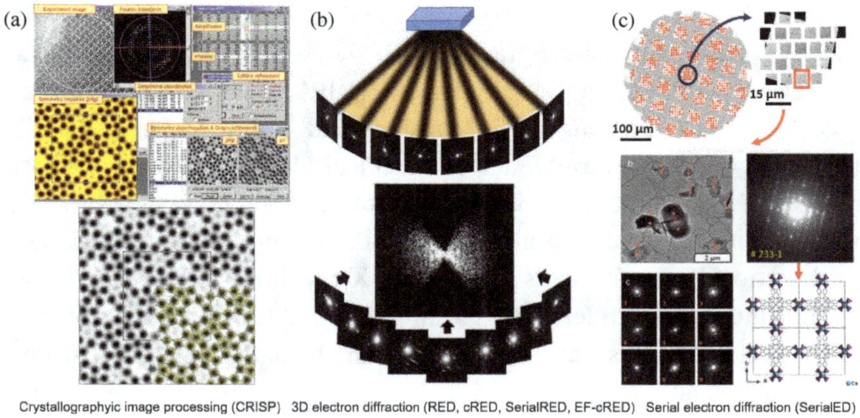

Fig. 1. Examples of electron crystallographic methods and software developed by our group. (a) Structure determination from HRTEM images by crystallographic image processing using the programme CRISP. (b) Rotation electron diffraction (RED) [5, 6], continuous rotation electron diffraction (cRED) [7], serial rotation electron diffraction (SerialRED) [9], and energy-filtered cRED (EF-cRED) [10] for 3D ED data collection and processing. (c) Serial electron diffraction (SerialED) for structure determination and phase analysis [8].

crystallographic image processing software CRISP for *ab initio* structure determination from HRTEM images (1992–2000, Fig. 1(a)) and demonstrated that complex structures could be solved from HRTEM images taken along different zone axes (1995–2012). We also developed the first software ELD for ED intensity quantification (1993) and unit cell determination and phase analysis (2004) and used them for phase identification and *ab initio* structure determination of new inorganic materials (1996–2004) [1].

Although we demonstrated *ab initio* crystal structure determination from nano- and micron-sized crystals by electron crystallography, the methods were both demanding and time-consuming, especially challenging for beam sensitive materials, such as zeolite, MOFs, and COFs. On the other hand, electron diffraction requires 100 times less electron dose than HRTEM imaging and has unique advantages for studying beam-sensitive materials. Since 2008, our group has developed several electron diffraction techniques, including rotation electron diffraction (RED) by combining fine beam tilt and rough goniometer tilt to obtain complete 3D ED data [5] and software RED for data collection and data processing [6]. In order to automate the ED data collection, we developed a software platform *Instamatic* and implemented continuous rotation electron diffraction (cRED) with crystal tracking [7]. We also developed serial (rotation) electron diffraction (SerialED and SerialRED) with fully automated crystal screening for single ED snapshots and high-throughput 3D ED data collection, respectively, and hierarchical cluster analysis for phase analysis [8, 9]. These make it possible not only for studying extremely beam-sensitive crystals but also for high-throughput phase analysis and for detection of minor phases invisible by X-ray diffraction. Recently, we implemented energy-filtered 3D ED in *Instamatic* [10]. The new electron diffraction techniques have made important breakthroughs in structure determination and phase analysis.

For more than 20 years, our group has also been working on the design and synthesis of novel nanoporous materials. We developed more than 80 open-framework germanates including the first synthesis of zeolite beta polymorph C (FOS-5) [11], a chiral zeolite SU-32 (STW) [12], and the first mesoporous oxide SU-M with gyroidal channels separated by crystalline walls [13]. Using 3D ED, my group determined more than 250 new structures, including zeolites, MOFs, and COFs (Fig. 2). We determined

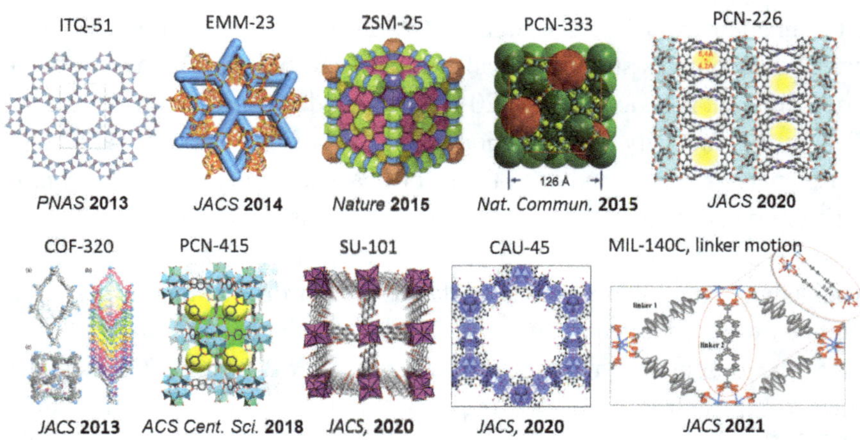

Fig. 2. Structures of novel nanoporous materials solved using 3D ED. (Top) Zeolites (16-ring ITQ-51 [18], 22-ring EMM-23 [14], 8-ring ZSM-25 [15]), Al-MOF PCN-333 [22], and Zr-MOF PCN-226 [25]. (Bottom) The first COF single crystal structure COF-320 with open and closed forms [21], PCN-415 MOF with bimetallic cluster [$Ti_8Zr_2O_{12}(COO)_{16}$] [23], Bi-MOF SU-101 [26], mesoporous Zr-MOF CAU-45 [24] where linker disorder was identified by 3D ED, and MIL-140C showing linker motions [28].

the structures of several zeolites that remained unsolved for decades [14, 15]. We developed a novel structure prediction approach that allowed targeted synthesis of new zeolites and discovered a family of zeolites with expanding complexity and embedded isoreticular structures [15, 16]. A large number of extra-large pore zeolites were also discovered and solved by electron crystallography [17–20]. We demonstrated the first single crystal structure determination of COFs by 3D ED [21] and solved the structures of many novel MOFs [22–25]. Detailed structural features such as hydrogen positions, disorders, and linker motions in MOFs could be identified [25–27]. Beyond the framework structures, 3D ED has also shown to be effective in studying host-guest interactions in MOF nanocrystals [28]. We demonstrated very recently high-throughput phase elucidation of highly complex polycrystalline zeolite materials using the automated SerialRED. Five different zeolite phases could be identified from a single synthesis product [29]. This includes phases with ultra-low contents or similar unit cell parameters that are unable to be identified using PXRD. SerialRED provides new opportunities for the exploration of complex material synthesis systems and the rapid development of polycrystalline materials.

How can emerging electron diffraction techniques accelerate the discovery of new materials?

Structural studies at atomic levels are indispensable to fundamentally understand physical and chemical properties, which is in turn important for the development of new materials. Because 3D ED is powerful for the structure determination of nanocrystals from samples containing many different phases, chemists can take advantage of this technique to explore new strategies that can produce multi-phasic samples and directly use SerialED and SerialRED instead of PXRD and SC-XRD for phase screening.

The 3D ED techniques have revolutionized crystallography and provided new opportunities for discovering novel structures and materials and exploring their properties and applications. As TEMs are widely available in laboratories around the world compared to synchrotron facilities, we foresee the significance of 3D ED for structural analysis will continue to increase, which will certainly accelerate the discovery of novel materials and research in the fields of porous crystalline materials.

Acknowledgements

This work was supported by the Swedish Research Council (2019–00815), the Knut and Alice Wallenberg Foundation (2018.0237), and SciLifeLab. I acknowledge the past and current students and postdocs in the group who made great contributions to the work and the collaborators who provided interesting samples and challenging problems.

References

[1] X. Zou, S. Hovmöller, and P. Oleynikov, *Electron Crystallography — Electron Microscopy and Electron Diffraction, IUCr Texts on Crystallography 16*, Oxford University Press (2011).
[2] M. Gemmi, E. Mugnaioli, T. E. Gorelik, U. Kolb, *et al.*, *ACS Cent. Sci.* **5**, 1315 (2019).
[3] Z. Huang, T. Willhammar, and X. Zou, *Chem. Sci.* **12**, 1206 (2021).
[4] Z. Huang, E. S. Grape, J. Li, A. K. Inge, and X. Zou, *Coord. Chem. Rev.* **427**, 213583 (2021).
[5] D. Zhang, S. Hovmöller, P. Oleynikov, and X. Zou, *Z. Kristallogr.* **225**, 94 (2010).
[6] W. Wan, J. L. Sun, J. Su, S. Hovmöller, and X. Zou, *J. App. Crystallogr*. **46**, 1863 (2013).
[7] M. O. Cichocka, J. Ångström, B. Wang, X. Zou, and S. Smeets, *J. Appl. Cryst.* **51**, 1652 (2018).
[8] S. Smeets, X. Zou, and W. Wan, *J. Appl. Cryst.* **51**, 1262 (2018).
[9] B. Wang, X, Zou, and S. Smeets, *IUCrJ*, **6**, 854 (2019).
[10] T. Yang, H. Xu, and X. Zou, *J. Appl. Cryst.* **55**, 1583 (2022).
[11] T. Conradsson, M. S. Dadachov, and X. Zou, *Micro Mesoporous Mater.* **41**, 183 (2000).
[12] X. Zou, T. Conradsson, M. Klingstedt, M. S. Dadachov, and M. O'Keeffe, *Nature* **437**, 716 (2005).
[13] L. Q. Tang, L. Shi, C. Bonneau, X. Zou, *et al.*, *Nat. Mater.* **7**, 381 (2008).
[14] T. Willhammar, A. W. Burton, Y. Yun, J. Sun, X. Zou, *et al.*, *J. Am. Chem. Soc.* **136**, 13570 (2014).

[15] P. Guo, J. Shin, S. B. Hong, P. A. Wright, X. Zou, *et al.*, *Nature* **524**, 74 (2015).
[16] J. Shin, H. Xu, P. A. Wight, X. Zou, S. B. Hong, *et al.*, *Angew. Chem. Int. Ed.* **55**, 4928 (2016).
[17] J. Sun, C. Bonneau, Á. Cantín, A. Corma, X. Zou, *et al.*, *Nature* **458**, 1154 (2009).
[18] R. Martínez-Franco, M. Moliner, Y. Yun, X. Zou, A. Corma, *et al.*, *Proc. Natl. Acad. Sci.* **110,** 3749 (2013).
[19] T. Willhammar, A. W. Burton, Y. Yun, J. Sun, X. Zou, *et al.*, *J. Am. Chem. Soc.* **136**, 13570 (2014).
[20] J. Jiang, Y. Yun, X. Zou, J. L. Jorda, A. Corma, *Chem. Sci.* **6**, 480 (2015).
[21] Y. Zhang, J. Su, H. Furukawa, X. Zou, O. Yaghi, *et. al.*, *J. Am. Chem. Soc.* **135,** 16336 (2013).
[22] D. Feng, T.-F. Liu, J. Su, X. Zou, H.-C. Zhou, *et. al.*, *Nat. Commun.* **6,** 5979 (2015).
[23] S. Yuan, J.-S. Qin, H. Xu, X. Zou, H.-C. Zhou, *et al.*, *ACS Cent. Sci.* **4,** 105 (2018).
[24] S. Leubner, V. E. G. Bengtsson, K. Synnatschke, X. Zou, N. Stock, *et al.*, *J. Am. Chem. Soc.* **142**, 15995 (2020).
[25] M. O. Cichocka, Z. Liang, X. Zou, H.-C. Zhou, Z. Huang, *et al.*, *J. Am. Chem. Soc.*, **142**, 15386 (2020).
[26] E. S. Grape, J. G. Flores, T. Hidalgo, I. A. Ibarra, A. K. Inge, *et al.*, *J. Am. Chem. Soc.*, **142**, 16795 (2020).
[27] Y. Luo, M. T. B. Clabbers, J. Qiao, Z. Yuan, X. Zou, *et. al.*, *J. Am. Chem. Soc.* **144**, 10817 (2022).
[28] L. Samperisi, A. Jaworski, K. P. Lillerud, X. Zou, Z. Huang, *et al.*, *J. Am. Chem. Soc.* **143**, 17947 (2021).
[29] B. Wang, T. Rhauderwiek, A. K. Inge, N. Stock, X. Zou, *et al.*, *Chem. Eur. J.* **66**, 17429 (2018).
[30] Y. Luo, B. Wang, S. Smeets, W. Yang, X. Zou, *et al.*, *Nat. Chem.* (2021) https://doi.org/10.1038/s41557-022-01131-8.

https://doi.org/10.1142/9789811282324_0018

Ab initio Free Energy Predictions with Chemical Accuracy: Adsorption and Catalysis in Nanoporous Materials

Joachim Sauer

Institut für Chemie, Humboldt-Universität zu Berlin, 10099 Berlin, Germany

Rational design of materials for adsorption and catalysis

Compared to solid-supported catalysts which have complex structures that are not ordered into unit cells, reticular materials like zeolites and metal–organic frameworks (MOFs) combine the advantages of reduced complexity and periodicity with a large structural variability and scalability. In zeolites, the networks of corner-sharing TO_4-tetrahedra (T = Si, Al^-) form a large variety of frameworks with different pore topologies and sizes [1]. The frameworks can have different Si/Al ratios and their negative charge can be compensated by different (metal) cations. Compensation with protons results in the large family of Brønsted acid catalysts. For the cosmos of one-, two-, and three-dimensional reticular structures, I refer to the contributions of Yaghi (3.1) and Fujita (3.5) in this chapter.

The design of materials is based on understanding gained from experiments and computation. An example is the development of the SAPO-34 catalyst for the UOP/HYDRO Methanol-to-Olefine process [2]. The small pore size leads to high selectivity for ethene and propene production, and the optimized acid function leads to much less formation of paraffinic by-products. In more recent work on MOFs [3], machine learning in combination with Monte Carlo simulations has been used to identify a top-performing material for H_2 storage within a database of more than 50,000 experimental structures. The material was synthesized and the measured isotherms agreed well with the predictions. The latter example demonstrates the increasing importance of data science as the fourth pillar of science in addition to experiment, computation, and theory.

The design of improved materials and — not less important — the understanding of existing ones require the *ab initio* prediction of adsorption constants with no other input than the positions of the atoms. Adsorption isotherms describe the adsorbed amount (given as loading Θ) as a function of the gas pressure p:

$$\Theta = K\,p/(K\,p + 1).$$

The adsorption (Henry) constant K is obtained from the Gibbs free energy of adsorption:

$$\Delta G_a = -RT \ln K.$$

To be useful, the Gibbs free energy predictions have to be chemically accurate (≈ 4 kJ/mol). For realistic models of molecule–surface interactions with hundreds of atoms in the simulation cell, this is a challenging problem of computational quantum chemistry.

Divide-and-conquer approach for *ab initio* calculation of Gibbs free energies

Free energy simulations require (i) a method to calculate the potential energy surface (PES) and (ii) a method to sample the PES. The computer time for both parts scales nonlinearly with the number of atoms in the simulation cell. To make accurate calculations for large simulation cells affordable, we use a divide-and-conquer approach. We have developed a methodology that relies on the Taylor expansion of the PES around equilibrium structures [4, 5].

(i) For calculating the PES, we use wave-function methods as high-level methods in hybrid QM:QM calculations (QM — quantum mechanics). Specifically, we use second-order Møller–Plesset Perturbation Theory (MP2). As a low-level QM method, we employ density functional theory with some account of dispersion (DFT+D), the workhorse in solid state and surface calculations.

(ii) We calculate *anharmonic* vibrational frequencies for each degree of freedom separately. The one-dimensional potentials are sampled in *curvilinear* coordinates.

Figure 1 summarizes the implemented methods compared to the current standard approach in computational adsorption/catalysis. For small alkanes in zeolite H-chabazite [5], we have demonstrated that both enthalpies and entropies of adsorption can be calculated with chemical accuracy. For this system, the computer time for a full MP2 calculation [6] is five orders of magnitude larger than for a DFT+D calculation (PBE+D2). Our hybrid MP2:(PBE+D2) calculation is only two orders of magnitude more expensive than the DFT+D calculation.

Ab initio isotherm predictions with chemical accuracy

For the adsorption of CO and N_2 in Mg-MOF-74, a comparison has been made with measured adsorption isotherms. Figure 4 in Ref. [7] shows the isotherms calculated in the harmonic approximation and with

$$K = (Q_a/Q_0)\exp(-\Delta E/RT) = \exp(-\Delta S/R)\exp(-\Delta E/RT)$$

$$k = (kT/h)(Q^{\neq}/Q_0)\exp(-\Delta E^{\neq}/RT) = (kT/h)\exp(-\Delta S^{\neq}/R)\exp(-\Delta E^{\neq}/RT)$$

$$Q = \Sigma_i \exp(-\varepsilon_i/k_BT)$$

$\Delta G =$	$-RT\ln(Q_1/Q_0) + \Delta E_{ZPV}$	$+\ \Delta E$
	Vibrational energies	Energy of stat. points
Standard	harmonic	DFT+D
This work	anharmonic	hybrid MP2:DFT+D + ΔCCSD(T)

Chemical accuracy (1 kcal/mol ≈ 4 kJ/mol)
periodic boundary conditions, 1000 atoms per unit cell

Fig. 1. Implemented methods for calculating equilibrium and rate constants compared to the standard approach for molecule — Surface interactions.

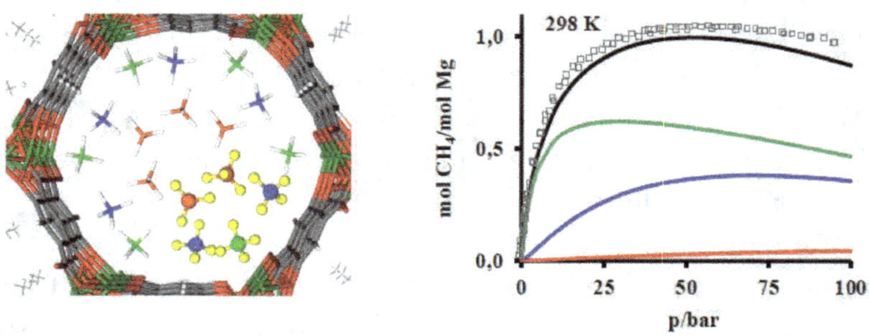

Fig. 2. CH_4 adsorption in Mg-MOF-74 at different sites (left), Mg^{2+} — green, linker — blue, second layer — red, and their contributions to the total (black line) excess adsorption isotherm (right) [8]. The experimentally determined availability (78%) of adsorption sites is assumed. Experimental data points [12] are given as squares.

anharmonic partition functions. The latter are in very good agreement with the experiment. Comparison with isotherms obtained after changing the Gibbs free energy (or energy) of adsorption by 1 and 4 kJ/mol shows that the accuracy of our anharmonic free energy calculations is better than 1 kJ/mol.

We have studied the adsorption of CH_4, CO_2, and H_2O in MOFs with regard to storage and separation of small energy-carrying molecules, carbon capture, and water harvesting, see Refs. [8–11]. Figure 2 shows

CH_4 molecules adsorbed at different sites in Mg-MOF-74 (which are coloured differently) and their contributions to the total (excess) adsorption isotherm [8]. This decomposition provides valuable information for a rational design of improved materials. However, this good agreement is not obtained for the ideal perfect Mg-MOF-74 crystal structure, but only if it is assumed that only 78% of the adsorption sites in the perfect structure are accessible for the adsorption of CH_4. This percentage has been determined experimentally (isosteric heats of adsorption as a function of loading) and explained with "a reduced rate of diffusion at such loadings" or with the fact that "the remaining metal sites are physically obstructed from access because of defects in the crystal structure" [12].

Outlook

Having a method for predicting adsorption isotherms with chemical accuracy for ideal MOF (or zeolite) structures, *deviations from the experiment may indicate imperfections of the sample or the measurements.* Together with progress in experimental structure characterization, this opens the door for understanding the structures of "real" materials which are employed for adsorption and catalysis. A recent example is the combined low-dose HRTEM and computational study on MOF UiO-66 [13] in which "… missing-linker defects were observed in various ... samples, including those that would have typically been assumed to be essentially defect-free, which underlines *that notionally perfect materials can contain defects invisible to most characterization techniques* that may influence observed variance in properties of MOFs such as gas uptake."

Imperfections of the materials may also explain the wide variation of experimental isotherms for H_2O adsorption in Mg-MOF-74, obtained in different laboratories or in the same laboratory using different synthesis procedures, see Ref. [14] for the original references. The similarly wide variation of the isotherms predicted with different simulation techniques [14] calls for chemically accurate *ab initio* calculations for this system which are in progress.

Not less important with respect to the rational design of improved materials for water harvesting is the 1:1 connection between points on the isotherm and the structures of the corresponding number of H_2O

molecules in the MOF pore which have been determined in parallel with single crystal-XRD experiments and quantum chemical calculations [11].

Acknowledgements

This work has been supported by the German Research Foundation (DFG) within the priority program 1570 "Porous media" and with a Reinhart Koselleck grant, as well as by the Fonds der Chemischen Industrie.

References

[1] C. Baerlocher and L. B. McCusker, *Database of Zeolite Structures*. www.iza-structure.org/databases (2021).
[2] B. Vora, J. Q. Chen, A. Bozzano, B. Glover, and P. Barger, *Catal. Today* **141**, 77 (2009).
[3] B. J. Bucior, N. S. Bobbitt, T. Islamoglu, S. Goswami, A. Gopalan, *et al.*, *Mol. Syst. Des. Eng.* **4**, 162 (2019).
[4] J. Sauer, *Acc. Chem. Res.* **52**, 3502 (2019).
[5] G. Piccini, M. Alessio, J. Sauer, Y. Zhi, Y. Liu, *et al.*, *J. Phys. Chem. C* **119**, 6128 (2015).
[6] F. Göltl, A. Gruneis, T. Bucko, and J. Hafner, *J. Chem. Phys.* **137**, 114111 (2012).
[7] A. Kundu, G. Piccini, K. Sillar, and J. Sauer, *J. Am. Chem. Soc.* **138**, 14047 (2016).
[8] K. Sillar and J. Sauer, *J. Am. Chem. Soc.* **134**, 18354 (2012).
[9] K. Sillar, A. Kundu, and J. Sauer, *J. Phys. Chem. C* **121**, 12789 (2017).
[10] A. Kundu, K. Sillar, and J. Sauer, *Chem. Sci.* **11**, 643 (2020).
[11] N. Hanikel, X. Pei, S. Chheda, H. Lyu, W. Jeong, *et al.*, *Science* **374**, 454 (2021).
[12] P. D. C. Dietzel, V. Besikiotis, and R. Blom, *J. Mater. Chem.* **19**, 7362 (2009).
[13] L. Liu, Z. Chen, J. Wang, D. Zhang, Y. Zhu, *et al.*, *Nat. Chem.* **11**, 622 (2019).
[14] E. O. Fetisov, M. S. Shah, J. R. Long, M. Tsapatsis, and J. I. Siepmann, *Chem. Commun.* **54**, 10816 (2018).

https://doi.org/10.1142/9789811282324_0019

Coordination Self-Assembly: Past, Present, and Future

Makoto Fujita
Department of Applied Chemistry, The University of Tokyo, Bunkyo-ku, Tokyo 113–8656, Japan

Short Summary: Since 1990, our group has been developing a new construction principle, featured as "coordination self-assembly", leading to the spontaneous formation of cyclic structures, catenanes, three-dimensional cages, and so on that are assembled from a large number of transition metal centers and simple coordinating organic molecules. A Pd(II)-cornered square complex served as the first example of coordination assembly that generated not only a framework but also a "space", a potential and excellent platform for creating new properties and functions. The cages can be gigantic and behave as large molecular containers, leading to new chemical properties and reactivity of the substrates entrapped in these cages. In 1994, the same principle for cavity construction was applied to the preparation of an infinite coordination network that showed molecular inclusion in the cavity. Recently, single-crystal-to-single-crystal guest-exchange observed in the porous coordination network led to a revolutionary method for determining X-ray structures of small molecules which were encapsulated in the crystalline self-assembled cages.

Over the last three decades, self-assembly based on coordination chemistry has made remarkable developments, and various types of well-defined nanostructures such as helices, macrocycles, cages, and so on have been constructed from a large number of transition metal centres and simple coordinating organic molecules. In 1990, our group reported a spontaneously assembled molecular square with 4 palladium(II) atoms at its corners [1]. This report served as one of the earliest examples of coordination self-assembly, a new principle of metal-guided synthesis, being antithetic to organic synthesis, of the structure of matter on the nanoscale with an enormous impact on the molecular sciences.

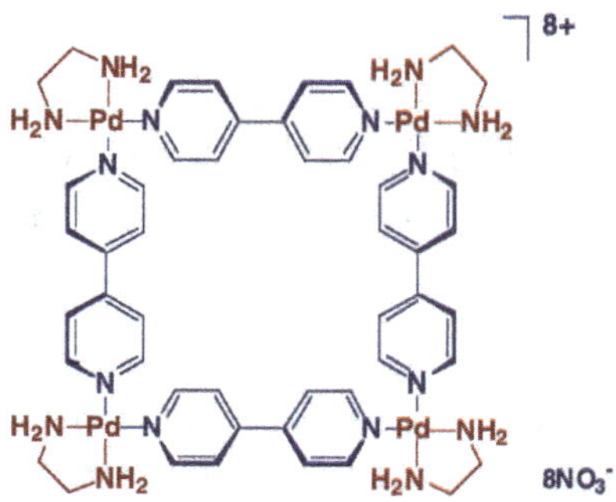

Even back in the 1980s, a few milestone works had already existed for molecular self-assembly around metal centres, for example, Cu(I)-templated catenane (Sauvage 1983) and double helical complexes (Lehn 1987). These earlier examples however did not show any particular functions or properties, in this regard comparable to most of the metal complexes formed under thermodynamic conditions. The square complex is distinctive from the others in that self-assembly generated not only frameworks but also cavities that are excellent platforms to create new properties and functions through molecular recognition therein. The square's two-dimensional (2D) framework was subsequently extended into three-dimensional (3D) frameworks like cages, capsules, bowls, tubes, spheres, and catenanes [2] (Fig. 1). Extremely simple procedures strikingly dominate over the tedious ones of previous covalent syntheses. The results enjoy its unique status for (i) the facile creation of large hydrophobic cavities, (ii) very strong binding of neutral molecules in the cavities, and (iii) unprecedented physical and chemical phenomena within the cavities.

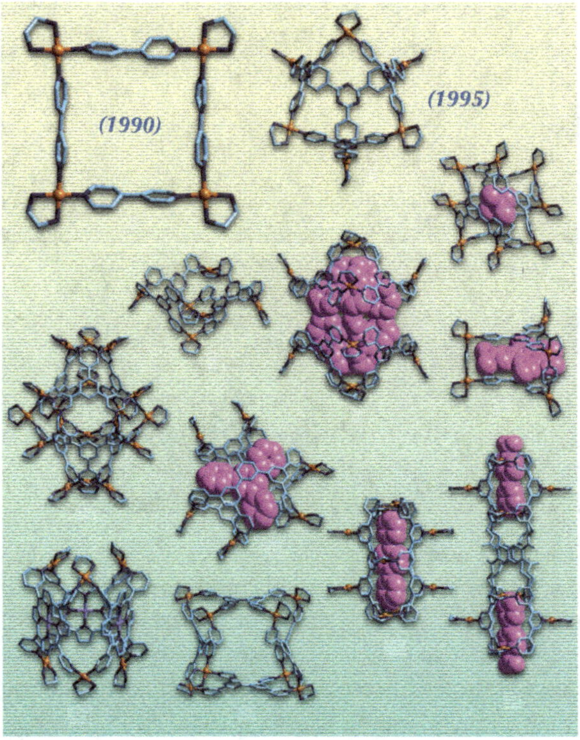

Fig. 1. Coordination self-assemblies with cavities.

In this chapter, among many contributions from our group to this field, several key self-assembled structures are disclosed emphasizing the past, the present, and the future of the field.

Molecular confinement effects in the self-assembled cages [3]

The cavities of self-assembled cages are extraordinarily large and are capable of binding neutral guests. Unique functions are created from the self-assembled architectures. In the M_6L_4 cage, four large guests (e.g., adamantane and carborane) are encapsulated. Two different guests can be accommodated in a pairwise selective fashion, facilitating reaction design in the

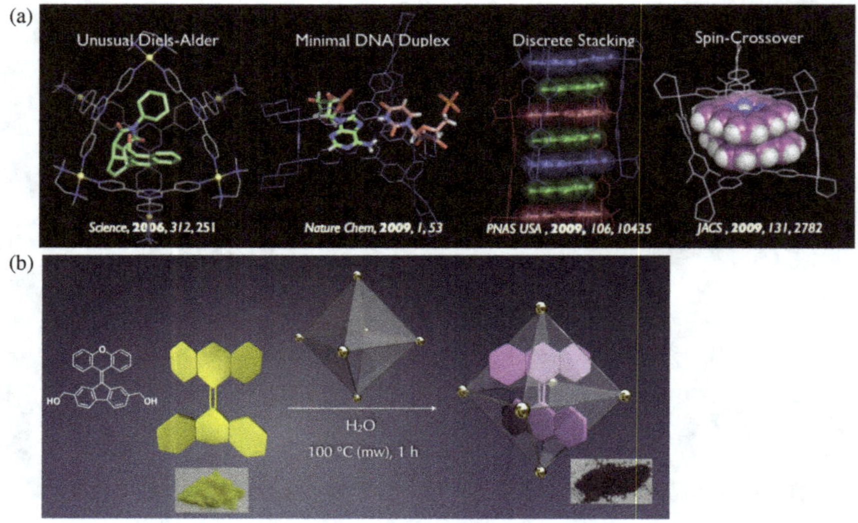

Fig. 2. (a) Examples of new reactions and properties developed in the self-assembled cavities. (b) "Twisting" manipulation of an over-crowded olefin to show biradical nature.

cavity. In the recognition of biomolecule fragments (oligopeptides or short nucleotides), their *in vivo* structures are reproduced in the cages in water.

Reactivity and catalysis controlled by cavities represent one of the most important functional properties of self-assembled hosts. The photodimerization of olefins as well as Diels–Alder reactions in the cage are featured by remarkable rate enhancement (>10^2 times), perfect regio- and stereo-selection, high pairwise selection, and even chiral induction up to 50% ee. Notably, unreactive aromatics, even naphthalene, smoothly undergo the Diels–Alder reaction. In a single crystalline state reaction, extremely unstable reaction intermediate generated in the cage can be examined by X-ray analysis (Fig. 2(a)).

Through confinement in the cavity, small molecules can be mechanically manipulated to alter their properties. For example, an over-crowed olefin in an anti-folded conformation was transformed into a twisted conformation revealing a biracial nature due to the poorer overlap of p-orbitals. An aromatic amide was bent into a non-planer conformation to be easily hydrolyzed due to the disruption of amide conjugation (Fig. 2(b)). Spin crossover, stable organic mixed valent states, metal–metal bonding, spin–spin interaction, and photo-induced guest–host electron/energy transfer have been observed through the accommodation of appropriately designed substrates.

Polyhedrons with protein size cavities [4]

Self-assembly of 5-nm sized cuboctahedral $M_{12}L_{24}$ completes and "EndoChemistry" at their endo-surface and inner space were developed (Fig. 3). Discrete inorganic synthesis was demonstrated by the precise endo-template synthesis of silica in the sphere. The sphere can encapsulate even a protein within the shell, providing potential methods for controlling protein functions and for protein structure determination.

Self-assembly of gigantic M_nL_{2n} complexes under mathematical restriction [5]

We also reported $M_{24}L_{48}$ rhombicuboctahedron and $M_{30}L_{60}$ icosidodecahedron, all of which belong to Archimedean solids. Recently, a non-Archimedean $M_{30}L_{60}$ polyhedron was reported, whose topology is

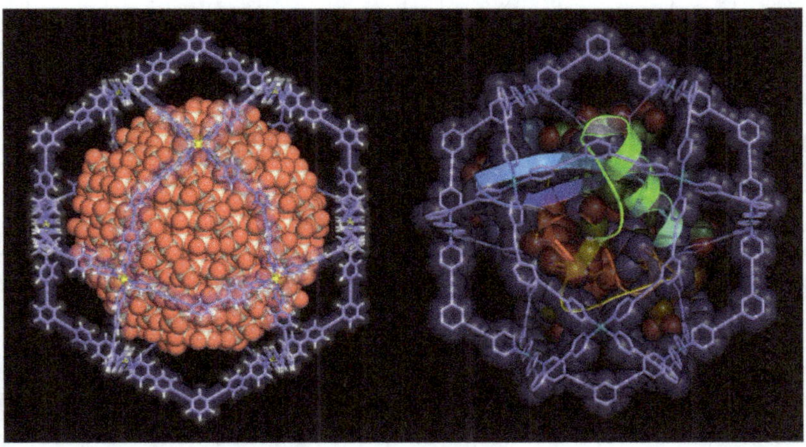

Fig. 3. Examples of Endo-Chemistry in the 5-nm sized $M_{12}L_{24}$ spheres. Left: Perfectly monodisperse silica nanoparticle (Mw/Mn = 1.004) prepared in the sphere. Right: A protein encapsulated by the sphere.

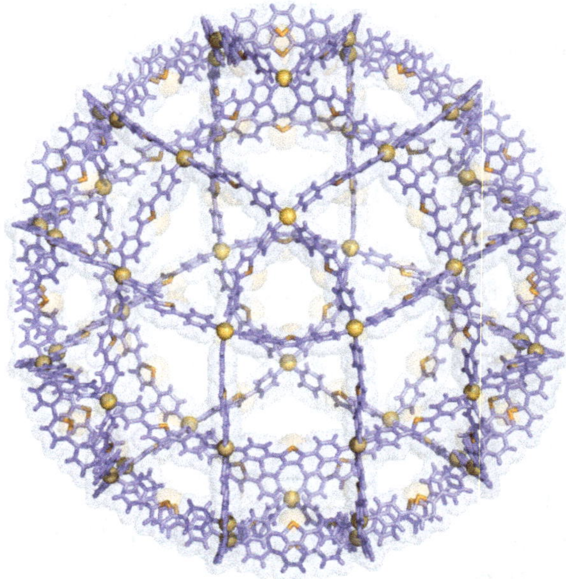

Fig. 4. X-ray structure of $M_{48}L_{96}$ complex.

mathematically rationalized based on a theory of tetravalent Goldberg polyhedra made up of squares and triangles. The self-assembly of $M_{48}L_{96}$ (Fig. 4) was achieved, which was predicted by the theory built by ourselves.

Porous coordination network and crystalline sponge method

Since 1994, coordination self-assembly has been applied to the construction of porous coordination networks [6], currently termed as MOFs. Only a few works were acknowledged in the 1980s with attention to the void of the structures (Robson, 1988). Our persistent endeavour in this field is to unambiguously confirm the events in the pore by X-ray analysis, based on the single-crystal-to-single-crystal guest exchange phenomena. A variety of organic transformations and even reaction snapshots have been observed.

Recently, these phenomena led to a revolutionary method for determining X-ray structures of compounds without sample crystallization

Fig. 5. Cartoon presentation of the crystalline sponge method.

(Fig. 5) [7]. The method has quickly arisen as a transformative technology and has been accelerating chemistry as well as many related disciplines not only in academia but also in industries.

Perspective

In traditional disciplines, organic chemists create new "shapes" (i.e., new frameworks, stereochemistries, and nanostructures), whereas inorganic chemists do new "states" (i.e., new spin state, oxidation state, and exited states). The scientific significance in coordination self-assembly is that a new interdiscipline in chemistry has been developed, in which inorganic chemistry creates new shapes.

References

[1] M. Fujita, J. Yazaki, and K. Ogura, *J. Am. Chem. Soc.* **112**, 5645 (1990).
[2] M. Fujita, F. Ibukuro, H. Hagihara, and K. Ogura, *Nature* **367**, 720 (1994).
[3] M. Fujita, D. Oguro, M. Miyazawa, H. Oka, K. Yamaguchi, and K. Ogura, *Nature* **378**, 469 (1995).
[4] M. Tominaga, K. Suzuki, M. Kawano, T. Kusukawa, T. Ozeki, S. Sakamoto, K. Yamaguchi, and M. Fujita, *Angew. Chem. Int. Ed.* **43**, 5621–5625 (2004).
[5] D. Fujita, Y. Ueda, S. Sato, N. Mizuno, T. Kumasaka, and M. Fujita, *Nature* **540**, 563–566 (2016).
[6] M. Fujita, Y. J. Kwon, S. Washizu, and K. Ogura, *J. Am. Chem. Soc.* **116**, 1151 (1994).
[7] Y. Inokuma, S. Yoshioka, J. Ariyoshi, T. Arai, Y. Hitora, K. Takada, S. Matsunaga, K. Rissanen, and M. Fujita, *Nature* **495**, 461 (2013).

https://doi.org/10.1142/9789811282324_0020

Colloidal Crystal Engineering with DNA: Repurposing the Blueprint for Life

Chad A. Mirkin[*,†,‡], Kyle J. Gibson[*,‡] and Kaitlin M. Landy[*,‡]

*Department of Chemistry, Northwestern University, Evanston, IL 60208, USA

†Department of Materials Science and Engineering, Northwestern University, Evanston, IL 60208, USA

‡International Institute for Nanotechnology, Northwestern University, Evanston, IL 60208, USA

Outcompeting nature using DNA as a structure-directing group

Nature has evolved to use DNA to store the genetic instructions for life, and all genetic data can be coded using only four nucleobases. When it comes to defining structure on the cellular level, however, nature has elected to employ polypeptides, as opposed to nucleic acids, as its building blocks of choice to construct functional entities, such as enzymes and tissues. In some respects, this is surprising given the complex relationship between amino acid sequence and protein folding as well as the myriad interfacial attractive and repulsive forces that govern protein–protein interactions and hierarchical structure formation. In our research, we have sought to highjack DNA for a different purpose: as both a structure-directing agent and structural material. Indeed, in addition to encoding genetic information, DNA also can be used as a bonding modality for dictating highly specific structural outcomes. Its sequence-specific interactions, programmable length with defined secondary structure, and tunable interaction strength make it a versatile synthon for interconnecting elementary non-nucleic acid building blocks and defining structure across multiple length scales. Oligonucleotides exhibit significantly longer persistence lengths and can be more deliberately and reliably tailored with respect to recognition properties than polypeptides, making them perhaps the materials of choice when attempting to create a new paradigm for bonding and structure design through a straightforward set of design rules. Indeed, a grand challenge for scientists is to outperform the structural control that nature has evolved over millions of years by utilizing DNA to both guide and control structural outcomes.

DNA as a versatile and unrivalled ligand for colloidal crystal engineering

Colloidal crystals are highly ordered arrays of nanoscale building blocks with properties that derive from their habits, crystal symmetries, lattice parameters, and constituent building blocks. They are found naturally, for example, in the form of opals or butterfly wings, where they are a source of structural colour. In addition, they have been prepared in many forms in the laboratory [1]. Approaches to prepare them have relied on electrostatic assembly, evaporation, and small molecule particle-interconnecting organic ligands [2]. However, these approaches do not offer the ability to programme structure outcome. The suite of interactions available through these approaches are limited with respect to length scale and lack orthogonality, which prevents the design and synthesis of targeted crystals. Therefore, a macromolecular ligand, such as DNA, that spans a greater length scale and that can be encoded with orthogonal interactions is needed.

To take advantage of DNA as a structure-directing material (a type of "source code"), we have explored colloidal crystals as a model system in which we can use hybrid DNA–nanoparticle conjugates as our building blocks. These DNA-functionalized nanoparticles are termed programmable atom equivalents (PAEs), as they represent the elementary blocks used to make all hierarchical materials in this field. Notably, unlike conventional atoms, the bonding characteristic of PAEs is independent of "atom identity" (nanoparticle core). This observation allows one to use DNA to create intuitive pathways based upon geometry-guided rules to new materials inspired by nature and ones with no equivalent in nature [3]. Colloidal crystals assembled using DNA utilize multivalent DNA hybridization interactions to drive cooperative binding between particles, enabling reliable, deliberate, and designed crystallization [4], Fig. 1. The guiding principle in this work is that the thermodynamically favoured structure will be the one that maximizes particle–particle contacts that lead to hybridization — this maximized bonding argument is known as the complementary contact model (CCM) [5].

The use of DNA as a ligand in colloidal crystal assembly goes beyond simply improving the colloidal stability of nanoparticles or providing a "glue" to hold the structure together [6]. This approach repurposes DNA from the blueprint of life to a blueprint for encoding a final crystal structure with precise control over design parameters, including lattice symmetry, nanoparticle spacing, and crystal habit. The outer DNA shell is primarily responsible for the process of crystallization, which means the chemical composition of the core material is entirely independent of the crystallization process. This allows any material to be incorporated into a crystal as long as it can be chemically modified with a dense shell of DNA [7]. In chemical terms, this is a seismic departure from traditional atom-based chemistry. For any given element, the identity of the atom is intrinsically linked to the bonding characteristics of that element; a fundamental tenet in chemistry is that two atoms cannot have identical bonding behaviour without also having identical nuclei.

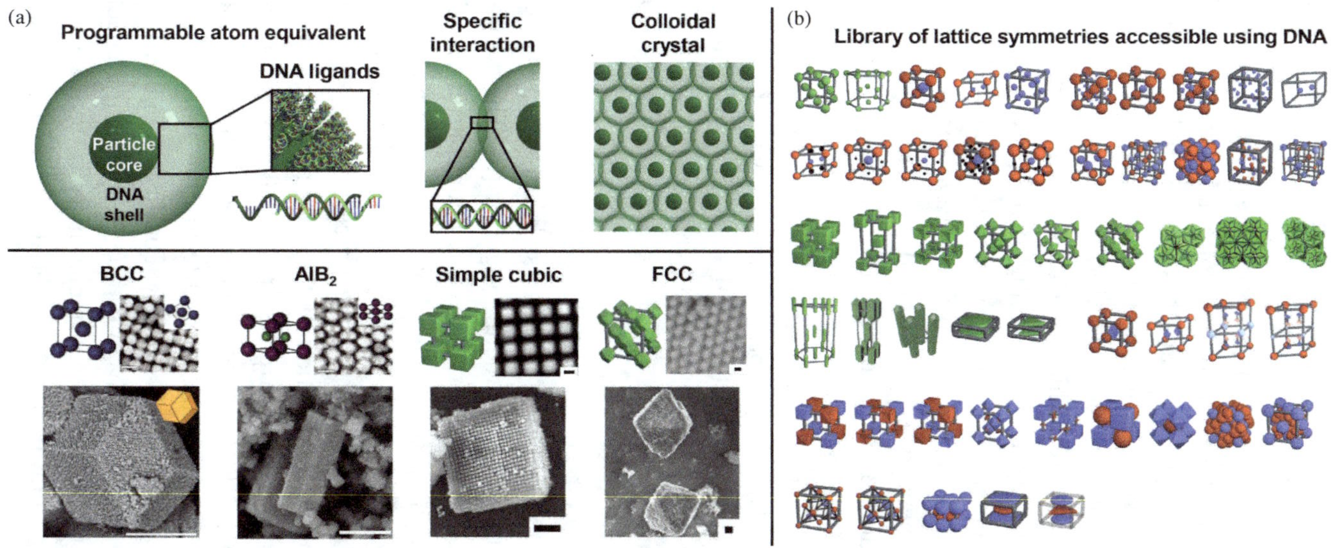

Fig. 1. a) Schematic describing PAEs and the crystallization process. (b) Electron microscopy characterization for select lattice symmetries and crystal habits. Scale bars: 50 nm (top) and 1 μm (bottom). (c) Unit cells from the library of over 70 synthesized to date.

Furthermore, the massive design space of DNA (both length and sequence) has led to thousands of crystals spanning over 70 different symmetries, all based upon the CCM [5, 8]. Accessible lattice symmetries range from relatively simple face or body-centred cubic lattices from the earliest demonstrations to complex lattices with large unit cells such as perovskites and clathrates, and several of these lattice symmetries have no mineral equivalent [9]. Introducing directional binding by using anisotropic nanoparticles with a shell of DNA of the appropriate length to retain the underlying nanoparticle shape enables particle assembly based on facet registration, which has dramatically expanded the scope of possible structures [10, 11]. Interestingly, when the particles are reduced in size, they will randomly diffuse through a lattice defined by larger complementary particles. These particles are referred to as electron equivalents and establish a new type of bonding in colloidal crystals akin to metallic bonding [12, 13]. In addition to providing exquisite control over lattice symmetries and parameters, this approach has provided a route to faceted macroscopic single crystals spanning eight different crystal habits to date [14–16]. Moreover, recent work on templating colloidal crystal growth on substrates has led to methods for device integration and controlling crystal position and orientation on a surface [17].

Colloidal crystals comprised of PAEs exhibit emergent properties, distinct from both the nanoparticle cores and DNA shells. For example, structures comprised of gold and silver particles can be designed to exhibit unusual catalytic behaviour [18] and photonic modes [19–21], and the stimuli-responsive nature of DNA can be used to dynamically tune such properties [22]. Additionally, it was recently discovered that DNA-mediated colloidal crystals exhibit a shape-memory property due, in part, to the hyperelastic nature of the DNA [23]. Upon drying, the removal of water around the DNA compresses it by ~90% and destroys the crystallinity of the sample. But upon rehydration, the crystals are restored to their original state, including intact habit. Despite the bonds responsible for maintaining the ordering of the crystal being fully disrupted, the DNA provides a memory capable of moving the particles in space to return them to their original positions in a type of self-healing process. This phenomenon is unlike any other in self-healing materials in that it restores all of the particle building blocks essentially to their original states, and it comes from the inherent ability of the DNA to form encoded bonds between the particles. Metaphorically, in this example, DNA serves as the blueprint, the construction worker, and the glue for creating such structures.

Challenges and opportunities in repurposing the blueprint of life

Historically, the focus in this field has been on exploring what is structurally possible and building a library of different lattices in an effort to understand the design rules that govern their formation. However, with the CCM and the data compiled by hundreds of research groups to date, we are approaching the point where many new crystalline materials can be targeted and reverse-engineered. A day is coming soon when computational programmes will take a target structure, with a given lattice structure and composition, and produce the conditions and particle and DNA building blocks required to make that structure. The phase space that can be explored is very rich and much more diverse than conventional chemistry. Because "atom identity" and bonding characteristics are independent of one another in this field, particles that normally would not interact can be forced to interact with one another. This observation alone allows one to dramatically expand the scope of possibilities beyond what we observe in nature. The arsenal of chemistry aimed at producing colloids comprised of unusual particle shapes and sizes creates tremendous opportunities in this regard. Take, for example, hollow particles — they could lead to a class of open architecture systems similar to MOFs and zeolites but spanning a larger and more difficult to access length scale. Indeed, the next developments in this field will focus on higher orders of structural control.

The building block parameter set is rich: particle composition, size, shape and structure (e.g., filled, hollow, alloy, and heterostructure), and nucleic acid length and sequence. Yet, the crystal product/target parameter set is even richer with an almost infinite number of features. Although we can reliably target many structures, we must gain additional understanding to exert greater synthetic control. All crystal features impact their properties, but all such features in this field are intrinsically linked and guided through the design of the DNA. Full control of crystalline structure will require the ability to decouple these features from one another. In addition, as the field moves towards functional colloidal crystals with properties by design, multicomponent architectures are of particular interest to create materials with properties that exceed what is possible with a single nanoparticle building block. Multicomponent architectures represent an ongoing challenge on the lattice level (i.e., increasing complexity required for each component to occupy a distinct lattice site) and on the mesoscale (i.e., core–shell, Janus, or other architectures).

Leveraging the unique structural control afforded by colloidal crystal engineering with DNA to create materials with functional properties is another ongoing challenge. Current studies on optically active colloidal crystals suggest that they may find uses as lenses, wavelength-dependent mirrors, and optical switches in a variety of nanoscale devices [24–26]. Light-matter interactions on this length scale lead to highly unusual properties, and the dynamic tunability of colloidal crystals engineered with DNA offers an attractive avenue for investigating long-range periodic structure–function relationships. Furthermore, this approach may enable access to negative refractive index materials, opening applications in cloaking.

Interfacing colloidal crystals with other materials is another promising direction that could yield structures with new properties. For example, by combining the ability to expand and compress polymeric matrices with the exotic structures attainable via colloidal crystal engineering with DNA, researchers should be able to realize all sorts of functional new materials with optical properties that are mechanically responsive. Finally, the approach of using DNA to control material architecture structure extends far beyond colloidal crystals. Proteins, polymers, hydrogels, and many other materials can have structural outcomes controlled by careful and strategic incorporation of oligonucleotides, and challenges include mapping out innovative ways to interface DNA with such materials.

Acknowledgements

This material is based upon work supported by the Air Force Office of Scientific Research under Awards FA9550-17-1-0348 and FA9550-22-1-0300.

References

[1] Z. Cai, Z. Li, S. Ravaine, M. He, and Y. Song, *et al.*, *Chem. Soc. Rev.* **50**, 5898 (2021).

[2] M. A. Boles, M. Engel, and D. V. Talapin. *Chem. Rev.* **116**, 11220 (2016).

[3] D. Samanta, W. Zhou, S. B. Ebrahimi, S. H. Petrosko, and C. A. Mirkin. *Adv. Mater.* **34**, 2107875 (2022).

[4] S. Y. Park, A. K. R. Lytton-Jean, B. Lee, S. Weigand, and G. C. Schatz, *et al.*, *Nature* **451**, 553 (2008).

[5] R. J. Macfarlane, B. Lee, M. R. Jones, N. Harris, and G. C. Schatz, *et al.*, *Science* **334**, 204 (2011).

[6] C. A. Mirkin, R. L. Letsinger, R. C. Mucic, and J. J. Storhoff. *Nature* **382**, 607 (1996).

[7] C. Zhang, R. J. Macfarlane, K. L. Young, C. H. J. Choi, and L. L. Hao, *et al.*, *Nat. Mater.* **12**, 741 (2013).

[8] C. R. Laramy, M. N. O'Brien, and C. A. Mirkin. *Nat. Rev. Mater.* **4**, 201 (2019).

[9] E. Auyeung, J. I. Cutler, R. J. Macfarlane, M. R. Jones, and J. Wu, *et al.*, *Nat. Nanotechnol.* **7**, 24 (2012).

[10] M. R. Jones, R. J. Macfarlane, B. Lee, J. Zhang, and K. L. Young, *et al.*, *Nat. Mater.* **9**, 913 (2010).

[11] H. Lin, S. Lee, L. Sun, M. Spellings, and M. Engel, *et al.*, *Science* **355**, 931 (2017).

[12] M. Girard, S. Wang, J. S. Du, A. Das, and Z. Huang, *et al.*, *Science* **364**, 1174 (2019).

[13] S. Wang, S. Lee, J. S. Du, B. E. Partridge, and H. F. Cheng, *et al.*, *Nat. Mater.* **21**, 580 (2022).

[14] E. Auyeung, T. I. N. G. Li, A. J. Senesi, A. L. Schmucker, and B. C. Pals, *et al.*, *Nature* **505**, 73 (2014).

[15] M. N. O'Brien, H.-X. Lin, M. Girard, M. Olvera de la Cruz, and C. A. Mirkin. *J. Am. Chem. Soc.* **138**, 14562 (2016).

[16] S. E. Seo, M. Girard, M. Olvera de la Cruz, and C. A. Mirkin. *Nat. Commun.* **9**, 4558 (2018).

[17] L. Sun, H. Lin, Y. Li, W. Zhou, and J. S. Du, *et al.*, *Adv. Mater.* **32**, 2005316 (2020).

[18] E. Auyeung, W. Morris, J. E. Mondloch, J. T. Hupp, and O. K. Farha, *et al.*, *J. Am. Chem. Soc.* **137**, 1658 (2015).

[19] K. L. Young, M. B. Ross, M. G. Blaber, M. Rycenga, and M. R. Jones, *et al.*, *Adv. Mater.* **26**, 653 (2014).

[20] D. J. Park, C. Zhang, J. C. Ku, Y. Zhou, and G. C. Schatz, *et al.*, *Proc. Natl. Acad. Sci.* **112**, 977 (2015).

[21] M. B. Ross, C. A. Mirkin, and G. C. Schatz. *J. Phys. Chem. C* **120**, 816 (2016).

[22] Q.-Y. Lin, J. A. Mason, Z. Li, W. Zhou, and M. N. O'Brien, *et al.*, *Science* **359**, 669 (2018).

[23] S. Lee, H. A. Calcaterra, S. Lee, W. Hadibrata, and B. Lee, *et al.*, *Nature* **610**, 674 (2022).

[24] D. J. Park, J. C. Ku, L. Sun, C. M. Lethicca, and N. P. Sternd, *et al., Proc. Natl. Acad. Sci.* **114**, 457 (2017).

[25] L. Sun, H. Lin, K. L. Kohlstedt, G. C. Schatz, and C. A. Mirkin, *et al.*, *Proc. Natl. Acad. Sci.* **115**, 7242 (2018).

[26] W. Zhou, Z. Liu, Z. Huang, H. Lin, and D. Samanta, *et al.*, *Proc. Natl. Acad. Sci.* **117**, 21052 (2020).

https://doi.org/10.1142/9789811282324_0021

Exploring the Matterverse Using Nanomaterial Megalibraries

Chad A. Mirkin*[†,‡], Peter T. Smith*[‡] and Carolin B. Wahl[†,‡]

*Department of Chemistry, Northwestern University, Evanston, IL 60208, USA
[†]Department of Materials Science and Engineering, Northwestern University, Evanston, IL 60208, USA
[‡]International Institute for Nanotechnology, Northwestern University, Evanston, IL 60208, USA

The discovery of new materials must address an enormous design challenge

Identifying new materials that exhibit desired properties is central to almost everything we do as a civilization. Imagine a world without catalysts, silicon, polymers, batteries, and construction materials. Although remarkable advances have been made in materials discovery, in many respects, we are moving through the possibilities at a glacial pace. Indeed, when just approaching the synthesis of new inorganic materials, one must consider an enormous parameter set, including, but not limited to, the elemental composition, oxidation states, size, shape, morphology, crystal structure, and support. This design space is massive and cannot be adequately surveyed using traditional synthesis or screening methods. For example, if we focus solely on metallic nanoparticle materials containing variations of the 61 metals with stable isotopes, and we target the synthesis of four-component nanoparticles, there are 520,000 possible combinations. If we then also consider size control from 1–50 nm in 1 nm increments, there are a total of 26,000,000 potential materials. Finally, if the four-component elemental composition is controlled in 5 atom% increments, we are confronted with 25,000,000,000 possible material targets. There are many more parameters besides composition and size that influence performance; therefore, the total possible number of chemically or physically distinct nanomaterials is near infinite, which defines what we term the "matterverse".

In the context of inorganic materials, combinatorial chemistry has advanced to the point of using hundreds to thousands of thin film features for screening properties [1–3]. While important advances, these approaches do not create features that are discrete in nature and of the appropriate length scales to uncover the effects of reduced dimensionality or structure/function relationships. Indeed, while multimetallic gradient thin film libraries are attractive because they are easily prepared, they lack control over most design parameters except elemental composition. In order to fully navigate the matterverse, one must (1) realize ultrahigh-throughput synthesis techniques that produce millions to billions of new materials per experiment with the chemical precision of traditional serial synthesis methods and (2) develop comparably fast characterization techniques to generate performance metrics, teaching us new lessons about materials design that can only be learned by exploring as much of the design space as possible.

Scanning probe nanolithographic techniques enable the synthesis of increasingly larger libraries of materials

Over the past two decades, we have developed several scanning probe lithography techniques that enable the high-throughput, on-chip synthesis of complex nanomaterials. This began by transitioning scanning probe microscopy (SPM) techniques from reading to writing tools, as in the case of dip-pen nanolithography (DPN) [4]. SPM-based lithography allows for positionally encoded deposition of attoliter-volume nanoreactors supporting specific chemical transformations for materials synthesis (Fig. 1(a)) [5]. The spatial confinement within a nanoreactor causes all material precursors to coalesce into a single nanostructure, affording unprecedented control over product composition and size on the single-particle scale [6–9].

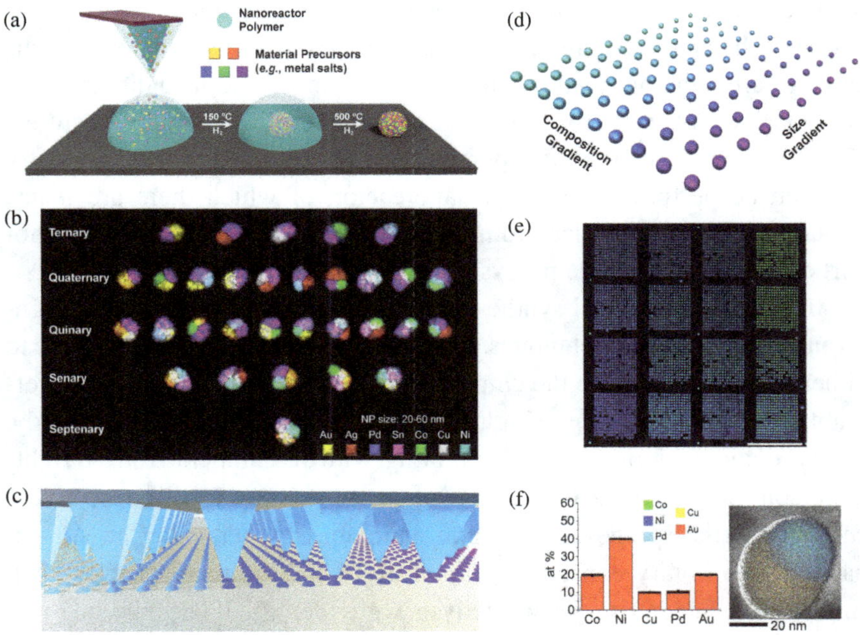

Fig. 1. (a) Scanning probe lithography of attoliter-sized nanoreactors containing materials precursors. For nanoparticle synthesis, thermal treatment induces particle formation and nanoreactor decomposition. (b) A library of polyelemental nanoparticles synthesized from nanoreactors showing elemental mapping from energy-dispersive X-ray spectroscopy (EDS) [10]. (c) Parallelized nanoreactor deposition where the composition and volume of the precursor at each pen is controlled. (d) A nanoparticle megalibrary schematic with both composition and size gradients. (e) A library of more than 4,000 $CsPb(Br_{1-x}Cl_x)_3$ perovskites with varied photoluminescence behaviour to optimize blue emission properties (scale bar = 100 μm) [11]. (f) AI-predicted synthesis of a five-component, single-interface nanoparticle, where the bar graph shows the measured composition by EDS and the black bars represent the AI-suggested ratios [12].

Remarkably, to date, a library of single heterostructured nanoparticles containing up to seven different elements has been synthesized by reductive, thermal annealing of patterned nanoreactors loaded with the requisite metal salts (Fig. 1(b)) [10].

By further transforming serial nanolithography methods into cantilever-free parallel writing tools with as many as 11 million tips, we now have ways to dramatically accelerate the pace of materials synthesis (Fig. 1(c)) [13]. For example, when precursors are distributed in predefined gradients across an array of pens, the composition and concentration at each pen are different, resulting in at least as many unique materials as there are pens on a single chip (i.e., a nanoparticle megalibrary — a positionally encoded chip with at least a million distinct features) (Fig. 1(d)). Importantly, this approach can be extended to any chemistry that can be performed within a nanoreactor, of which there are many, including reactions that lead to multicomponent metallic and ionic nanoparticles as well as halide perovskites [6–9].

To be useful, parallel synthesis must be paired with sufficiently high-throughput screening techniques. Spectroscopic methods that allow one to either rapidly interrogate the chip in a site-specific manner or, more preferably, collectively during a single analysis step are desirable. Accordingly, in one demonstration, we (collaborating with researchers from Wright-Patterson Air Force Base) used Raman spectroscopy to discover a new catalyst for the synthesis of carbon nanotubes by screening an Au–Cu nanoparticle library containing a total of 14 million nanoparticles [14]. Importantly, hundreds of new catalysts were identified, but only one was a global maximum performance-wise (Au_3Cu), which highlights the need to search a large parameter space to identify the best material for an intended use. We have also synthesized a library of 4,624 mixed halide perovskites, each subtly different in chemical composition, which was screened for blue photoluminescence (Fig. 1(e)) [11]. From this work, $CsPb(Br_{0.6}Cl_{0.4})_3$ was identified as the highest intensity blue emitter. We now can control the precursor composition and volume at each pen tip, allowing us to pattern and synthesize at least 100,000 unique materials per square centimetre in a single experiment using commercial tools. Moreover, we are making rapid progress towards the synthesis of megalibrary chips with greater than 200,000,000 distinct particles.

Given the extraordinary complexity of the matterverse, it is imperative to take a systematic approach to megalibrary design by predicting composition spaces that are likely to exhibit properties of interest. To that end, we are interfacing the megalibrary platform with artificial intelligence (AI), combining our ability to generate large volumes of first-party data with the predictive capabilities of AI. Megalibraries provide a path to synthesize large portions of the matterverse and acquire massive, high-quality structural and performance datasets, which then can be mined for optimized materials using AI. Thus far, in collaboration with Toyota Research Institute, we have integrated AI with libraries of heterostructured polyelemental nanostructures to successfully predict new element mixtures for biphasic nanoparticles, resulting in the most chemically complex biphasic nanoparticles ever made: $Au_{20}Pd_{10}Cu_{10}Ni_{40}Co_{20}$ (AuPdCu-NiCo) and $Au_{10}Ag_{10}Pd_{10}Cu_{10}Ni_{40}Co_{20}$ (AuAgPdCu-PdCuNiCo) (Fig. 1(f)) [link] [12]. Building on this foundation, we are working to integrate the multimodal structural and functional data streams produced by our high-throughput characterization tools with suitable models to harness both the predictive and analytical power of AI and allow us to comprehensively survey the matterverse.

Megalibraries and AI will accelerate materials discovery across chemistry fields

The megalibrary platform, enabled by the remarkable rise of scanning-probe nanolithography, is primed to revolutionize many aspects of materials discovery. By providing access to unexplored synthetic avenues, megalibraries offer opportunities beyond high-throughput synthesis to include screening and massive data analytics with implications far beyond the development of new screening tools and automated instrumentation. Any sufficiently automated experimentation can quickly scale beyond conventional data processing, which requires us to rethink how to acquire, store, and analyse data. This paradigm shift began a long time ago — one is already hard-pressed to find an analytical instrument that does not generate data in a digital format, and it does so at an ever-increasing rate. The integration of AI provides a way to manage these massive datasets by mining large databases, analysing multimodal experimental results, and making predictions about new materials and their properties to inform future experiments [15]. Excitingly, this will enable a transition from automated to fully autonomous workflows, essentially digitizing a component of the workforce. In so doing, AI integration with megalibraries will change the way we design and conduct experiments, allowing us to focus on discoveries that will tackle some of the greatest challenges we face, be it materials for fuel cell catalysts, CO_2 reduction to fuels, energy storage and conversion, chemical synthesis, or whatever the future holds — we will find important answers by mining the matterverse.

Acknowledgements

This material is based upon work supported by the Sherman Fairchild Foundation, Inc. and the Toyota Research Institute, Inc.

References

[1] E. Danielson, J. H. Golden, E. W. McFarland, C. M. Reaves, and W. H. Weinberg, *et al.*, *Nature* **389**, 944 (1997).

[2] P. J. McGinn, *ACS Comb. Sci.* **21**, 501 (2019).

[3] L. Yang, J. A. Haber, Z. Armstrong, S. J. Yang, and K. Kan, *et al.*, *Proc. Natl. Acad. Sci. USA* **118** (2021).

[4] R. D. Piner, J. Zhu, F. Xu, S. Hong, and C. A. Mirkin, *Science* **283**, 661 (1999).

[5] J. Chai, F. Huo, Z. Zheng, L. R. Giam, and W. Shim, *et al.*, *Proc. Natl. Acad. Sci. USA* **107**, 20202 (2010).

[6] G. Liu, D. J. Eichelsdoerfer, B. Rasin, Y. Zhou, and K. A. Brown, *et al.*, *Proc. Natl. Acad. Sci. USA* **110**, 887 (2013).

[7] P. C. Chen, X. Liu, J. L. Hedrick, Z. Xie, and S. Wang, *et al.*, *Science* **352**, 1565 (2016).

[8] J. S. Du, D. Shin, T. K. Stanev, C. Musumeci, and Z. Xie, *et al.*, *Sci. Adv.* **6**, eabc4959 (2020).

[9] D. D. Xu, C. B. Wahl, J. S. Du, S. Irgen-Gioro, and E. A. Weiss, *et al.*, *J. Phys. Chem. C* **125**, 26125 (2021).

[10] P. C. Chen, M. Liu, J. S. Du, B. Meckes, and S. Wang, *et al.*, *Science* **363**, 959 (2019).
[11] M. Lai, D. Shin, L. Jibril, and C. A. Mirkin, *J. Am. Chem. Soc.* **144**, 13823 (2022).
[12] C. B. Wahl, M. Aykol, J. H. Swisher, J. H. Montoya, and S. K. Suram, *et al.*, *Sci. Adv.* **7**, eabj5505 (2021).
[13] F. Huo, Z. Zheng, G. Zheng, L. R. Giam, and H. Zhang, *et al.*, *Science* **321**, 1658 (2008).
[14] E. J. Kluender, J. L. Hedrick, K. A. Brown, R. Rao, and B. Meckes, *et al.*, *Proc. Natl. Acad. Sci. USA* **116**, 40 (2019).
[15] E. Stach, B. DeCost, A. G. Kusne, J. Hattrick-Simpers, and K. A. Brown, *et al.*, *Matter* **4**, 2702 (2021).

https://doi.org/10.1142/9789811282324_0022

The Design of Self-Assembling Protein Cages and Other Reticular Protein Materials

Roger Castells-Graells, Joshua Laniado,
Kyle Meador and Todd O. Yeates

*Department of Chemistry and Biochemistry,
University of California, Los Angeles, 90095 CA, USA*

Outlook on novel material design from a biomolecular perspective

The last two decades have brought exciting developments in the design of molecular materials. This is particularly true for subfields where macromolecules — nucleic acids and proteins — have been exploited as the building blocks. Macromolecules offer exceptional versatility in terms of design and production, along with diverse chemical functionality. They also bring high levels of complexity, with attendant opportunities and design challenges. For materials based on DNA or RNA, the widely understood rules of Watson–Crick base-pairing provide general strategies for controlling molecular associations and geometric shape. The unique building opportunities presented by DNA were of course appreciated first by Ned Seeman, who pioneered the area of DNA nanotechnology. Protein molecules, on the other hand, offer numerous contrasts to nucleic acids as molecular building materials. They offer rich chemical and functional diversity, and indeed nature evolved myriad sophisticated self-assembling architectures based on protein molecules. But the rules that dictate how linear amino acid sequences will fold up in three dimensions, and then potentially assemble into larger structures, are much more complex. Accordingly, building geometrically regular materials from protein molecules has required the development of new principles and molecular strategies. A range of engineering ideas have been explored in the last 20 years, with diverse goals and outcomes. Here we will emphasize some of the most influential developments that have come from focusing on the use of symmetry to create extraordinary, self-assembling protein materials, with emerging applications.

Symmetry principles bring about a new era of protein-based reticular materials

Just over 20 years ago, motivated in part by realizations that came from an esoteric puzzle in macromolecular crystallography [1], we asked how it might be possible to create complex three-dimensional assemblies from simpler protein molecules. Numerous preceding efforts in protein design had explored the formation of relatively simple units, e.g., dimers or trimers of helical polypeptides, or extended linear filaments. Based on parallels to protein crystals and mathematical connections to symmetry groups, we observed that much more sophisticated architectures — including cubic- and icosahedral-shaped cages as well as extended materials — might be made to self-assemble into atomically predictable structures if two different kinds of oligomeric protein units could be brought together *in precisely defined orientations*. That idea and molecular engineering methods for achieving it were articulated by graduate students Jennifer Padilla and Christos Colovos in 2001 [2]. In addition to laying out a set of rules for creating a first set of geometrically regular (or reticular) materials, that study provided the first experimental demonstration of a designed protein cage. By genetically fusing a natural protein dimer to a natural protein trimer (using a continuous alpha helical connection) with their symmetry axes in a precisely defined configuration, a self-assembling symmetry T cage was created, comprising 12 identical subunits, roughly 600 kDa in size [2, 3] (Fig. 1). The approach was subsequently extended to create a distinct 24-subunit cube from different components [4].

Following Padilla's pioneering demonstration, numerous engineering variations have been explored for building novel, self-assembling protein architectures. A key innovation, which rapidly expanded the design possibilities, came from work by King and Baker *et al.* where, instead of genetically fusing together oligomeric protein components, oligomeric units were brought together in defined orientations by introducing complementary amino acid surface mutations (chosen based on computational methods) to

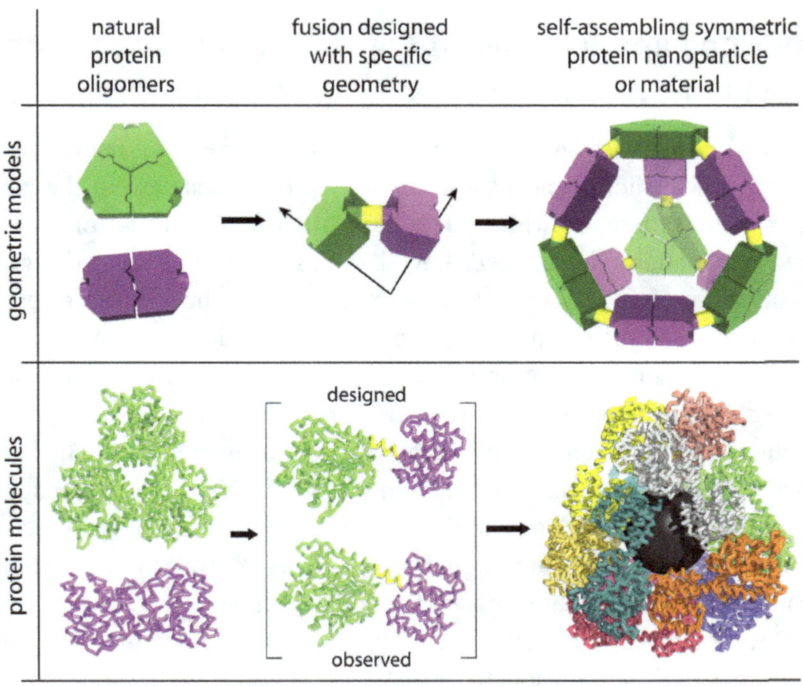

Fig. 1. Illustration of the use of symmetry combinations, as introduced by Padilla *et al.* [2], to create protein cages and other reticular protein materials. Figure adapted with permission from Lai *et al.* [3].

hold the components together non-covalently [5, 6]. The computational interface design approach avoids the limitations inherent to the original fusion method, though at the expense of greater design difficulty and often modest experimental success rates; it has since led to numerous novel protein cages and materials. In addition to these two basic approaches to building geometrically regular protein materials, many other research groups have shown success with methods that deviate in one way or another, e.g., by introducing metal ligands or coiled-coil peptide elements, or by using natural protein cages as starting points for diversification [7–12]. Across the wide range of design strategies and variations that have been explored, a general finding is that those that have adhered most strictly to principles of symmetry have led to the most predictable and geometrically regular protein materials.

Emerging applications and opportunities for reticular protein materials

Designed protein assemblies of the type described here are beginning to find uses in medicine and nanotechnology. Many ideas are in the early stages of development, while a few have already proven useful in real-world applications. King *et al.* have demonstrated the use of designed icosahedral protein cages as vaccines, by decorating the surfaces of novel capsids with carefully situated viral antigens [13]. Liu *et al.* have demonstrated the use of designed protein cages as modular scaffolds for rigidly binding diverse cargo proteins, with the purpose of allowing the cryo-EM structure determination of proteins otherwise too small (i.e., smaller than about 50 kDa) for that powerful technique [14, 15]. Numerous other applications will come to fruition in the next few years. An exciting direction for new work concerns the design of dynamically controllable assemblies, such as architectures that respond to specific cellular cues [16–18].

Recent advances in theory and computing are also notable. The number of distinct design architectures that have been realized experimentally with protein molecules is relatively small compared to the full range of possibilities. In recent theoretical work, motivated by the better-developed understanding of MOF architectures, we articulated the complete space of three-dimensional *symmetry combination materials* (SCMs) that can be created through the combination of two component protein oligomers [19]. The geometric rules for filling out that design space should guide many new studies. Improved computational methods for designing protein assemblies are also developing. New machine learning methods promise to make important advances along that front [20].

Acknowledgements

The authors recognize support from the US Department of Energy, the US National Science Foundation, and the US National Institutes of Health.

References

[1] S. W. Wukovitz and T. O. Yeates, *Nat. Struct. Biol.* **2**, 1062 (1995).
[2] J. E. Padilla, C. Colovos, and T. O. Yeates, *Proc. Natl. Acad. Sci. USA* **98**, 2217 (2001).
[3] Y.-T. Lai, D. Cascio, and T. O. Yeates, *Science* **336**, 1129 (2012).
[4] Y.-T. Lai, E. Reading, G. L. Hura, K.-L. Tsai, A. Laganowsky, *et al.*, *Nat. Chem.* **6**, 1065 (2014).
[5] N. P. King, W. Sheffler, M. R. Sawaya, B. S. Vollmar, J. P. Sumida, *et al.*, *Science* **336**, 1171 (2012).
[6] N. P. King, J. B. Bale, W. Sheffler, D. E. McNamara, S. Gonen, *et al.*, *Nature* **510**, 103 (2014).
[7] A. D. Malay, N. Miyazaki, A. Biela, S. Chakraborti, K. Majsterkiewicz, *et al.*, *Nature* **569**, 438 (2019).
[8] A. Sciore, M. Su, P. Koldewey, J. D. Eschweiler, K. A. Diffley, *et al.*, *Proc. Natl. Acad. Sci. USA* **113**, 8681 (2016).
[9] J. M. Fletcher, R. L. Harniman, F. R. H. Barnes, A. L. Boyle, A. Collins, *et al.*, *Science* **340**, 595 (2013).
[10] D. J. E. Huard, K. M. Kane, and F. A. Tezcan, *Nat. Chem. Biol.* **9**, 169 (2013).
[11] Y. Azuma, T. G. W. Edwardson, and D. Hilvert, *Chem. Soc. Rev.* **47**, 3543 (2018).
[12] F. Lapenta, J. Aupič, Ž. Strmšek, and R. Jerala, *Chem. Soc. Rev.* **47**, 3530 (2018).

[13] J. Marcandalli, B. Fiala, S. Ols, M. Perotti, W. de van der Schueren, *et al.*, *Cell* **176**, 1420 (2019).
[14] Y. Liu, S. Gonen, T. Gonen, and T. O. Yeates, *Proc. Natl. Acad. Sci. USA* **115**, 3362 (2018).
[15] Y. Liu, D. T. Huynh, and T. O. Yeates, *Nat. Commun.* **10**, 1864 (2019).
[16] J. E. Miller, Y. Srinivasan, N. P. Dharmaraj, A. Liu, P. L. Nguyen, *et al.*, *J. Am. Chem. Soc.* **144**, 12681 (2022).
[17] C. Gu, T. Zhang, C. Lv, Y. Liu, Y. Wang, *et al.*, *ACS Nano* **14**, 17080 (2020).
[18] Y. J. Kang, D. C. Park, H.-H. Shin, J. Park, and S. Kang, *Biomacromol.* **13**, 4057 (2012).
[19] J. Laniado and T. O. Yeates, *Proc. Natl. Acad. Sci. USA* **117**, 31817 (2020).
[20] J. Dauparas, I. Anishchenko, N. Bennett, H. Bai, R. J. Ragotte, *et al.*, *Science,* doi: 10.1126/science.add2187 (2022).

https://doi.org/10.1142/9789811282324_0023

Discussions of Session 3 — Reticular Chemistry and New Materials

Chair: Omar Yaghi
Auditors: C. Aprile*, G. Van Assche†

*Group of Applied Materials Chemistry (CMA), Unit of Nanomaterials Chemistry (CNANO), Department of Chemistry, Namur Institute of Structured Matter (NISM), Université de Namur, Rue de Bruxelles 61, 5000 Namur, Belgium
†Physical Chemistry and Polymer Science, Department of Materials and Chemistry, Vrije Universiteit Brussel, Pleinlaan 2, 1050 Brussels, Belgium

Omar Yaghi
I think we will just open the presentations for discussion. Questions from any of you about anything that you have in mind?

Kurt Wüthrich
Well, the word *complexity* has been very frequently used since yesterday. Should we not try to define what we really mean with complexity? Is it size? Is it lack of symmetry? Or is it too many panels on a single slide? Is it dynamics, where we add at least a fourth dimension to describing

structure, I mean time-dependent fluctuations. Is it a lack of periodicity? How do we define complexity? The word was used so many times in different connections since yesterday, that I think it would be very nice if we could come to some conclusion on what we want to say with complexity.

Omar Yaghi

Nature of course is the best teacher and, I think, *complexity*, for me, means the presence of a variety of interactions — weak interactions, hydrogen bonding, hydrophobic interactions, polar interactions — within an entity. But also, there is a landscape of energy, different energy levels that electrons can travel through. And also, there is a gradient, not just a gradient of composition, although that is a very important one, the gradient of composition, a gradient of interaction, a gradient of energy levels. And of course, there is also hierarchy in terms of structure, there is a hierarchical arrangement of molecules or compartments. All of those come together. And, of course, you want this to have a purpose. And, hopefully, directionality. I think all those things define *complexity*. My comment yesterday about supramolecular chemistry is that you can create complex systems very easily. But to control them, and to have directionality and purpose, is much harder. I think you saw today that we worked very hard, all of us, at making ordered systems. And now, because you can modify those ordered systems in different ways, covalently and non-covalently, now you can begin to build *complexity* without falling into chaos, which I called goo. This is the ideal area of growth, the marriage of supramolecular chemistry to what reticular chemistry was.

Makoto Fujita:

It is exactly the same question as I had yesterday. We should discuss what are *complex systems* or *complexity*. First, let us think of assembly. Weak interactions work between or among any molecules. But there are many, many miserable assemblies. Then we should not say this is the assembly. Only when the ordered structure is formed, then we can say that this is self-assembly or assembly. So, the order of the structure is necessary. And, in the same time and in a similar way, in complex systems, by networking there are many chemical events. Some order should generate, something ordered should be generated. In my opinion, the generation of constancy or, in biology, *homeostaticity*: by the networking of many chemical events, the system will try to keep their constancy in the system. Maybe most people are interested in the life system. The most, the best complex system is life. And homeostaticity could be a target of the complex system. So, we should aim at self-constancy systems.

Joachim Sauer:

First of all, complexity is not disorder. And if we talk about complexity, there are different dimensions. Lots have already been said about complexity of structures. There are different organizations at different levels, and there may be a hierarchy. There is also complexity in the dynamic behavior, if different parts of the system have different time scales, are doing things at different time scales. And there is complexity in the reactivity: you may have a system where you may have just one or two important steps, but then you may have, in heterogeneous catalysis, reaction systems where you have possibly 130 steps to take into account, including heat and mass transport. So, there are different dimensions. And when it comes to, you have asked this question, *symmetry*, also the absence of symmetry is not complexity. It would be that you have in a structure, at different levels, different symmetries that you make. You have building units that have some symmetry, and then you have an organization, which may have a different symmetry or no symmetry.

Arne Thomas:

After all these very elaborated answers from my colleagues, we probably also have to confess and be honest that *complexity* is often just another term for "we don't understand". When I read "there is a complex reaction mechanism", "there is a complex structure" or something, very often this just means "I don't understand this structure" or "I don't understand this complex mechanism". This is just something we have to be careful with. Always this "complexity" sounds better than "I don't get it". I do think to understand complexity, one has first to reduce complexity. This is why we all try to make more defined materials to have a chance to understand these materials better. And it would be for me a less complex material, though it might be a multifunctional material, but as long as I can understand what happens in this material, I say then this is not complex anymore for me.

Xiaodong Zou:
Regarding complexity, I agree with Arne Thomas that you do not understand, but we should actually make use of these opportunities. We usually think about simplified things, like you want to look at one material, pure phases, and simple structures, but the systems are complex, in the form of, like, when you talk about the functionality. It can be used in different ways of generating materials in the complex system, or in trying to understand this complex system, and in being able to have characterization tools, to be able to develop, to study, this complex system. So that changes the *complexity*, until something we really understand maybe becomes simple.

Chad Mirkin:
I do not have a lot to add. I think it is a term. First of all, I did not use it, I want to say that, and I do not think you heard me use it. But second, I think it is very similar to the whole evolution of the term *self-assembly*, which meant many things to many people. The substance of it, is that complex multicomponent systems often have properties that go beyond subunit-based systems. And that is an important pursuit, understanding primary structure, secondary structure, tertiary structure, and then function that correlates with that structural understanding, that structural control. To me, that is the ultimate goal in all this: how do you begin to zip together subunits to create structures that are "complex", but more importantly, have properties that you cannot realize from the subunits themselves. I think there are many examples that we heard today and yesterday, frankly, and there will be many more from that type of pursuit.

Kurt Wüthrich:
I would like to follow up with a specific question to Professor Fujita. You showed to us that with a single parameter, you could go to higher order structures of a given type. Now does this mean that you get more complexity, or, because you now handle this with a single parameter, it is actually a simple system?

Makoto Fujita:
In my understanding, the complex system is on the next hierarchy of the ordering. Weak interaction of molecules can induce their self-assembly events. In a similar way, weak interaction of chemical events, so chemical

reactions, can induce their self-ordering of the system. It is not there are events among the molecules. So, I do not like to use the term complexity or complex system for making a structure from molecules only. It is not the assembly of molecules, but the assembly of chemical events.

Omar Yaghi:

I think *complexity* is not necessarily or should not be confused with *complicated.* That is why my description originally is a structural based description. I do like the idea that complexity produces a function that is much larger than the sum of the parts, and I think that is important. So to me, glass is not complex. There is no recognition, there are no weak interactions, things like that, that you can talk about in terms of function and things like this, but hemoglobin is complex.

Ben Feringa:

Another aspect of *complexity*: you guys make all these fantastic well-organized structures. When I talk with my colleagues in solid state physics about properties, it is always about step edges, defects sites, all these things, which determine the properties ultimately to a large extent. Could you comment on that? Because that adds another dimension of complexity to the system: to go to real functions.

Omar Yaghi:

I think every system is complex at some level. I mean a lot of the structures that you saw, if you look in detail, these are average crystal structures, and there would be defects. If you look at them in a TEM, you can see sometimes missing linkers, you can see rings that are supposed to be hexane and some of them have five membered rings, some of them have seven member rings distributed here and there. So to me, if you look at all these systems in great detail, and move away from the average structure, they are very complex systems. Which reminds me to also say that complex systems also have perturbations along the landscape of the composition.

Arne Thomas:

I think also important is, of course, to talk about the difference with solid-state chemistry structures, where, as you said, point defects, step defects, and so on, have a huge influence on the bulk properties of such materials. Most of the materials we were talking about are actually molecular entities, which are connected by strong bonds. It is actually then the molecules that count, and not so much the bulk property. If you think about hydrogen storage or CO_2 separation, or what I have shown, there is actually a catalyst sitting there. If there is a defect, I have the feeling, it is not making such a big effect like in the solid state structure, where you look at the bulk property band structure, and so on and so forth. We see that, when we really get out of crystallinity, that then the properties of such materials change. So, we should not go away too much from the ordered structure, then something changes, but as we are actually talking most of the time about the function of the molecular entities in these materials, it is ok, I think, and there are some termination defects or something.

Ben Feringa:
I am happy to hear that. Because all the time with these band gaps, the energy levels for conductivity but also for catalysis and so on, and also we know from zeolites: that plays a crucial role. So, engineering these kinds of sites might be another challenge for the field.

Chad Mirkin:
But it is more complex than that and more difficult than that. Take the synthesis of the particles that I talked about. We can beat our chest and we can make 225 million distinct particles on a surface, which is true, in a four by four centimeter area. But when you get down to the individual particles, no two particles are exactly the same, the orientation of the particles, or even the composition differs. So the way we get to that is actually to use numbers, to use redundancy, to use the ability to make many things that are effectively the same, and look at all of them in one shot. Let us not miss one because of some of aberrations you are talking about, and then focusing on the actual structures that are giving us the activity that we see. But it is a difficult problem, because these systems are multiparameter, and their complexity makes them interesting, but also makes it very difficult to really pin them down.

James Liao:

If we try to define complexity, perhaps we can borrow the definition from mathematicians, where complexity is a property of a non-linear system. If you have a linear system, then typically you can predict everything, from the beginning to the end. It is completely predictable. If you have a non-linear behavior, nonlinear structure, sometimes you get to this bifurcation behavior, and all kinds of these strange attractors, strange behavior, emerge. So that is one line of origin, from where the word complexity came from. But now we use that word in the chemistry context, where things are more static, that is not that dynamic, but I am sure we can find something that is analogous to the dynamic complexity that mathematicians use.

Joachim Sauer:

Coming back to Ben Feringa's question about the role of defects and steps. We have already discussed, and I have mentioned it in my presentation, that you may have imperfections, and they may play a big role. I have briefly shown largely varying isotherms measured for the same nominal material, which change loading per metal site between two and four molecules per site. And if you look in details, this has to do with the synthesis process, so the role of steps and these things. So when it comes to electronic properties, then of course we would have the same problems as they have. If we have transition metal ions, the electronic structure maybe different if we have defects in the system. And if we go to diffusion, for example, then the question appears what happens at the surface of the crystallite. From zeolites, we know very well that there is this discussion of a possible barrier and the effect on the intracrystalline diffusion, in-out diffusion, and then the transport in the mesopores of the particles, which is finally in the interest of the user of these materials.

Xiaodong Zou:

In terms of defects, I would say that all these deviations from the periodicity of structures — for example like multivariant structures where you do a functionalization so that you have different functions, and also the metals can be changed — generate multiple components in your structure. I think this actually will be a very positive way of making new functions into the material. Of course, the characterization will be challenging, and it needs to be developed.

Ben Feringa:

I had no intention to make a negative comment, I just wanted to say that defects and all these kinds of things are intriguing and offer all kinds of opportunities, and Nature probably knows how to handle it. I again refer to a cell. There are many things that a cell knows how to control, and in catalysis, for instance, this plays a key role. It offers also a lot of opportunities for us, to introduce function and so on.

Donald Hilvert:

You have shown all these beautiful structures, and it looks like the chemical instruction to create them is fairly well understood. How reliable is your ability to impart specific functions on these structures, and is it always mediated by adding specific metal ions, are these new coordination sites, or additional functionalities?

Omar Yaghi:

I can give you the example of water, because we ultimately figured out exactly the absorptive sites of each water molecule in a structure, exactly how the water is sitting, exactly the hydrogen donor and hydrogen acceptor in the structure, and exactly the geometry of how it is sitting. Now you can design, incorporate linkers that will enhance or strengthen the interaction or weaken the interaction, depending on what you really want, and at what humidity you want to take up the water. So there is a lot of design going on there. I think in terms of designing the MOF itself, because of the multi-metallic nature of the sum of the building units in MOFs, there is always uncertainty whether you are going to get a six-connected multimetallic unit or a four-connected one. But once you make one member of a family, you can develop the rest of it, change the matrix functionality and everything like that. You can know a lot about the system. With COFs, that is completely different, because the COFs, like what Arne Thomas was talking about, their building units do not change during the reaction.

Only the linkage between them is what forms, and so they remain the same. So, in fact, they are made by design, because the building units remain intact during the reaction. So, there is a lot of design in COFs, there is a lot of design once you understand the system in MOFs, and especially if you are lucky to get the absorptive sites of gases, then there is a tremendous amount. That is why in my talk yesterday, I said we chemists are surgeons, because you could instead of a pyrazole dicarboxylate linker for a harvesting MOF, you could introduce an oxygen in a five-membered ring, a furan dicarboxylate, in a structure exactly the same as the one with the pyrazole, thereby weakening the interaction between water and the framework. And you could also mix those two together and modulate the hydrophobicity, so you can do a lot.

Chad Mirkin:

It is a really interesting question, because I did not make it clear. A distinction between how molecular architectures are engineered and how colloidal crystals with DNA are engineered, is that while we talk about the particle building blocks' atom equivalence, there is a fundamental difference, and that is that the bonding is independent of atom identity, and that is just not true in conventional chemistry. That actually dramatically simplifies the design space, because it now distils down to geometric arguments, and how things fit together to maximize duplex formation or hybridization. And what that allows you to do now, is engineer structures that we do not even see in Nature. I mean, we have now six that have no mineral equivalent. It allows you to dial in bond length, based upon the length of oligonucleotide. It allows you to dial in crystal habit, which you cannot do with conventional systems. And the reason that is so important is — connecting with what you said, that it feels we have now the design space down — that it is down. It is getting to the point where there will be computational programs that if you want to build a particular structure out of particles, the program will spit out the particle building blocks, the oligonucleotide sequences, required to get there. And then you can flip it around and do exactly what you were talking about, which is go to the physical scientists and say: "What property would you like? What structure should exhibit that property theoretically?", and I will make it based upon these design rules. And that is what I think is so exciting and what

I tried to share in the last slide, that we are getting to the point where we can design these types of materials properties not found in Nature, and properties that you design ahead of time and do not discover by happenstance.

Nicolas Giuseppone:
Related to the question of Ben Feringa and defects in your materials, but not related to the functions, just to the way you form your materials: Professor Yaghi, you mentioned that you use for your reticular approach less kinetically labile interactions than in supramolecular chemistry. So I am wondering how much relies on the correction of defects when you form your self-assemblies, or relies on the directional kinetic progression, crystallization of your material? And I would have the same question for the discrete MOF of Professor Fujita: how labile are they? Because you have shown this magic ring catenation, and I am wondering, looking at that, if in your MOFs, for instance Professor Yaghi, you can have reconfigurable structures. Can we for instance imagine catenation of 3D MOFs and materials brought from this approach?

Omar Yaghi:
Catenation of framework is a very interesting example, because when you make a MOF, let us say based on a primitive cubic structure, it naturally will want to interpenetrate. If the pores are large enough then it will self-interpenetrate. We think that this is happening at the very initial nucleation of the framework. Remember, what you are doing in the crystallization is you are trying to balance two things: you are trying to balance the self-correction with the kinetics of the reaction, to get you to form a crystal that you could see or you could analyze. So, along the way, depending on your reaction conditions, you could make a defective structure. And sometimes, we happen to look at it in TEM, sometimes we will see those defects in some of the granules that we are looking at. There are systems that no matter how large the pore gets, they will not interpenetrate. Mathematically it is just not possible because the synthesis codes for a certain connectivity, certain coordination, and in order to have that interpenetration you need a different coordination, and therefore it does not

happen. So I would say that a lot of the defects have to do with how you make the material. Sometimes, in some cases, some people have been able to introduce defects in a systematic way in a crystal, where they yank out one of the linkers at periodic intervals and then insert another linker with a different functionality at those sites, and this is proven through single crystal to single crystal to single crystal transformation. So it is done on the same crystal.

Makoto Fujita:

Even in the solid state, we sometimes observe the "magic rings" type behavior, I mean there is a kind of phase transition from the interpenetrated framework to the non-interpenetrated framework in the single crystal fashion. There are several examples. That means in microscopic views, even in the crystalline state, it behaves like in solution, it behaves like a liquid. So, although it is a crystal at macroscopic levels, it can still behave like a solution. So interpenetrated frameworks can be transformed into the non-interpenetrated framework.

Peter Palese:

We have heard about these wonderful structures in molecular entities. Are any of those new compounds toxic in humans or to humans? Do they cause cancer? Is there any concern for people working with them? Is that something which has come up?

Chad Mirkin:

I can give you an answer you probably will not expect, but the building blocks themselves, in the case of our particles we called them programmable atom equivalents, the other term for them is a spherical nucleic acid, they are actually used as drugs, and they are in six clinical trials, and in certain cases to cure different forms of cancer by gene regulation type pathways. That is probably not what you expected, but the interesting thing is that they have the property that they will go through your skin, so there is a concern working with them. We have to utilize a lot of protocols to make sure that we minimize contact, but we use that to actually create pathways to drugs.

Omar Yaghi:
There is an article in the Wall Street Journal today about Wenbin Lin's work at the University of Chicago on using MOFs for cancer therapeutics.

Karla Kirkegaard:
I was very intrigued by the topological consequences of some of your synthesis, and I was thinking about topology from biology. Sometimes, I guess, the interpenetration involved free ends, but sometimes you can get changes in topology that have to do with the reversibility of the bonds that you make. You break and rejoin them on the other side of what you pass through. I was wondering if you think the topological changes that surprise us have to do mainly with interpenetration, or sometimes it is the reversibility of the bonds that you make?

Makoto Fujita:
Nature uses many strategies for making interlocking structures. Sometimes it is very similar to ours: small fragments are preorganized, then enzymes will link the precursors to make highly complicated structures. And sometimes, just a single protein strand can fold and spontaneously knot. And previously, the number of knotted proteins was not so large, but in recent years, many, many knotted proteins were found. So, Nature has many strategies to make knots.

Karla Kirkegaard:
But in your case, you think it is a condensation of small pieces?

Makoto Fujita:
In our case, yes.

Omar Yaghi:
It would be interesting in his systems, because they are very dynamic, to take two entities that he made of different sizes, and try to not link them, but more, like join or melt them together, fuse them together, into some regularly shaped thing, and then in a reversible way. I do not know, have you tried that?

Makoto Fujita:
No.

Clare Grey:
It is really a sort of challenge to the panel and particularly to Chad Mirkin and others. A lot of these processes are very close to room temperature, and as an inorganic chemist you are really often restricted to binary oxides. I think the organic PVs in the lead-based halide perovskites are a beautiful example of where they can be processed to room temperature, and you control things. But I just wonder what the panel thinks about now trying to go up in temperature and explore a much wider range of materials and properties with the same sorts of unique, and very careful and precise approach? I know in metal-organic frameworks you use hydrothermal processes, but still you are restricted in the classes of materials you can do, for example, by doing these room temperature methods. I just wonder where the next directions are in terms of expanding the types of materials by putting temperature, and controlling of kinetics of reactions, to then get different classes in materials?

Chad Mirkin:
I may have left something out that was key. In the making of the particles in the chip, that is done at 500 to a thousand degrees. So those are high temperature materials. We can make high entropy alloys and lots of structures. You have seen quenching protocols, they can be employed. So those are materials that in fact are very compatible with high temperature. We can also make low temperature materials, like the perovskites, that is crystallization type of processes as opposed to a thermal degradation process. In the DNA based structures, those are inherently materials that are stable between room temperature and 220°C, depending upon the solvents you use and how you make them. But once made, they can and have been moved into solid state supports, silica for example, so that you can control architecture at low temperature and then move to supports that allow you to work at high temperature, to create all sorts of interesting catalytic architecture, for example.

Omar Yaghi:

I think in terms of the MOFs and COFs, there is a range of temperature from room temperature all the way to 300–400°C for the triazine frameworks. You want to stay cool enough, so that you do not burn off the C-H's that you need to introduce functionality down the road, right? Otherwise, we will finish off with graphite, which is not functionalizable or not as easily functionalizable. So I think you have to keep that in mind. There is a lot that you can do within 400°C.

Arne Thomas:

I agree, of course, that when you are working with organic materials or metal organic materials, you are restricted in temperature at some point. Already when you go to 500°C to make a synthesis, you see a lot of carbonization going on and then you are back to a carbon, which is a complicated material. I think, actually, the challenge we have is not so much going high with the temperature, but finding other covalent bonds where we can induce reversibility. The first COFs, were boroxine COFs from Omar Yaghi. If you put water on them, they dissolve again. Then came the imine networks, which are much more stable. And now we made vinyl-bonded bonded COFs, I do not want to go into detail, but these are C-C bonded, entirely C-C bonded COFs. No one thought ten years ago that it could ever be possible to make such a bond in a reversible fashion. And there are much more out there. This will be the way to go, I think. Temperature does not help so much there, but other tricks, so to say, to induce the reversibility. So far there has not been made any COF which is, for example, by transition metal coupled C-C coupling reactions. This is so far not possible. Any chemical reaction, any covalent bond is reversible of course, in principle. When you do Le Chatelier and actually get out, for example, some bromide as a potassium salt, then you shift the equilibrium so much to one side that you actually restrict reversibility. And this is something we all think about, I guess, because new linkages are always something which is interesting for us, as we add another point of functionality to a network.

Xiaodong Zou:

We have made use of temperature, also concentration, to modify MOFs because of the bond reversibility, so that you can extend, exchange the linkers with different functionalities for the post synthesis processes. And also to expand the pores to larger ones, to change linker with a longer linker. And also to change topology, making use of this reversible self-assembled bonding. So in this, temperature is a parameter to play on.

Omar Yaghi:

I just want to say that if you want something higher than 400°C then you just work with carbonitride, but you will not get the functionalization that you would want, because now you are in the inorganic regime.

Gerald Joyce:

I have a question about informational complexity. You were talking about multivariate MOFs and differential occupancy within the lattice, and you raise this really intriguing possibility that there could be a linear ordering with regard to what the occupancy is. And of course, that information only has value if it could be propagated, if it could be read to an output channel. So, the question is: are there nearest neighbor effects, or what is the potential for nearest neighbor effects if you have a linear ordering? If it is periodic, that is boring. If it is a spin glass, that is boring. But if it is aperiodic and can be propagated, read out through nearest neighbor effects, then that is informationally interesting. So, what are the prospects for informational complexity of those systems?

Omar Yaghi:

What I love about those systems is that you are taking a very repetitive crystal and you are superimposing unique sequences of functionalities, of organic functionalities, or even you could do it with the metals, along the entire crystal. So now, if you are a substrate moving along the crystal, you are experiencing different microenvironments all the way from the beginning to the end. And we do not know anything about what is that sequence, but we know what the functionalities are, we know where they are, we know their ratio, we can vary all those components, so it is not a completely unknown system. So I think this is a computational problem. Let us say that I have a property, a catalytic reaction, that is sensitive to input and output. You can put in 10000 things and then look at how the product varies, and then work backwards to get to what potentially that sequence might be. I have no doubt that you can come up with that. We do not have enzymes, like biologists, that can sequence that for us, but this is our poor way of doing it. The problem is that we, as chemists, are very uncomfortable with this. We like to make things, then characterize them, study their properties, and then push them to application. And what I am proposing, is that we skip the characterization and just use the system like a black box in terms of the sequence, and then let the computation and the machine learning give us some guesses as to what is going on. But we do see aggregation, as I mentioned, with NMR, you can do Rotational-Echo DOuble-Resonance (REDOR) NMR on these. It does not give you information on the molecular level, but in terms of the nano regime, you do see aggregation of one functionality relative to another. And in terms of the functionalities interacting, if they are fat enough, they do interact, and that does influence where they sit in the sequence.

Bert Weckhuysen:
We talk about thermal stability, but I think there is also something else important in, for example, catalysis. Are you stable in a solvent, water? What when there is, like in electrocatalysis, electrons and charging, pH? All these things when we would move to photo-, photoelectro-, or electrocatalysis, what would then happen with material? In zeolite or other heterogenous catalysis, what we sometimes have, is that the inorganic structure, even as perfect it is, will become partially in solution. Something is happening at the interface between the solution and the catalytic surface, and then after reaction it precipitates again. So we sometimes make the analogy, it was already made between other materials, zeolite is one example of it. What about metal organic frameworks, COF materials, etc, and do you see similar things?

Arne Thomas:
Probably, one general comment on stability, and actually Omar Yaghi can tell much more about this, I think, he has experience of 20 years. The people come and say: “But your MOFs are not stable, right, and your COFs are not stable”. I think there is misunderstanding in this because the inorganic chemists were the first ones who were actually interested in MOFs, of course, because there is crystallography and something in there, and they compared materials all the time to mesoporous silicas, silica

alumina, and so on. When we think about stability, we have of course, and there you are totally right, to think of what kind of conditions we have. I can just give you an example. I have one material which is made, you just look at it with wet eyes and it will flow away because it is absolutely not stable under humidity. But it is an anionic framework, and you can put it into an application where water is not present, like in a lithium ion battery, for example, in a membrane. And there, it is super stable and can do lithium channeling through this membrane. That is just one example. Of course, you think about catalysis, I fully understand, and there we will not do steam reforming with MOFs, I guess, and also not with COFs. If there is high temperature and water involved under these conditions, they are just not made for this. In electrocatalysis and photocatalysis, there is always the issue of photooxidation, for example, bleaching, which is happening. For example, these acridine dyes are known to photo-bleach with some time. We are not really good in making long-term tests: our long-term tests always mean 20 hours, and then we analyze the materials, and if it looks like before, everything is good. So therefore, I cannot give you a definite answer, but I think there are enough materials reported, for example Omar Yaghi talked about the new material for CO_2 separation, which is made in kilogram, ton scales, now in industry. They tested certainly for a long time.

Bert Weckhuysen:

I want to bring this to the positive side because, sometimes, what you have in heterogeneous catalysis is that you bring something in solution, and it is actually having an additional effect on the reaction mechanism. So, you could have like a heterogeneous-mediated catalysis, but it is still something at the near-surface that is happening. I think maybe these materials could also bring that to life, that is actually a positive thing of leaching.

Omar Yaghi:

I think the stability is a positive thing. It is just that Arne Thomas did not mention that, when you have olefin backbone COFs, they are stable in strong acid, strong base. It is not a secret, it is published. And even in corrosive environments, like n-butyllithium. It is a carbon-carbon bonded network, so the question of stability is an old question, it is no longer there. We got MOFs that are cycling a hundred thousand cycles of water uptake and release in the middle of the desert, with absolutely no impact on the MOF. When you look deeply into what is happening in that MOF, there are water molecules that are seeds, and they stay in the MOF, and in a way they protect the MOF. We do not take them out, they just stay in there. They only represent less than 5% of the uptake. In terms of COFs, the stability issues are not there at all. You can cook them up in strong acid, strong base, nothing happens to these olefin COFs that we reported two years ago.

Arne Thomas:
Still, there are some materials where it could happen. Yesterday there was a question, I am not sure who asked, which sounded a bit like: what is better, homogeneous or heterogeneous catalysis? I have the feeling this is the answer what we have here, because we have molecular catalysis and an open framework structure. Some of them, for example, are anionic solids with immobilized cationic molecular catalysts. And then you cannot actually differentiate between homogeneous and heterogeneous catalysis, because the catalyst might work in solution, but if you filter the solid material, also the catalyst is separated. This might be exactly the effect you are referring to, and it happens in some of these materials too.

Omar Yaghi:
I think you also have to think about the applications, because you do not want everything to be a rock. You have to think about what kind of application you need, because in some cases you want the MOF to break apart. In fact, with the pesticide application, you want to encapsulate the pesticide in a MOF that, when exposed to the atmosphere, slowly releases the pesticide. So, it just depends on what kind of application you have.

Henry Snaith:
It is a very fascinating concept for the billions of material-manufacturing discovery processes. Two questions related to that. One is sort of a practical question: how do you get the different materials and compositions on the tips, before you put them into whatever vessel or whatever substrate you are synthesizing them on, or growing them on? How do you mix the

compositions to address all those? The second question is, there is a real challenge when trying to discover new materials. Firstly, you have got compositional space, but then you have got how you have synthesized it, the quality of the material. And you can synthesize things badly and then they can appear badly. And that is related to also how you assess the material, and how you are characterizing these billion samples on a 2 by 2 cm substrate. Is this aspirational or do you have methods that you characterize these materials with already?

Chad Mirkin:

To me, there are three silos that are being built. One is synthesis: what can we make? Two is screening: what are the structures that are active, what are they actually doing? And the third is data collection and using that to train machine-learning and artificial intelligence. So we kind of work on building-up all three of those up in parallel. Remember that everything is position encoded. So, each feature is a discrete material, that is a big difference with the inkjet type synthesis, which people often refer to as *spray and pray*. In thin films, where you got a lot of heterogeneity, every site is a single particle of a fixed composition. How do you get that? The commercial print heads have 160000 elastomeric tips, all nanoscopic in terms of sharpness. They are uniform, that is something that we worked out over the last decade in half. We spray a precursor on those arrays, that varies in composition, from left to right. So, when you do that, every tip is printing a reactor that has a different set of precursors. These elastomeric tips are compressible. In the instruments, if you tilt them, the left side comes down first and makes big features, because the tips compress. The middle makes some medium features, and the right side makes very tiny features. So, in one shot, you can not only make 160000 different features with respect to composition, but different with respect to size. That takes a fraction of a second, to make the precursors. Do it over-and-over-and-over-again. That is how you get to millions and billions very quickly. And then, everything is processed in one shot and converted into particles. So, all the different reactors have different volumes and different numbers in ratios of elements, but they got fixed positions. And now, you screen, and you collect your information. You look at where do we have activity — we are choosing to use electrocatalysis as the main driver with calorimetric indicators to tell us where we have hotspots. And then we go in and do exactly what you said. If this is a hotspot, we go in and structurally characterize that and understand whether it is a discrete particle, or was it an error — because that would be a problem. And if it is a discrete particle, we are now at 95% being particles that are fully intact. What is the composition? How is it presented? Making it many times over to see if we can make it over-and-over-and-over-again to create structures that exhibit qualitatively and quantitatively similar properties. But it is a challenge, and it is one that we have been working on, but there is a lot going on. There is a company called Stoicheia that we started, and a big investment is being made in this space, for obvious reasons. This is a way to look at materials really, really, fast, and we think it is going to yield some winners pretty early in the process.

Omar Yaghi:
I just have a follow up from Henry Snaith's question. Are you making compounds? What I mean by that, a compound has a well-defined composition. The things that you are making, are they compounds? And if they are, what are the chemical formulas?

Chad Mirkin:
It depends upon the materials that we are looking. In case of the halide perovskites, we are definitely making compounds, with well-defined compositions, ratios, and structures. They are characterized at the X-ray level. In the case of the particles, the heterostructures, those are alloys and heterostructures that consist of different domains that have been merged together. And in many respects, it is kind of interesting they even form, because you could imagine a situation where instead of forming one particle with four different phases all merged together, they could form four different particles. But the conditions of the experiment typically drive all of them together, as I showed in the 7-element library.

Omar Yaghi:
If these alloys, are not compounds, then how can they be reproduced? Or do you mean you do know the exact ratios of those? I am having trouble understanding how these compositions can be reproduced, like the alloys that you just mentioned.

Chad Mirkin:

The thing about molecular chemistry is that I can make a mole of molecules, where in principle every molecule is the same. With particles above a certain length scale, I cannot make a mole of particles where two particles are the same. So, there is an average structure, defined by composition and general phase, and that is how they are characterized, identified, and reproduced. But I would have to tell you that, for the vast majority of the catalysis, that is what we work with, what we identify, that is what we run with. The molecular approaches try to replace that, to use precision to exceed that capability. But in many cases it does not work, because you cannot work under the conditions that are relevant. Or it does work, and you find routes to go down that particular path.

Omar Yaghi:

I guess my question is, if they are not compounds, then how do you calculate the yield at the end?

Chad Mirkin:

You define an acceptable set of parameters. Just like in any heterogeneous material, you define what is an acceptable product, based upon general size. What is the dispersity, greater or less than 10%? What is the elemental ratio? How far does it deviate from that? You define, as the user, what is acceptable. And you can drive yourself nuts with it. If you drive it down to an individual atom, you know, you will never get to an ability to make structures of interest. But the good news in this space: for a lot of the materials that are very useful, average structure and control over many of these parameters, within the set of possibilities, gives you the ability to make materials exhibit generally reproducible properties and properties can be used, especially in the area of catalysis.

Nicholas Turner:

Continuing the application theme, several of the speakers discussed the encapsulation of proteins, including enzymes, in MOFs, COFs, metal peptide frameworks, which I found very interesting. So I started to think about what would be the sort of advantages or also some of the challenges in putting enzymes inside some of these frameworks. Stability, I can imagine a protein would be stabilized by encapsulation, that would be advantage. Porosity, I think is interesting, because when you immobilize enzymes, they are often mass transport limited depending upon the immobilization and encapsulation method used, so that could be interesting. Separation is an obvious one, you might want to have enhanced separation technology. Leaching or lack of leaching, which was mentioned earlier, is very important. If you make pharmaceuticals and if you use a biocatalyst in the last or penultimate step, it is very important that you establish rigorously that there is no protein component in the API, for obvious reasons. I wondered just to finish off, whether you could encapsulate two different enzymes within a MOF? I do not quite understand how you do this, but I would be very interested to know. If that is possible, then one can imagine building encapsulated cascades, potentially. I would appreciate any feedback or comments on any of those questions.

Makoto Fujita:

The encapsulation or spatial isolation of proteins is a very interesting and challenging topic in both MOF chemistry and in our cage chemistry. So, let us think about the instability of proteins. Proteins often easily denature, unfold. But most of the degradation steps initiate through the aggregation of proteins. Finally, it precipitates. It is always a major pathway of the degradation of proteins. Once the protein is isolated, so physically isolated, they have no opportunities to aggregate with each other, then the protein will be remarkably stabilized. Maybe the same phenomena were observed both in MOF chemistry and in the cage chemistry. Regarding a previous question, there are several examples of wrapping proteins by polymer matrices. But the proteins are too much biased and enzymatic activities significantly changed, reduced. But so far as we experienced, its chemical nature, including the enzymatic activity, does not change while it becomes very stable because their aggregation is completely prevented.

Xiaodong Zou:

It is possible to immobilize more than one enzyme in the pores. There are MOF materials with different types of cages, and you can first immobilize the larger enzymes, and then the next step is to immobilize the smaller enzymes into the smaller pores.

Chad Mirkin:

In the case of DNA, it is actually trivial and done. You can take almost any protein and modify the periphery with oligonucleotides, and use the same design rules and program crystallization. In fact, you can get higher quality crystals than you get with the particles, because the subunits are molecularly pure.

This idea of building cascades: we have done as many as five different proteins and multiple different enzymes within one structure, and that is the exact direction that this is all headed in. Controlling protein crystallization, where you do not pray for a particular crystal, but you guide its outcome based upon these design rules.

Yamuna Krishnan:

I was interested in why you chose to describe the DNA duplex connecting two entities as a bond, rather than as a linker, because in my view, bonds are a subset of linkers, that it is every bond is a linkage, but every linkage need not be a bond. So I was just wondering, if you could expand on that?

Chad Mirkin:

It is a really good question. We wrote a review on the whole area because we cannot view each other going down this path with very different sets of principles, and he asked the same thing. I said the reason is, when we make these constructs, density is critical. We do not use a single oligonucleotide as a linker. If you recall, Paul Alivisatos did work where he kind of aligned particles on a DNA template. I call that linking, to me that is a labeling, it is very similar to labeling with a fluorophore. In this particular case, the density is critical. We load up the oligonucleotide on the surface of the particle, so the oligonucleotides were forced to stand upright and adopt the directionality imposed by the particle core. That leads to true bonding characteristics. That allows you to get valency, when you go from a sphere to a cube, to get an octahedral type of arrangement, trigonal prismatic. It allows you to dictate architecture in a way where you can predict ahead-of-time bond lengths. If the oligonucleotides are floppy and allowed to all move and sample space in different directions, then you have no predictability in terms of bond length. Here, if I want to adjust bond length, every base I will add 2.6 Angströms of distance between the particles. It is perfectly linear, over really large distances. So, I truly believe in the context of these types of constructs where the oligonucleotide are loaded, the appropriate analysis here is a type of bonding. It is a bonding that we take advantage of, and this idea that bonding is independent of atom identity gives you the design space that I have tried to articulate.

Yamuna Krishnan:

Prior to today, my understanding of a bond was something where the length, the angle, and in fact the covalent character, was set by the quantum mechanical descriptions of the atoms involved on either side of that bond. So, you are saying that we should now kind of free our mind of that construct, of how we perceive bonds?

Chad Mirkin:

Let us say, expand it. What do you think of the mechanical bond? There are many types of bonding. We are expanding the definition of bonding and how you ultimately use it.

Omar Yaghi:

I think the question is: is a DNA linker a bond? I think chemists are very fussy about what a bond is. And even the mechanical bond is an entanglement, and I agree it does not have a quantum mechanical description. So I think that the source of this question is: should a DNA strip be called a bond because it links two things together? How about we ask Professor Sauer?

Joachim Sauer:

First of all, I do not understand this. Can you explain me how is the DNA interacting with these things you are connecting?

Chad Mirkin:

In this type of situation, you have particles, they are not atoms. We call them atom equivalents and if you want you can call the bonds between them bonding equivalents. These particles are loaded up with oligonucleotides with sticky ends, so stretches of bases, single strands that can define the recognition properties of the particle with respect to other particles. There is directionality imposed on the bond identity, the oligonucleotide in this case, because of the density of the oligonucleotides on the particle surface. And when you use these types of constructs in this way, you can begin to build matter, as I have said, that spans an incredible amount of phase space, where you can control lattice parameter, which we typically associate with bonding, with sub-nanometer resolution, over very large length scales, and you can realize crystal symmetries that have mineral equivalents and ones that do not. Finally, you can drive to specific crystal habits, because thermodynamically what is favored in this particular case, is to maximize bonding or pairing, oligonucleotide pairing. And so those principles allow us to make the equivalent of molecules and materials, crystalline materials, in a way that people have not done before.

Joachim Sauer:
I know that I am entering difficult territory, but would it not be a supramolecular arrangement?

Chad Mirkin:
Not a doubt. It depends on the way you look at it. We used the chemistry metaphor in this case, the model, as a strawman, to say what we can do with particles as atoms and oligonucleotides as bonds that we do with conventional chemistry, where do we fall short, and where can we go beyond. And that has been kind of the framework of developing this field for the last 30 years, and it has proven pretty useful.

Omar Yaghi:
I think that the source of the confusion is that the chemical bond derives its attributes from the atoms that it links, and I think that the people who are sticklers for this, probably are saying: "Hey look, there is a strip of DNA, and it does not matter what stuff you have linking, on either end". I think that is probably the source of the question.

Chad Mirkin:
That is why I started up by saying it is the advantage that it is independent of atom identity, so nobody is calling this an electron bond.

Omar Yaghi:
I think then it should be called a linker, because I could call terephthalic acid that is linking two clusters, a bond.

Chad Mirkin:
If you get directionality, yes.

Omar Yaghi:
It certainly does.

Kurt Wüthrich:
I have a comment to the question by Nicolas Turner. In my field, we have used inverted micelles to suspend functional enzymes in aqueous solution in a low-viscosity organic solvent, just to improve the quality of the NMR spectra. In spite of the very specialized use, I think it relates here to the kind of encaging that you all are talking about.

Arne Thomas:
In short, this is exactly what you said, you suggested. Everything was totally right about what are the advantages of immobilizing enzymes into MOFs or COFs materials. Then you said there might be also coupling, and also this has been done, using two different enzymes, as Xiaodong Zou said. But I think you can have a little bit more fantasy there and think of why just coupling two enzymes. You can also probably couple an artificial catalyst with an enzyme. And then, because we can compartmentalize the whole thing — Omar Yaghi got this idea of actually separating gas molecules — we can probably also separate solvent streams, so that you have one compartment where an enzyme can work with an organic catalyst that should work in an organic solvent that would normally kill the enzyme. We can actually have different compartments of a COF or MOF do different kind of catalysis and then transfer products to make a real cascade. Because this is a big problem, yesterday cascade catalysis was mentioned. For two catalysts working together for cascade catalysis, always all conditions have to be exactly working together: temperature, solvent, and so on. I have at least the hope, this is not done so far, but I have the hope that one day we will be able to do this in one material.

Ben Feringa:
I hear a lot of interesting discussions about templating and precise organization. As far as I know, for decades, people have tried to imprint in polymers, sticklers, to do polymer imprinting, like for instance a transition state for a catalytic reaction. As far as I know, they have not met with much success at all. Should we revisit this field, using your approaches, and getting real imprinted, for selectivity, for separation technology, for catalysis, all these kind of things, using your approach?

Omar Yaghi:
I will give you a free idea: I think you can evolve. The multivariable systems are really fascinating, because when you take a dumb crystal that is not multivariable, and put it in a soup of linkers that are differently functionalized, you may be able to evolve a structure around your substrate. And after you have evolved it several times, several generations, you may have the best imprint for that substrate. I think that is one way to do it. I do not like the term templating from MOFs and COFs, because they are really being put together by those linkages, they are very directional, and the stuff in the middle does not seem to matter; it is just space-filling rather than templating. I like to think of templating as something substrate specific for the assembly of a specific MOF, and we do not see that.

Ben Feringa:
That is what I would like to hear from you, because in the traditional polymer synthesis, there was hardly any success with this approach because apparently the dynamics, the reconfiguration, the adaptive behavior was not there, but here there seems to be opportunities.

Omar Yaghi:
Yes, because the backbone is rigid, and then functionalities are the ones that are shaping.

Makoto Fujita:
We have once tried the guest-induced assembly of its own optimal receptor structure from a mixture of several different ligands. There is a complex mixture of the metal and just several ligands, but by adding specific

template molecules, its optimal receptor structure would be predominantly assembled. We also tried to template with transition-state-like molecules, but unfortunately, we have not obtained good results. We actually tried to do that imprinting in our self-assembly system.

Stefan Lutz:
I want to follow up on Nicolas Turner's question about proteins in MOFs and COFs. Two parts. I would be curious about electron transfer. If you encapsulate an enzyme or protein in your type of arrangements, you showed the very nice example with the nickel catalyst, could this also work for the entire protein? And then thinking along the lines of just the last comment, talking about the evolvable container, but of course the protein can also be evolved to work optimally in that container, so here you are now looking at a system that can be improved in two dimensions. I would love to hear your thoughts.

Arne Thomas:
The first answer is yes, this we know already. I will not tell you too much about it, but it is working. You can encapsulate enzymes and you make an electron transfer, photo-catalytically or electro-catalytically. This is both working. It is a bit of a question, and now we come to the difficulties, and you know these of course. The question is how the electron transfer occurs. It is a little bit the orientation of protein and so on. So, there are many things happening in three-dimensional pores and the question is: from where is the electron coming? I cannot answer for the moment, but I can tell you it is working.

Regarding the evolvable container, this is actually the same answer more or less, as the question is: how it orients? I think we have many opportunities, because we can change, and this is always what I say with this multifunction and multivariate thing. You cannot change only such that we optimize electron transport, but also the polarity, for example, this might be superb. This is extremely important, I think, to evolve the protein there. Should the protein really stick to the surface, because it is hydrophilic and has hydroxy functionality? Should it be a little bit more apolar, so that it stays more within the cavity, but not attached to the support too much? These are things we are trying at this moment to find out. The last example, you have shown there was something where we can gradually change polarity of this type of COF materials, and this is something that I think will be very interesting, to see what the protein in such pores is doing. Again, with this notation I made in the beginning: do not change three things at once, because then you do not know what the effect is. Try actually to stick to one material, but gradually change one parameter, like polarity, and see what protein is doing in there.

Karthish Manthiram:
I think it is brilliant to see structure at such a fine scale. Xiaodong Zou, as you described it, it is very remarkable to be able to pick up these minor phases that we would otherwise be blind to. And I think, as you implied, these minor phases in many cases are probably, or could be, responsible for catalysis. I just want to get your take on to what extent do you foresee this possibility to be able to map catalytic activity at that fine length scale, at this tens of nanometer length scale, so you can now correlate these maps of catalytic activity with the phase behavior, this heterogeneity in phase behavior that you see, and bring those two together?

Chad Mirkin:
I think that's where it is definitely ultimately headed to. But you know you have to start a new area. Start talking about things as DNA bonds. You have to think about how you do navigate in that area in a way that allows you to do biggest things first. So you are talking about something that is really, really important, and will undoubtedly be an issue of interest, but it is not the big ticket item from the start. The question is: can you use this to find things, first of all, that do what we need, that we are all excited about, and then use redundancy and the ability to duplicate many times over to explore that to the level that you need to get to the point where you can scale up and use it for that particular purpose? If that demands that we understand that architecture down to the individual atom level, then we got a much higher challenge or higher bar for us. In many cases, it will not, I am convinced of that, but we will eventually use this as a tool to get better and better doing this. I have one student who is taking it down to making particles that are 1 nm in size. It is a countable number of atoms in the cluster. It changes how we characterize. We are having to use

aberration-corrected electron microscopes to characterize these types of structures. Challenges in terms of characterization go way up. This is all part of the whole evolution, I think, of this particular area, but initially we are going in with the brute force approach. Let us take what we can control today, elemental composition, size, define parameters and acceptable variances, build those libraries, and see if we can find things that do what we cannot with conventional materials. So far, we found a few, and we are going to find a lot more.

Xiaodong Zou:
I would like to say that now, with electron crystallography, it is possible to really both see the materials and also determine the structure, and understand the detailed structure at the atomic level. And also most importantly, you can study the heterogeneities, which means that you can have a complex material or a material with different components, and it is possible to study it. So, I would like to think of how that would change the current way of making materials. Because you are thinking about the simple components and trying to get a big crystal and pure materials, but what if you put many components and molecules together and see what you get? And also, being able to look at the evolution of materials in different environments, and the catalysis, while we can both see and know the structure with atomic precision.

Chad Mirkin:
I just think it has become a necessity that people like you come up with high-throughput ways of characterization. Characterizing one versus 225 million are two different issues. So, figuring out how you get structural information on the things that matter is going to be a big part of how this either moves or does not move going forward.

Xiaodong Zou:
This is a challenge, and we are moving towards that field.

Matthew Kanan:
My question is with respect to catalysis. I guess the way I see what you are able to do, is you can control the functionality in the individual spaces inside these materials pores, and then you can create a distribution of

combinations of those functionalities, which you are calling multivariate materials. I am trying to understand the added capabilities there a little bit better. Obviously, the functionality within a certain space is what makes zeolites special. They have very acidic sites, they have pore sizes that interact with hydrocarbon substrates, and they are thermally stable, and that is the magic of zeolites. Your new materials, you are, sort of, in a different temperature regime, and you can decorate, I suppose, with more functionality. But is there an example of a catalytic transformation that you can perform, based on your ability to position three or four different functional groups that are important, that cannot be performed with a molecular catalyst that is tailored to that specific transformation?

And the second question is that, I guess, I still do not quite see the power of it. If I have a substrate coming in and it sees one environment, that maybe catalyzes a reaction of that substrate at a certain configuration, and it is going to diffuse over to the next pore that has a different configuration. I do not understand how that works, unless it happens to be that ABCD vs DCBA, that those two configurations give you sequentially two transformations, which seems to be rather unusual circumstances. I am missing the power of having that variability. You both sort of alluded to this sort of cascades and these networks, and I guess it would just help to have an example to understand how one can do that, as opposed to lots of great technologies for immobilizing enzymes and solid supports to be amenable to high-throughput synthesis. There are ways of encapsulating and protecting things. I am just trying to understand the power of ordering in these three-dimensional spaces, or sampling large heterogeneous mixtures of configurations of functionality a little bit better.

Omar Yaghi:

Molecular chemists spend their lifetime trying to organize ligands around a metal for a specific transformation. There has been a lot of time perfecting that catalytic pocket, if you want to call it that. What my thinking from multivariate was is to introduce those ligands that a molecular chemist would, although the molecular chemist has to also do protections and things like that, that you do not have to do in the MOF because the MOF is a protecting group. Let us take the TEV protease enzyme example that I gave. We just introduced the amino acids that we thought approximate the pocket into the pore, and we build a short peptide of four amino acids, one amino acid at a time. So the pore is really internally heterogeneous and my thinking is that the reason the MOF, with all that lack of preplanning, still broke that specific amide bond that the enzymes did, is because it has billions and billions of microenvironments and a certain, perhaps small, fraction of them is fertile for that particular selective bond cleavage. The rest of them are inert to that transformation. The problem of course of this approach is that, while a molecular chemist does things with precision around the metal, in this one you are throwing everything in there and saying: "Hey, instead of making one complex, I will make billions of microenvironments, billions of complexes and just let the substrate sort of find its way through". The problem, ultimately, that you have to reckon with is, what are those environments that are fertile? And the second thing, will the substrate after transformation be able to diffuse in and out easily, with all that stuff that is happening in the pore? So, there is a lot of stuff you have to do to make sure that it follows the logic of molecular chemistry. But I feel that it is an opportunity to sample a combinatorial set up space within the crystal. This was my idea about the fact that we do not know what these sequences are, we need to figure out a way of doing that.

Matthew Kanan:
Maybe I just misunderstood. The combinatorial space is the different functionalized linkers. How many different functionalities are you putting in one of these?

Omar Yaghi:
I think there are seven reactions that are leading to a short peptide of four amino acids, and each transformation is not 100% complete. So, you have really a forest of one amino acid, two amino acids, three amino acids, four amino acids, and it is just everything is in there. It is not a clean environment, but nevertheless it did these transformations.

Matthew Kanan:
And based on the kinetics, can you estimate different possibilities of what fractions of those sites are active?

Omar Yaghi:
Hard to say. We just know that there was, I think, 70% yield in terms of the cleavage.

Karen Goldberg:
Have you ever in these MOFs used the ligand in a metal-ligand cooperation, in terms of using the metal center and the ligand within the MOF? So, using the metal as a catalytic center? Because you are using ligands, but a site of ligand could cooperate with the metal in terms of doing a bond activation.

Arne Thomas:
This has been done many times. You can actually make MOFs with open sites. In some MOFs, some middle sites are fully coordinated and they are not active in catalysis anymore, but you can create MOFs with open metal sites, and then the metal site itself can be catalytically active, and you can use the linker for another incorporation of metals. This has been done, even though I do not have the application in mind, which kind of catalysis it was.

Omar Yaghi:
Ok, we have come to the end of the discussion. We had a great discussion, thank you, thanks everybody.

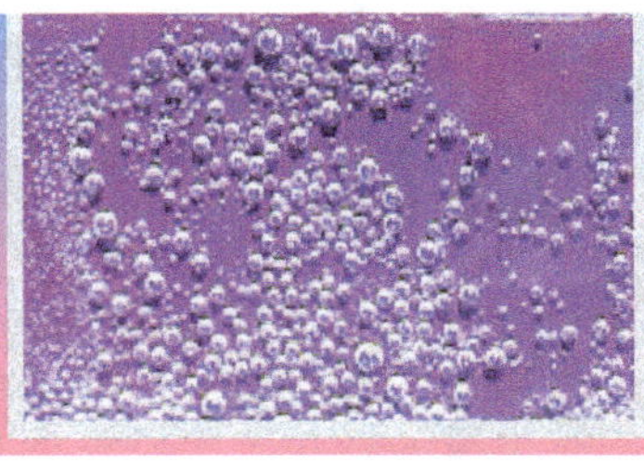

Image courtesy of Daniel Nocera

Session 4

New Chemistry for Renewable Energy

https://doi.org/10.1142/9789811282324_0024

Sustainable Energy Production and Manufacturing Using Only Sunlight, Air, and Water

Daniel G. Nocera

Harvard University, Cambridge, 02128 MA, USA

Energy demand in the 21st century will roughly double by mid-century owing to rising living standards in emerging economies and to a growing world population, primarily located in the Global South. Most of the increased demand will come from 3 billion people currently without access to reliable energy and 3 billion new inhabitants to our planet who will mostly reside in the Global South. For this reason, this global cohort will have a major impact on climate change over the approaching decades. Providing these 6 billion people with carbon-neutral alternatives to meet their societal energy needs is a requisite to mitigating climate change. To meet the increased energy demand of these 6 billion additional energy users will require invention, development, and deployment of carbon-neutral energy and energy-intensive processes on a scale commensurate with, or larger than, the entire present-day energy supply from all sources combined.

Nowhere has science wandered away more from the world it needs to serve most than in the field of energy. Addressing the energy needs of the underserved requires us to rethink how energy systems are implemented.

Energy systems of today are heavily engineered, and they are designed to rely on a centralized infrastructure. In addressing the underserved, the dogma is to translate science discoveries and attendant technologies that have worked in the developed world to the emerging world (i.e., pound the square peg into the round hole). However, climate change is outpacing the accrual of wealth needed to realize centralized infrastructures in the developing world. A new science needs to be discovered. Critical to this rethinking is a move towards a more low-cost and distributed energy ecosystem where both the production and delivery of energy are better aligned with the context of the underdeveloped regions of the world. Specifically, what new science will engender distributed energy systems that use sustainable inputs that are available to all? We have addressed this question with a series of science advances that underpin distributed and sustainable systems for biomass, fuels, and fertilizer production, and the synthesis of complex molecules such as vitamins using the inputs of only solar energy, air, and any water source.

Self-healing catalysis: Solar water splitting using any water source

All sustainable energy conversion schemes eventually rely on water as the source for proton- and electron-reducing equivalents. Because the coupling of the electron and proton is demanded for high energy efficiency, we were led to create the field of proton-coupled electron transfer (PCET) at a mechanistic level [1] with the first measurements that temporally resolved the movement of an electron coupled to a proton, followed by our development of the first theory of PCET [2]. Controlling PCET is at the heart of efficiently harnessing the protons and electrons of the water-splitting reaction, in which the energy of a solar light input is stored in the rearranged bonds of two water molecules to produce hydrogen ($\times 2$, $4H^+/4e^-$) and oxygen [3].

The implementation of water splitting for distributed energy systems in the developing world presents major challenges and thus an imperative for new science. Foremost is the criterion to develop catalysts that operate in any water source. Most water-splitting catalysts require a concentrated base where the protons that are released from the oxygen evolution reaction (OER) are promptly neutralized. In natural waters, however, the concentration of hydroxide is small and the strongest base is the oxide, which readily reacts with acid, thus leading to its corrosion by its dissolution. To overcome corrosion, we developed cobalt/nickel-phosphate/borate (Co/Ni-P_i/B_i) catalysts [4, 5] that are self-healing [6, 7] wherein an equilibrium for catalyst self-assembly is established that lies energetically within that for the OER of water-splitting catalysis (Fig. 1). Because these self-healing catalysts continually renew themselves during OER, they can operate in neutral and natural waters, including wastewater [8, 9]. The ability to perform OER under neutral conditions has several advantages. First, a major milestone is achieved with regard to responding to the underserved as solar water splitting is not reliant on a specific water source and it may be found locally. Second, the catalysts operating under mild environmental conditions facilitate their interface with materials such as Si, which is unstable in corrosive environments. Finally, the ability to operate water splitting from buffered water allows for the integration of water-splitting catalysis with bioorganisms.

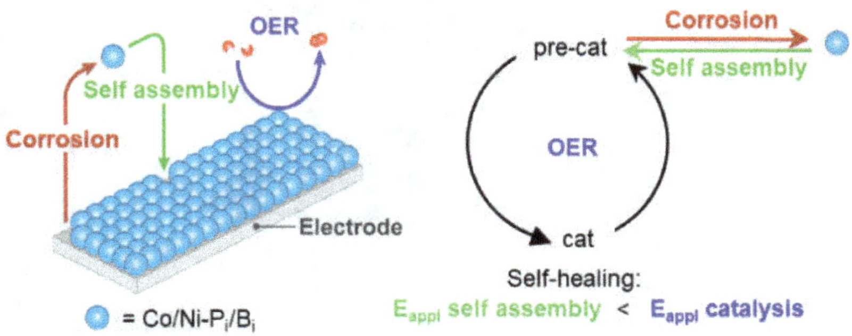

Fig. 1. Self-healing OER Catalysis. Self-healing is achieved if the potential for self-assembly of the Co/Ni-P_i/B_i OER catalyst (blue spheres) is less than the potential for water splitting.

Artificial leaf: Emulating the solar energy reaction of photosynthesis

The ability to interface water-splitting catalysts with Si under non-corrosive conditions led to the development of the Artificial Leaf [10, 11]. The Artificial Leaf, which is the first wireless water-splitting device, comprises a silicon wafer coated on one side with the Co–P_i OER catalyst and the other side with a NiMoZn alloy or a cobalt–phosphide (Co–P) hydrogen evolution reaction (HER) catalyst (Fig. 2). The Artificial Leaf is quite remarkable. When immersed in any water source and exposed to sunlight, oxygen bubbles from the OER side of the wafer and hydrogen bubbles from the HER side of the wafer at a rate that is commensurate with the solar flux impinging on the silicon. The Artificial Leaf captures the direct solar process of photosynthesis — the use of sunlight to split water into hydrogen and oxygen from neutral water, at atmospheric pressure and room temperature. However, inorganic water splitting systems such as the Artificial Leaf can absorb light and separate charge at efficiencies [12]

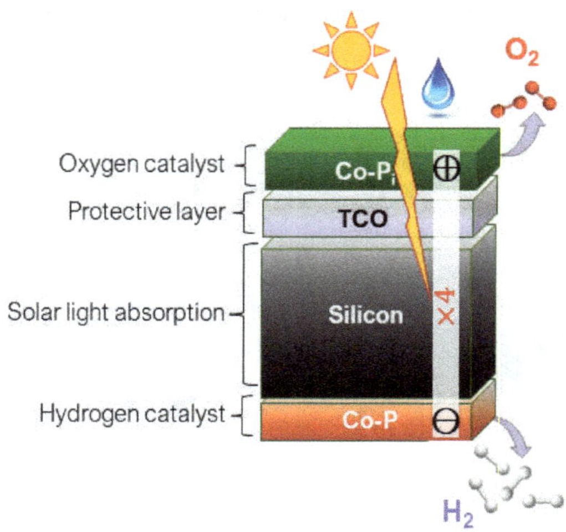

Fig. 2. The Artificial Leaf. The p-side of an Si is coated with a transparent conducting oxide (TCO) protective layer and a Co–P_i OER catalyst. The n-side is coated with a Co–P HER catalyst. Similar to the Kok cycle in Photosystem II, the absorption of four photons to produce oxidizing and reducing equivalents is translated into energy storge via the water-splitting reaction using natural water or wastewater.

much greater than Photosystems II and I [13] (PSII and PSI, respectively) and hence solar-to-fuels (hydrogen–oxygen) efficiencies are much greater than that achieved by the photosynthetic membrane for water splitting. Here again, the Artificial Leaf achieves an important milestone with regard to the undeserved — a minimally engineered device, Si with simple coatings, can produce oxygen and hydrogen from natural water and wastewater sources with sunlight as an input.

Bionic Leaf-C: Distributed carbon fixation to biomass and fuels

The Bionic Leaf-**C** was created to use the hydrogen produced from the catalysts of the Artificial Leaf to power the cellular biosynthetic machinery of bacterial microorganisms to convert carbon dioxide from air into biomass and liquid fuels [14]. In this approach, the entire PSI and PSII water-splitting assemblies of the photosynthetic membrane are replaced by catalysts of the Artificial Leaf. The hydrogen from water splitting is coupled to the production of NADPH within the cell via hydrogenases; the NADPH in turn drives cellular energy production (ATP) and is also the reductant for the fixation of carbon dioxide from air (Fig. 3). A solar-to-biomass yield of 10.8% is achieved with a typical Si photovoltaic efficiency of 20%. The biomass efficiency is authentic as it accounts for limitations arising from H_2 generation from water splitting, H_2 solubility, and the energy efficiency associated with biomass maintenance and growth. Moreover, Bionic Leaf-**C** operating in air is only 2.7 times lower in efficiency than that when it operates in pure CO_2, despite a CO_2 concentration difference between air and pure CO_2 of 2500. The high product yield for the Bionic Leaf-**C** in air demonstrates the advantages offered by biological carbon-concentrating mechanisms to alleviate mass transport limitations, absorb solubilized CO_2, and maintain high intracellular CO_2 concentrations. The carbon flux within the Bionic Leaf-**C** may be redirected from cell growth (i.e., biomass) to metabolically engineered pathways placed in the bioorganism to result in the synthesis of liquid alcohols, namely isopropanol (C_3), isobutanol (C_4), and isopentanol (C_5) at solar-to-liquid fuel efficiencies of 6–8%, depending on the liquid fuel.

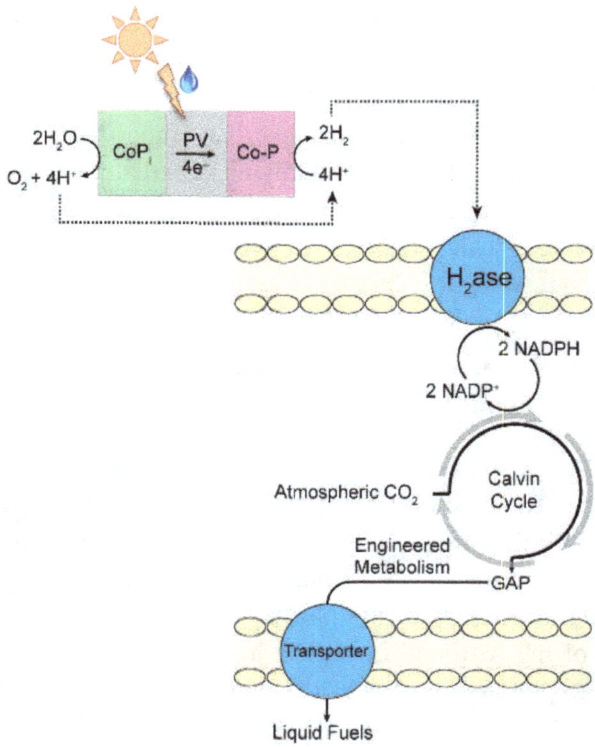

Fig. 3. Bionic Leaf-**C**. H_2 produced from the catalysts of the Artificial Leaf is used to power cellular biomachinery for CO_2 fixation along engineered metabolic pathways to make liquid fuels.

Thus, the Bionic Leaf-**C** is 10 times more efficient than natural photosynthesis for the best growing biomass (e.g., 1% for soy, switch grass) [15] and, with regard to liquid fuels, is 50–100 times more efficient than liquid fuels provided from natural biomass. The Bionic Leaf-C in effect is a distributed and sustainable surrogate for Fischer–Tropsch synthesis using only air, water, and sunlight as its inputs.

Bionic leaf-N: Distributed nitrogen fixation to fertilizer

Extending the hybrid inorganic–biological (HIB) approach, we have realized a renewable and distributed Haber–Bosch process with the creation of the Bionic Leaf-**N** [16]. The Bionic Leaf-**N** uses the autotropic and nitrogen-fixing microorganism, *Xanthobacter autotrophicus* (*X.a.*), which has parallel C- and N-fixing pathways and may be directly grown from hydrogen (Fig. 4). *X.a.*, as all nitrogen-fixing bioorganisms, down-regulates nitrogen fixation owing to the enormous energy intensity required to convert nitrogen to ammonia (nitrogenase requires 16 ATPs per nitrogen molecule). To overcome this roadblock, we have combined the hydrogen produced from the catalysts of the Artificial Leaf with CO_2 from air to produce polyhydroxybutyrate (PHB), which is stored internally within the bacteria. The PHB biopolymer provides the bacteria with an energy supply, as it is a nascent source of intracellular energy (ATP) and hydrogen (NADPH/H^+). With its own internal energy and hydrogen supplies, down-regulation of nitrogen fixation in *X.a.*-PHB is mitigated and nitrogen fixation in the form of ammonia and "N" biomass occurs unencumbered as reflected by exceptionally high nitrogen conversion activity (turnover frequency of 1.9×10^4 s^{-1} per bacterial cell and turnover number of ~9×10^9 bacterial $cell^{-1}$). The ammonia produced directly or from biomass breakdown can diffuse across the *X.a.* membrane and be

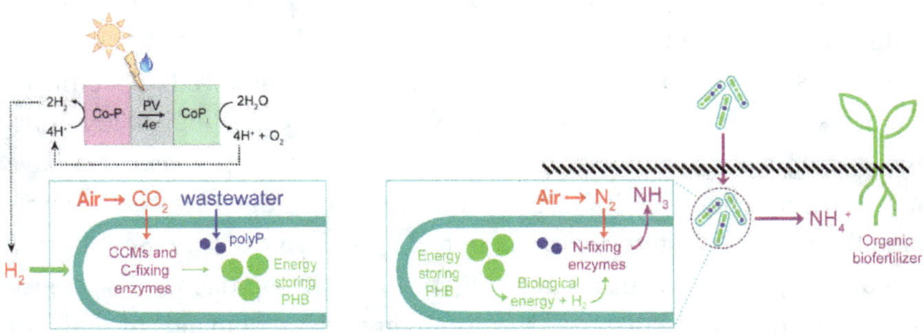

Fig. 4. The Bionic Leaf-**N**. (a) The Bionic-Leaf N combines H_2 from solar water splitting with CO_2 to produce polyhydroxybutyrate (PHB) and then (b) draws on the PHB as internal energy and hydrogen supply to power the nitrogen fixation cycle to produce ammonia and solid N biomass. Once PHB is produced, a solar source is no longer needed, and the organism can be introduced into soil and perform fertilization. When grown in the presence of wastewater, the organism will also sequester P in the form of polyphosphate (polyP). The process allows for the sustainable and distributed production of fertilizer from sunlight, atmospheric CO_2 and N_2, and wastewater.

used as a living biofertilizer. The Bionic Leaf-**N** is of significant consequence to mitigating CO_2 release into the atmosphere as the industrial synthesis of ammonia results in CO_2 emissions that are greater than for any other chemical-making reaction [17]. As an example, in a large 400-acre farm trial for leafy vegetables, 130 lbs of N is needed per acre for crop growth. By replacing 90% of the chemical fertilizer (UAN = urea ammonium nitrate, 5.98 lb CO_2 produced per lb N for UAN) with the Bionic Leaf-**N**, 141 metric tons of carbon dioxide was prevented from being released into the atmosphere [1]. Additionally, the Bionic Leaf-**N** sequesters CO_2 from air in the form of PHB (–1.105 lb CO_2/lb N) and thus the use of Bionic Leaf-N as a living biofertilizer is a carbon-negative process. Beyond mitigating the carbon budget associated with crop growth, the Bionic Leaf-**N** has an important future role in sustainable farming as the living biofertilizer avoids nitrogen runoff, and it is also able to fix P from wastewater [18] to allow for cyclic and renewable rotation of the biogenic elements of C, N, and P.

Sustainable manufacturing

The HIB approaches reveal a path forward to creating a distributed and sustainable manufacturing industry. The powerful tool of synthetic biology allows the HIB approach to be generalized to a renewable chemicals synthesis platform, depending on the biomachinery to which water splitting is coupled. As an example of the power of the approach, we have metabolically engineered *X.a.* to fix carbon and nitrogen from air to produce the complex heterocycle, vitamin B2 — if you will, vitamins from thin air [19]. For solar manufacturing by the HIB approach, new science needs to be developed that meets the high throughput demands of scale for manufacturing processes. High mass fluxes of reactants and products are needed for manufacturing and thus catalysts with high turnover rates and accelerated metabolic pathways of bioorganisms will be needed. The HIB approach need not be confined to the cell. In this case, outside the cellular confines, enzymes will have to function with stability over long periods of time under manufacturing relevant conditions. Embodied in the catalysis and reactor designs are many challenges including the need to furnish products selectively, isolate those products, and capture CO_2 at low energy intensities. Ultimately, the science that underpins solar manufacturing must be cost-effective over long-term operation.

Concluding comments: The sustainocene

The Holocene ("recent whole") period began 12,000 years ago with the introduction of human activity on our planet. From that time until circa 1800 CE, the presence of humans did not alter the natural systems of this world. Since 1800 CE, human activity has pushed this planet from the Holocene into what has been termed the Anthropocene period, an epoch when human interference with earth systems threatens the capacity of our biosphere to sustain life. The science and technology that created the Anthropocene now needs to be addressed with a science and technology that creates a new epoch — the "Sustainocene" — a period of ecological sustainability, environmental integrity, and the elimination of gross societal imbalances in poverty by providing a global population with access to energy, food, and water. The connection of the Sustainocene with the need for new science for the poor is plainly evident from the Kaya identity:

$$CO_2e = N \cdot (GDP/N) \cdot (E/GDP) \cdot (CO_2e/E),$$

where CO_2e is the global CO_2 emissions, N is the population, GDP/N is the gross domestic product per capita, E/GDP is the energy needed to maintain GDP, and CO_2e/E is the carbon footprint of the energy supply. For climate change to be stabilized, it is essential that CO_2e/E be minimized, i.e., that the new energy demand, E, for the underserved be sustainable owing to the addition of 6 billion to N in the Kaya identity. The advances described herein establish that the Sustainocene is possible. However, more scientists need to embrace a science targeted to the poor where they will realize a plethora of new challenges and a fertile space for new discoveries. In doing so, they will set out on the quickest path to stabilizing climate change and they will be the leaders of a historic change to a global chemical and energy infrastructure, one powered by the sun, air, and water.

References

[1] D. G. Nocera, *J. Am. Chem. Soc.* **2022**, 144, 1069–1081.

[2] R. I. Cukier and D. G. Nocera, *Annu. Rev. Phys. Chem.* **1998**, 49, 337–369.

[3] N. S. Lewis and D. G. Nocera, *Proc. Natl. Acad. Sci. USA* **2006**, 103, 15729–15735.

[4] M. W. Kanan and D. G. Nocera, *Science* **2008**, 321, 1072–1075.

[5] M. Dincă, Y. Surendranath, and D. G. Nocera, *Proc. Natl. Acad. Sci. USA* **2010**, 107, 10337–10341.

[6] D. A. Lutterman, Y. Surendranath, and D. G. Nocera, *J. Am. Chem. Soc.* **2009**, 131, 3838–3839.

[7] C. Costentin and D. G. Nocera, *Proc. Natl. Acad. Sci. USA* **2017**, 114, 13380–13384.

[8] A. S. Esswein, Y. Surendranath, S. Y. Reece, and D. G. Nocera, *Energy Environ. Sci.* **2011**, 4, 499–504.

[9] T. P. Keane and D. G. Nocera, *ACS Omega* **2019**, 4, 12860–12864.

[10] S. Y. Reece, J. A. Hamel, K. Sung, T. D. Jarvi, A. J. Esswein, *et al.*, *Science* **2011**, 334, 645–648.

[11] D. G. Nocera, *Acc. Chem. Res.* **2012**, 45, 767–776.

[12] M. T. Spitler, M. A. Modestino, T. G. Deutsch, C. X. Xiang, J. R. Durrant, *et al.*, *Sustain. Energy Fuels* **2020**, 4, 985–995.

[13] R. E. Blankenship, D. M. Tiede, J. Barber, G. W. Brudvig, G. Fleming, *et al.*, *Science* **2011**, 332, 805–809.

[14] C. Liu, B. C. Colón, M. Ziesack, P. A. Silver, and D. G. Nocera, *Science* **2016**, 352, 1210–1213.

[15] D. K. Dogutan and D. G. Nocera, *Acc. Chem. Res.* **2019**, 52, 3143–3148.

[16] C. Liu, K. K. Sakimoto, B. C. Colón, P. A. Silver, and D. G. Nocera, *Proc. Natl. Acad. Sci. USA* **2017**, 114, 6450–6455.

[17] Ammonia: Zero-Carbon Fertiliser, Fuel and Energy Store, The Royal Society, London, UK (2016).

[18] V. S. Mandala, D. M. Loh, M. B. Geeson, S. M. Shepard, I. V. Sergeyev, *et al.*, *J. Am. Chem. Soc.* **2020**, 142, 18407–18421.

[19] R. S. Sherbo, P. A. Silver, and D. G. Nocera, *Proc. Natl. Acad. Sci. USA* **2022**, 119, e2210538119.

https://doi.org/10.1142/9789811282324_0025

Novel Materials for Photovoltaic Solar Energy Materials

Henry J. Snaith

Clarendon Laboratory, Parks Road, Oxford OX13PU, UK

My view of the present state of technology and research on renewable energy

The continued development of renewable energy is clearly critical for eliminating the use of fossil fuels. There is no possibility of a sustainable, stable future for humanity or the broader environment and ecosystem on Earth without making this transition to renewables over the next few decades. The renewable power generation technologies are numerous, and many will play a role in our future power generation network, but there exist some clear front runners. Hydroelectric power (HEP) is well established, but the capacity is geographically dependent and the large dams required for HEP can have severe consequences to the local environment. Wind power is widespread and will continue to expand and generate a significant fraction of our future power demands, although the predominant requirement to create off-shore wind farms, rather than on-shore, has and will continue to slow down the price reduction. At some point however, the potential for larger and larger off-shore wind turbines, enabled by advanced composites and optimized designs, will lead to off-shore wind being more economic in the future. Solar power is the most abundant power source and even though the levels of irradiance vary, it is ubiquitous across the globe. The primary industrial method for harnessing solar energy is via photovoltaic solar cells, which convert sunlight directly into electricity. Here, silicon dominates, and advances in silicon technology have been made by moving from doped positive(p)-negative(n) junctions, to well-passivated heterojunctions, with record cell efficiencies having reached between 25 and 27%, and commercial modules at around the 21–22% level. The continued scaling of manufacturing and deployment has also led to continuous price drops in deployed PV, with the lowest cost solar farms having price purchase agreements at around 1$c/kWh (10$/MWh). With solar PV, the key driver to cost reduction is now increased efficiency, rather than reduced manufacturing costs. The latter has already been successfully achieved by the existing PV industry. Here, a general approach to deliver higher efficiency than Si is to move towards more advanced concepts. These include more exotic concepts such as hot-carrier collection and photon-multiplication or up-conversion and also more near-term practically realizable concepts such as concentrator PV and multi-junction PV [1]. Concerning the latter, metal-halide perovskites have proven to be extremely effective and record efficiency perovskite-on-silicon tandem cells have now surpassed 31% [2–4]. Power generation is one piece of the puzzle, and the combination of wind and solar works well in most locations in the world to create a relatively balanced power generation profile. Yet the power generation profile is variable day and night and seasonal depending upon geographic location. There is therefore a major requirement to bring online large-scale power storage to go hand-in-hand with the ramping up of the renewable power generation capacity. The advancements in li-ion battery storage for electric vehicles (EV) are making battery storage affordable for domestic use. However, it is questionable whether the technology developed for EVs will be ideal for grid storage. It is likely that certain elements, such as Co, will need to be replaced. Furthermore, batteries have been optimized for lightweight high-power density. For grid storage, weight and arguably round-cycle efficiency are less critical, which may open more diverse possibilities for developing new battery technologies. Hydrogen generation using electroliers is likely to become a major contributor to grid-storage capacity, where the hydrogen can then be stored and used to regenerate electricity in a converted gas-fired power station or through a fuel cell. Although the round-cycle efficiency of electrolysis and subsequent combustion of the hydrogen are low, the continued drop in the price of electricity generation from solar and wind makes the absolute efficiency of this process less critical, as opposed to the practicality and cost of electrolysis and storage. Beyond hydrogen, a range of other synthetic fuels, including ammonia, are likely to also play an increasingly important role.

My recent research contributions to field of photovoltaic solar energy conversion

My research focuses on PV technology, specifically investigating and advancing novel materials and device concepts. Over the last decade, we have been central pioneers in the development of metal-halide perovskite PV, having made some of the key early discoveries of the outstanding optoelectronic properties of lead-halide perovskites, and demonstrating that these materials can be integrated into a simple thin-film device structure which delivers extremely high efficiency [4–7]. Metal-halide perovskites have an ABX_3 stoichiometry, where A is an organic or alkali metal cation, B is a divalent metal cation (Pb, Sn, or Ge), and X is a halide anion. Motivated by the requirement to deliver higher efficiency PV technology to make a positive impact upon the industrialized solar energy generation, we have focused our efforts on enabling and delivering multi-junction perovskite PV cells, both as stand-alone thin-film technologies and in combination with silicon [8–11]. In Fig. 1, we show a cross-sectional electron microscope image of an "all-perovskite" triple junction [12].

A key challenge for the multi-junction perovskites has been to enable high open-circuit voltage from the wide gap perovskites and large "deficit" in voltage existed between the band gap energy and the open-circuit voltage. In order to obtain wide band gaps, we need to form alloys of I/Br at the X-site. Early observations saw that under illumination the halides tended to segregate into at least two different domains. This resulted in the

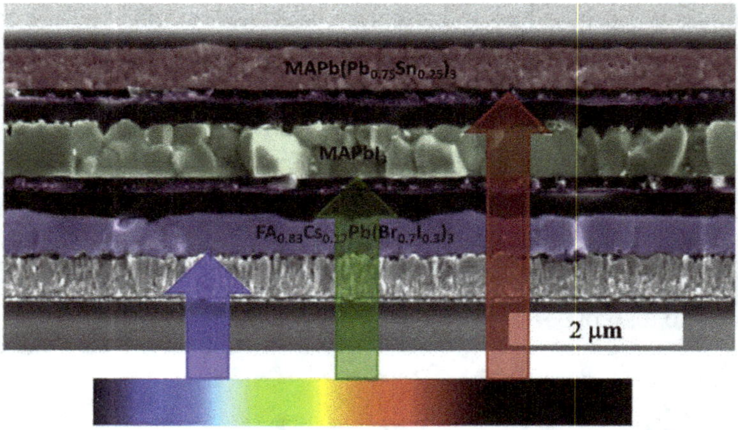

Fig. 1. Cross-sectional electron microscope image of an all-perovskite triple junction cell. Reproduced with permission from reference [12].

formation of lower band gap I-rich domains within the perovskite and subsequent emission of light through the low band gap domains [13]. This was thought to be the cause of the voltage issue with wide band gap perovskite cells. We performed a combined optical and thermodynamic simulation to quantify the impact of halide segregation upon open-circuit voltage deficit, and surprisingly we found that the halide segregation alone could only account for a small fraction of the voltage deficit [14]. We thus identified that suppressing trap-assisted charge carrier and recombination at the perovskite absorber/charge extraction layer heterojunction was the area that required the most effort to improve. Since then, we have made significant progress in delivering higher voltage wide gap perovskite PV cells by focusing upon both internal and interfacial passivation of the absorber layers [15].

A central challenge for metal-halide perovskites is to enable long-term operational stability commensurate with Si PV. Early compositions employed methylammonium as the organic cation, which suffers from poor thermal and chemical stability [16]. Moving towards formamidinium (FA) as the organic cation lead to a substantial enhancement in the thermal stability [17], and alloying FA with caesium, in addition to I with Br lead to highly stable perovskite thin films with tunable and stable band gaps [11]. With these adaptations to the perovskite, the "weakest link" ceased to be the organic–inorganic nature of the perovskite and became the photochemical degradation of the metal-halide framework. Here, under the combined stresses of light and temperature, I_2 gas is generated, which is considered to originate from holes being trapped on interstitial iodide [18, 19] [ref two past papers]. We found that crystalizing the perovskite in the presence of quaternary ammonium salts greatly suppressed this degradation leading to significant enhancements in the long-term stability [20, 21].

Outlook to future developments of research on novel materials for PV

As for perovskites, the following key challenges remain: (i) obtaining very high voltages for the wide gap cells, but the road map to achieve this is very clear, via controlled crystallization and growth to achieve homogeneous alloyed compositions with low defect densities, coupled with bulk and interface passivation. This may include molecular passivation or the inclusion of lower-dimension perovskite phases forming heterojunctions with the bulk 3D ABX_3 perovskite material. (ii) The second main challenge is long-term stability. Unlike efficiency, this cannot be quantified unambiguously in a lab and will require a combination of laboratory-based stress tests and real-world outdoor testing. According to our present knowledge, understanding the photochemical degradation process in metal-halide perovskites, and how and why certain additives suppress this is central to both delivering a highly stable technology, and having the confidence that it will last for 25 years in the field. In addition to the academic works, there is much industrial activity advancing the perovskite PV technology, which has a central focus on long-term stability. As scientists, we are best placed to stress these materials and devices via diverse methods so that we can discover and understand instabilities from a fundamental basis.

Beyond lead-halide perovskites, there are many possible new materials which may deliver complementary functionality or enhanced properties for a broad range of PV applications. Specifically, the lower bap perovskite which is composed of an alloy of Pb and Sn does not have the same stability as the wider bap neat-Pb perovskites. There is therefore an opportunity to discover a new compound which has a much lower band gap to enable more efficient and stable thin-film multi-junction cells. Here, AgBi compounds (halides and chalcogenides) have attracted recent interest, but nothing has yet become anywhere near as promising as the metal-halide perovskites. A notable importance for new compounds is to ensure that they are sustainable for TW scale PV production. Unfortunately, both Ag and Bi are materials that the present PV industry is trying to reduce the use of due to their scarcity of supply. Hence, this has to be considered when embarking on a large program to discover new compounds for PV.

The central target for novel materials has to be to deliver higher efficiency than existing technologies, and the multi-junction cells are a means to do this. Since tandems with perovskites have progressed so quickly to becoming industrialized, we may see reinvigorated efforts towards realizing other more exotic higher efficiency concepts being developed in the near future, including the conceptualization and demonstration of yet to be conceived ideas.

Acknowledgements

We have received funding from the Engineering and Physical Sciences Research Council (EPSRC), UK, the European Commission through the FP7, Horizon 200 and Horizon Europe frameworks, the Office of Naval Research (ONR and ONR-G) and the Air Force Office of Scientific Research (AFOSR) (USA), the Royal Society Wolfson Fund, Oxford PV, and the Binks Trust.

References

[1] M. A. Green, *Phys. E: Low-dimensional Sys. Nanostruct.* **14**, 65 (2002), https://doi.org/10.1016/S1386-9477(02)00361-2.

[2] NREL Transforming Energy, https://www.nrel.gov/pv/cell-efficiency.html (Accessed August 14, 2020).

[3] A. Kojima, K. Teshima, Y. Shirai, and T. Miyasaka, *J. Am. Chem. Soc.* **131**, 6050 (2009), doi:10.1021/ja809598r.

[4] M. M. Lee, J. Teuscher, T. Miyasaka, T. N. Murakami, and H. J. Snaith, *Science* **338**, 643 (2012), doi:10.1126/science.1228604.
[5] M. Z. Liu, M. B. Johnston, and H. J. Snaith, *Nature* **501**, 395 (2013), doi:10.1038/nature12509.
[6] S. D. Stranks, *et al.*, *Science* **342**, 341 (2013), doi:10.1126/science.1243982.
[7] J. M. Ball, M. M. Lee, A. Hey, and H. J. Snaith, *Energy Environ. Sci.* **6**, 1739 (2013), doi:10.1039/c3ee40810h.
[8] K. A. Bush, *et al.*, *Nat. Energy* **2** (2017), doi:10.1038/nenergy.2017.9.
[9] G. E. Eperon, M. T. Horantner, and H. J. Snaith, *Nat. Rev. Chem.* **1** (2017), doi:10.1038/s41570-017-0095.
[10] G. E. Eperon, *et al.*, *Science* **354**, 861 (2016), doi: 10.1126/science.aaf9717.
[11] D. P. McMeekin, *et al.*, *Science* **351**, 151 (2016), doi:10.1126/science.aad5845.
[12] D. P. McMeekin, *et al.*, *Joule* **3**, 387 (2019), https://doi.org/10.1016/j.joule.2019.01.007.
[13] E. T. Hoke, *et al.*, *Chem. Sci.* **6**, 613 (2015), doi:10.1039/C4SC03141E.
[14] S. Mahesh, *et al.*, *Energy Environ. Sci.* **13**, 258 (2020), doi:10.1039/C9EE02162K.
[15] R. D. J. Oliver, *et al.*, *Energy Environ. Sci.* **15**, 714 (2022), doi:10.1039/D1EE02650J.
[16] S. N. Habisreutinger, *et al.*, *Nano Lett.* **14**, 5561 (2014), doi:10.1021/nl501982b.
[17] G. E. Eperon, *et al.*, *Energy Environ. Sci.* **7**, 982 (2014), doi:10.1039/c3ee43822h.
[18] S. G. Motti, *et al.*, *Nat. Photonics* **13**, 532 (2019), doi:10.1038/s41566-019-0435-1.
[19] G. Y. Kim, *et al.*, *Nat. Mater.* **17**, 445 (2018), doi:10.1038/s41563-018-0038-0.
[20] S. Bai, *et al.*, *Nature* **571**, 245 (2019), doi:10.1038/s41586-019-1357-2.
[21] Y.-H. Lin, *et al.*, *Science* **369**, 96 (2020), doi:10.1126/science.aba1628.

https://doi.org/10.1142/9789811282324_0026

Towards a Zero-Carbon World: New Batteries and Chemistry Challenges

Clare Grey

Yusuf Hamied Department of Chemistry, University of Cambridge, Cambridge CB2 8BJ, UK

Climate change and batteries

More powerful, longer-lasting, faster-charging, and cheaper batteries — made from increasingly more sustainable resources — will play a critical role in the transition to a Net Zero world. While they are currently "fit for purpose" for many types of electric vehicles (EVs), the batteries are still too expensive for many and rely on critical or scarce metals (e.g., Co and, increasingly, Ni), often with environmental and social concerns around their extraction. Developing even larger batteries that are suitable for heavier vehicles, and for storing electricity to deal with the use of intermittent renewable sources of energy on the grid, remains a massive challenge — one where no clearcut technology winners exist. Lithium-ion batteries (LIBs) remain the most viable short-term technologies for these applications, with mass production resulting in significant price reductions and making it difficult for other (possibly more sustainable) technologies to compete, at least in some markets. Yet, with the massive number of giga-factories being built globally, the sustainability of resources will increasingly become even more of a concern. Fierce global competition must also mean that new manufacturing methods must be developed for the new players to develop a competitive advantage (and indeed even stay in the race).

Challenges

Having progressed rapidly over three decades, battery technology now faces a series of challenges in making the advances required for long-term grid storage or heavy vehicle and aviation applications. These include the following:

Cost: In both automotive and grid sectors, cost remains a major challenge. Much of the cost arises from the minerals themselves (mainly those used for the cathode, such as cobalt and nickel) and then the processing and synthesis of the materials.

Materials: Battery development is also challenging as there are few elements that are usable in electrodes that are cheap enough while also (i) having oxidation/redox chemistry within an appropriate voltage window and (ii) being stable when oxidized/reduced and lithium is extracted/reinserted over multiple cycles.

Degradation: Degradation at the cathode side is generally caused by structural changes that reduce capacity and can increase impedance. These phenomena increase as more lithium is extracted and the material is pushed to higher voltage during cycling. The highly oxidized cathodes react and oxidize the electrolyte. On the anode side, batteries form a protective or passivation layer known as solid electrolyte interphase (SEI). It is fundamental to performance as it prevents degradation of the electrolyte as well as conducting ions to the anode and insulating it. However, over time, growth of the SEI consumes active lithium and electrolyte materials, affecting both capacity and power density.

Energy density: Energy density is important in creating batteries that are compact and can power the device over long time periods. In some applications, volumetric energy density is key, but in aeroplanes, in particular, higher gravimetric energy density is important, while in grid-scale batteries, both volumetric and gravimetric densities are less critical but a major factor in cost. Current generations of LIBs (based on layered oxides and graphitic structures) are hitting the theoretical limits of battery chemistry, i.e., there are no more Li ions to be removed from the cathode material, and no space to insert Li ions into the anode, before Li metal plates on it. This makes enhancements to energy density more and more challenging, without consequences for battery degradation and/or safety. Use of silicon instead of graphite could theoretically increase energy density by approximately 40% at the cell level. However, when using silicon, the first cycle leads to an increased irreversible loss of capacity (over that seen for graphite) and more capacity is lost on subsequent cycles, primarily due to continued electrolyte degradation.

Battery research: Current and future research directions

Research into battery technology focuses on materials, engineering, and more generally, the performance of battery systems. In terms of the materials, research has (at least) two strands: On one side, research focuses on improving LIBs as briefly outlined in the following. The second considers a myriad of different "beyond-Li" chemistries, which include batteries that are not based on Li, along with batteries that still contain Li but may, for example, contain a Li metal anode — to increase energy density — or a solid-state electrolyte (as in a solid-state battery or "SSB"). SSBs in principle will be safer if the organic electrolyte is completely eliminated and, to have higher energy densities, if Li metal is used instead of graphite at the anode. Sodium- and magnesium-ion batteries (SIBs and MIBs, respectively) offer routes towards increased sustainability but come with new challenges including low energy density (SIBs) and poor performance (MIBs) due to very low rates and/or low voltages and degradation (Mg). Redox flow batteries provide a route towards extremely high capacity and scalable batteries for use on the grid but cost and lifetime remain serious challenges.

Advances in the short term involving current generation LIBs must come via increasing lifetime by reducing degradation, reducing cost, removing cobalt (e.g., by finding new cathodes), and decreasing charging times. Innovation in this field can be hampered by the need to keep costs down, essentially ruling out the use of large parts of the periodic table. Yet radically, new approaches may not — at least in their first generation — be "cheap". Research is often closely aligned with mission-oriented agencies or institutions, or commercial stakeholders, and there is an inherent tension between short-term wins and long-term step-change research. Close collaboration with relevant agencies/industries is extremely beneficial, by helping researchers to identify and understand the most pressing problems, for speeding up research that has clear targets, and for helping to progress research results more quickly from the lab to industry. However, the community needs to ensure that fundamental science is still strongly encouraged — even in this commercially driven field if the really hard problems — where there is no clear route or pathway to solve them — stand any chance of being addressed. Furthermore, there are many exciting avenues in this field that require new experimental methods and theory developments and use analytical techniques that are at the forefront of science. The field is massive and a short review cannot discuss them all, however, some illustrative examples are given in the following:

Metastability: This is a materials challenge that needs to be understood and controlled. To illustrate, a battery active material (a cathode (anode) or positive (negative) electrode material) is typically in its thermodynamic ground state when it is synthesized. Yet, when Li^+ ions are inserted or removed, the material is generally metastable towards decomposition. The kinetics and processes involving the transformation of the metastable state towards equilibrium need to be understood and then controlled — to prevent, for example, oxygen release from the cathode (with safety and degradation consequences, e.g., in Ni-rich cathodes, this results in an increase in cell impedance) or the continued reaction of the anode with the electrolyte, consuming Li and increasing cell resistance. In a recent elegant example, a combination of theory and experiment was used to show that the degradation that is induced by cycling to high voltages is exacerbated by the structural relaxation and rearrangements that occur at lower voltages. Continued reactions with the electrolytes are more severe in higher density anode materials such as silicon and lithium because fresh surface is continually exposed, either as the material expands and contracts on lithium insertion and removal (Si) or on Li plating.

Control of electron/ion transfer processes: Progress in, for example, Mg batteries requires an understanding of how a doubly charged cation is desolvated and inserted into a cathode material, as two electrons are — in theory — simultaneously removed from the cathode. At the anode, the process involves the reduction of the cation via the transfer of two electrons from the Mg metal, a process also coupled with the desolvation reaction. While in some systems these processes are understood, they are far from being controlled and optimized — something that would lead to faster-charging batteries.

The development of **Li-metal batteries** comes with its own set of questions and challenges. For example, how do the surfaces on which the Li metal plates affect the plating mechanism and Li corrosion? How can smooth Li metal deposition be ensured? One new approach to examining Li metal–SEI interfaces is shown in Fig. 1, where microwaves, applied at the electron spin frequency (to partially saturate the Li metal electron spin resonance), result in an enhancement of the ^{7}Li metal signals via an Overhauser effect [1]. This method can in principle be used to identify species at the buried SEI–metal (Li) interface (LiF in the example given here).

Theoretical studies of increasingly large systems — likely aided by machine-learning approaches — will be required to help unravel the often, competing processes. However, it is not only the complexity of the

(a) (b) (c)

Fig. 1. Li metal dynamic nuclear polarization (DNP) of a lithium metal electrode showing (a) the enhancement of the Li metal signal and SEI on the application of the microwave (mw on vs. off). (b) An SEM image of Li metal microstructure formed in an electrolyte containing a fluorinated ethylene additive. (c) The LiF at the Li–SEI interface can be detected on the detection of a microwave signal (red = mw on; blue = mw off) [1].

systems but the various timescales over which processes occur that are huge challenges, requiring considerable method development.

The shape and structure of a material, or its morphology, can be as important as its chemical composition. New morphologies have the potential to improve batteries by, for example, preventing particle cracking. For example, single crystal micron-sized particles may perform better (as cathode materials) than conventional polycrystalline NMC agglomerates that are subject to intergranular cracking, resulting in extended battery lifetime.

Coatings and additives: One challenge is to design bespoke coatings that can withstand expansion/contraction. Another is to identify appropriate additives that react electrochemically or chemically to passivate the surfaces — via a less empirical approach than often used currently.

Electro-catalysis at interfaces: A better understanding of how surfaces catalyse electrolyte–electrode degradation reactions must be developed. For example, why does the electrolyte ethylene carbonate react more readily with Ni-rich electrodes than linear chain carbonates? Why are lithiated silicides more reactive than lithiated graphites? What surface and interfacial structures prevent such degradation reactions?

The development of new analytical tools that track the myriad of degradation processes on the fly is critical — including those at liquid–solid interfaces and in the electrolyte — to provide real-time information on degradation, and mechanistic insight, to understand the state of the cells and ultimately inform how the cells are used (to minimize degradation). Synchrotron-based experiments provide exciting opportunities to interrogate batteries with ever-faster and higher resolution techniques but simpler metrology that could be used more widely is also vital. To illustrate, Fig. 2 shows a recent technology advance for batteries, interferometric scattering microscopy (iSCAT), which has allowed, for example, different phase transition mechanisms to be determined, lithium-ion diffusion coefficients to be extracted (with continuum modelling) and particle cracking to be visualized in real time [2].

Recycling: New and cheaper routes for recycling used materials are needed together with the development of battery materials that are inherently easier to recycle.

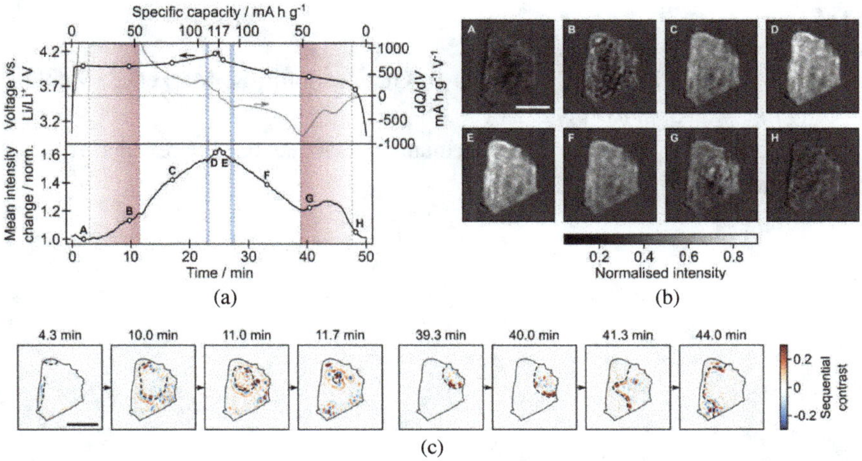

Fig. 2. Optical imaging of $LiCoO_2$ particles showing (a) the voltage profile (top) and the optical intensity (below), extracted from the single particle seen in (b). The images in (b) are extracted at points A–H on the cycling curve. (c) The images taken from the pink-shaded region in (a), illustrating the progression of the insulator-to-metal phase transition that occurs upon removing Li from the structure. Adapted from reference [2].

Sustainability: We must pursue to develop chemistries with materials that are abundant and sustainable, to include a greater emphasis on sodium-ion batteries with earth-abundant electrode materials and new, more sustainable LIB cathodes. While lithium-ion phosphate will play an important role in some batteries, higher energy-density materials are still needed.

In conclusion, research challenges span the synthesis of new materials, new characterization approaches, and new processing approaches — and demand interdisciplinary teams and longer-term strategies and research programmes.

Acknowledgements

The EU ERC is thanked for an Advanced Fellowship to CPG (BATNMR).

References

[1] M. A. Hope, B. L. D. Rinkel, A. B. Gunnarsdóttir, K. Märker, S. Menkin, *et al.*, *Nat. Comms.* **11**, 2224 (2020).

[2] A. J. Merryweather, C. Schnedermann, Q. Jacquet, C. P Grey, and A. Rao, *Nature* **594**, 522–528 (2021).

https://doi.org/10.1142/9789811282324_0027

Renewable Energy Storage in Hydrogen and Synthetic Hydrocarbons

Andreas Züttel

Laboratory of Materials for Renewable Energy (LMER),
Institute of Chemical Sciences and Engineering (ISIC),
École Polytechnique Fédérale de Lausanne,
EPFL Valais/Wallis, 1950 Sion Switzerland
Empa Materials Science and Technology,
8600 Dübendorf, Switzerland

Abstract

Renewable energy appears in the form of heat, electricity or biomass. The chemical challenge in the renewable energy economy is to provide the energy in a storable and usable form at the place and time the energy is required. Electricity and heat allow to split water into hydrogen and oxygen, hydrogen can be stored as compressed gas, liquid hydrogen or in metal hydrides. Furthermore, hydrogen reduces CO_2 to hydrocarbons, e.g., synthetic oil, kerosene or diesel, in order to produce the CO_2 neutral fuels. Beside the technical requirements to produce the renewable electricity and heat, the seasonal storage and the production of aviation fuels are the main challenges of an energy economy entirely based on renewable energy.

Renewable energy

Fossil fuels and materials on Earth in general are a finite resource and the disposal of waste into the air, into water, and on land has an impact on our environment on a global level [1] (chemical pollution, CO_2 concentration in the atmosphere, and radioactive waste).

The combustion of fossil fuels (currently burning fossil fuels +9.4 $GtC{\cdot}y^{-1}$ and deforestation +1.6 $GtC{\cdot}y^{-1}$) leads to an increase in the CO_2 concentration in the atmosphere [2] (Fig. 1). Fortunately, natural sinks (ocean: −2.5 $GtC{\cdot}y^{-1}$ (±5%) and forests: −3.4 $GtC{\cdot}y^{-1}$ (±20%)) lead to the reabsorption of CO_2 of −5.9 $GtC{\cdot}y^{-1}$ (at 410 ppm) leaving +5.1 $GtC{\cdot}y^{-1}$ (±50%) increase in the atmosphere. The reserves of fossil fuels are

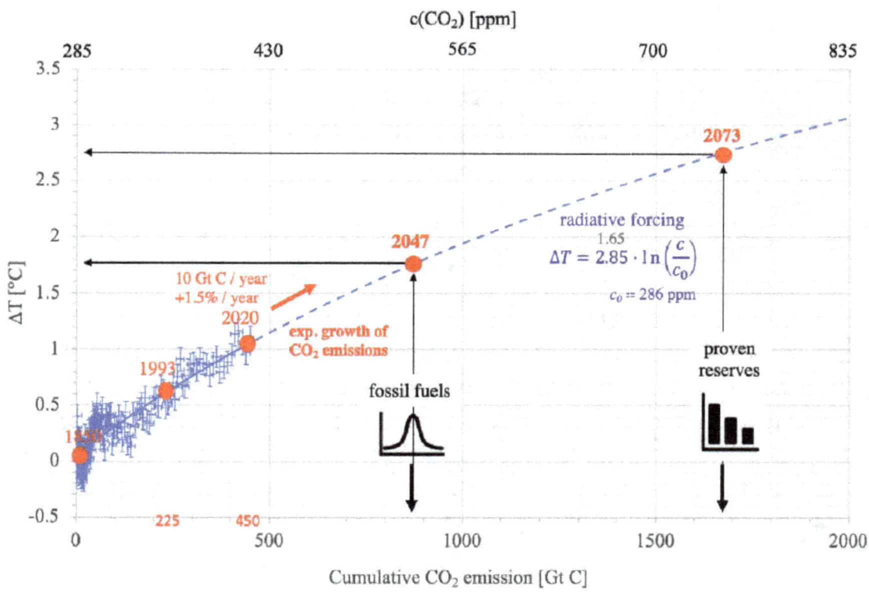

Fig. 1. Observed and extrapolated temperature increase as a function of the cumulated CO_2 emissions. The CO_2 emissions grow by 1.5% every year and the cumulated CO_2 emissions double every 27 years.

limited [3] and are estimated to be between 900 and 1750 GtC [4]. The anthropic CO_2 emissions [5] grow by 1.5% every year and the cumulated CO_2 emissions double every 27 years. The reserves of fossil fuels are therefore completely consumed between 2047 and 2073. Global renewable energy production is growing more than exponentially [6] and will reach the world energy demand in 2035 (Fig. 2).

However, renewable energy appears in the form of heat, electricity, or biomass. The chemical challenge in the renewable energy economy is to provide the energy in a storable and usable form at the place and time the energy is required.

The two main challenges in the transition from the current fossil-based energy economy to a future renewable energy-based one is the seasonal storage of renewable energy and the synthetic fuels for aviation because of the large amount of energy to be stored and the high energy density required, respectively. The electricity produced from renewable energy,

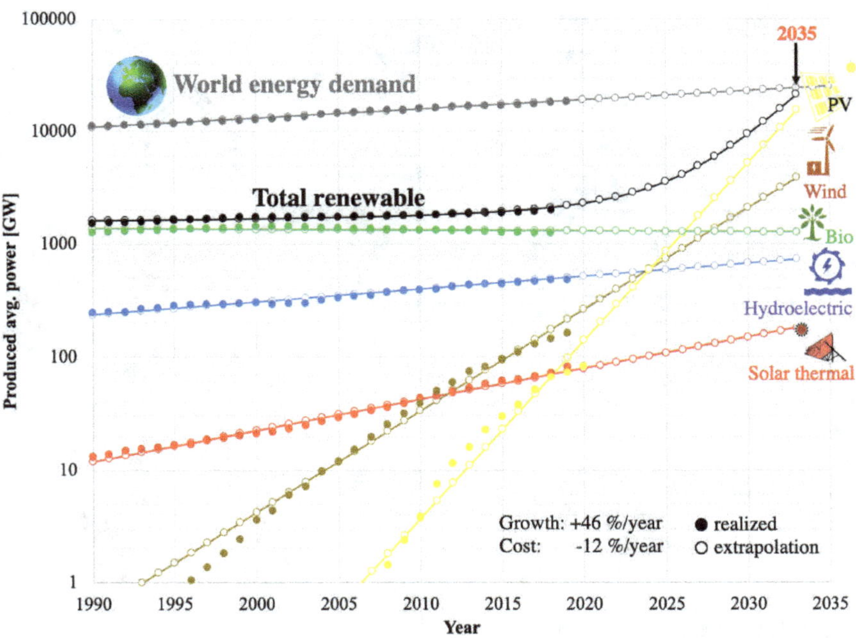

Fig. 2. The development of renewable energy in comparison with the world energy demand.

e.g., photovoltaics (Fig. 3), is stored for short term locally in batteries or over the seasons in hydroelectric storage lakes. The applications, e.g., mobility and heating, are electrified, which reduces the overall energy demand by approximately 33%. However, the seasonal storage of electricity requires large hydroelectric installations, which are not everywhere possible, and battery-electric aviation is limited to small aeroplanes for short distances. Even if the seasonal storage is lower in some regions closer to the equator or because the renewable energy is imported in the form of hydrogen, storage for the energy demand of approximately 4 months is required for energy security and redundancy of the energy production chain.

Chemical energy carriers

Electricity allows the splitting of water into hydrogen and oxygen, a technology that is called power to X (P2X). Electrolysis [7] exhibits today an

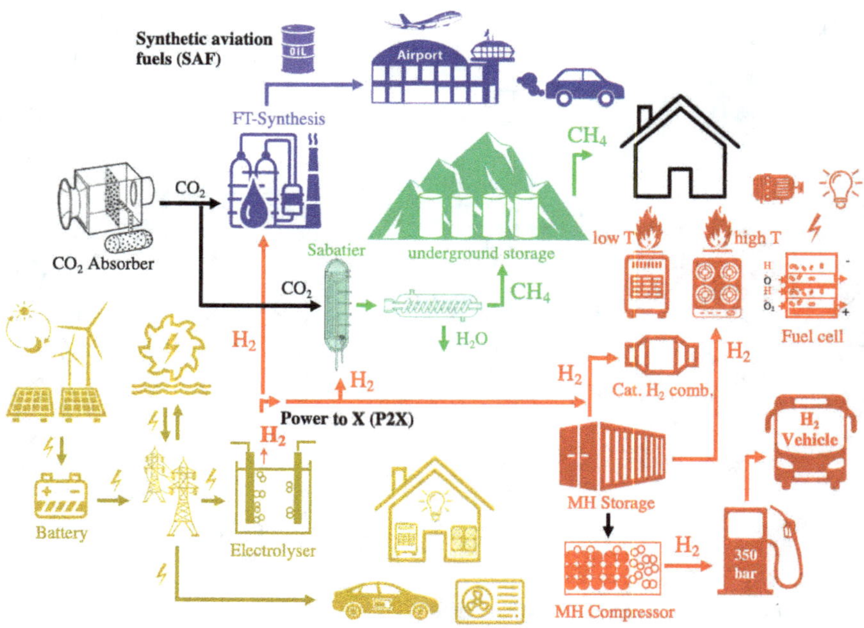

Fig. 3. Renewable energy conversion chain for electricity (yellow), hydrogen (red), synthetic methane (green), and synthetic liquid hydrocarbons (blue). The electrolysis of water allows the transition from power to chemical energy carrier (P2X) and the CO_2 air capture allows the production of synthetic methane and synthetic aviation fuels (kerosene and diesel).

efficiency of <60% depending on the type, i.e., PEM, alkaline, or SOEC, and on the size or power of the electrolyzer. Electrodes of modern electrolyzers work with a power density of >30kW·m^{-2}. The CAPEX of electrolyzers is >2000 €/kW$_{HHV}$ (HHV: Higher heating value of the hydrogen = 39.4 kWh/kg H_2) leading to an energy cost in hydrogen of +6 €cts/kWh.

The research topics in electrolysis focus on the increase in efficiency by reducing the overpotentials on the electrodes with improved electrocatalytic materials, reducing the ohmic losses in the gas separation membrane, and developing new water-splitting technologies which use the solar energy in photons directly to dissociate water by photoelectrochemical process. Hydrogen is stored as compressed gas or as liquid at low temperature (−252°C) or absorbed in metal hydrides.

Today, hydrogen storage reaches approximately 15% of the volumetric energy density (0.4–20 kWh/kg, 0.5–1.3 kWh/l) of liquid hydrocarbons, e.g., oil. The storage of hydrogen consumes between 15 and 30% of the energy in the hydrogen and CAPEX of the storage is between 9 and 2500 €/kgH_2 (+0.4€cts/kWh to +0.12€/kWh). Hydrogen is used in catalytic combustion for heat production or fuel cell to produce electricity with an efficiency of approximately 50%. An energy system based on hydrogen requires the infrastructure for production, storage, and distribution of hydrogen and the adaptation of the applications to hydrogen, e.g., hydrogen combustion and fuel cells. Alternatively, hydrogen can be produced as an energy carrier for transport and storage and used in a combined cycle power plant to produce electricity with an efficiency of 50% while the applications are electrified.

Hydrogen together with a carbon source, e.g., CO_2 from air capture or biomass, allows the synthesize of CO_2 neutral hydrocarbons. While selective reactions to methane, methanol, and dimethyl ether are known, the Fischer–Tropsch reaction from syngas (CO and H_2), the product of the reversed water gas shift reaction (RWGS) $CO_2 + H_2 \rightleftarrows CO + H_2O$, leads to a large variety of products described by the Anderson–Schulz–Flory distribution [8]. The products are refined and the undesired products are recycled in the process. The cost of synthetic oil is dominated by the electricity cost and the cost of the CO_2, which is currently between 160 and 830 €/t C, while the biomass is at 800 €/t C. Synthetic hydrocarbons can directly replace fossil fuels in all applications and the storage and

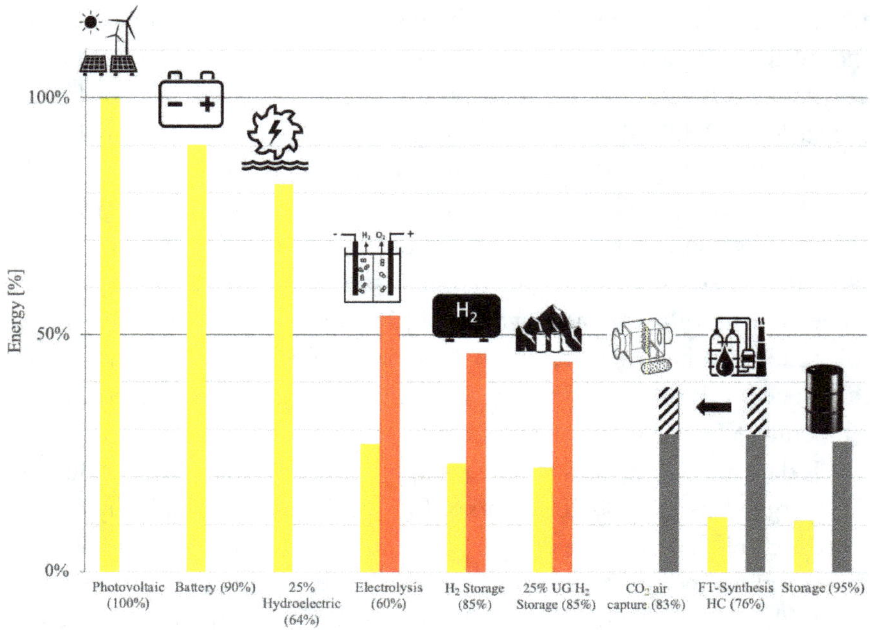

Fig. 4. Conversion efficiencies, from PV and wind to hydrogen and finally hydrocarbons. Energy in electricity (yellow), hydrogen (red), and hydrocarbons (grey).

distribution of the synthetic fuel is an established technology and exists already.

Due to the thermodynamic limits and technical conversion limitations described as maximum conversion efficiencies (Fig. 4), energy conversion and storage leads to a partial loss of energy (exothermic reactions) and a reduction of the finally available energy in the product. Therefore, the choice of energy carrier depends on the energy density required in the application as well as the storage options. The conversion efficiency, the investment for the conversion device, and the storage system determine the economics (Fig. 5) of the energy system.

Conclusions

The transition from an energy economy mainly based on fossil fuels to an energy economy based on renewable energy requires chemical technology

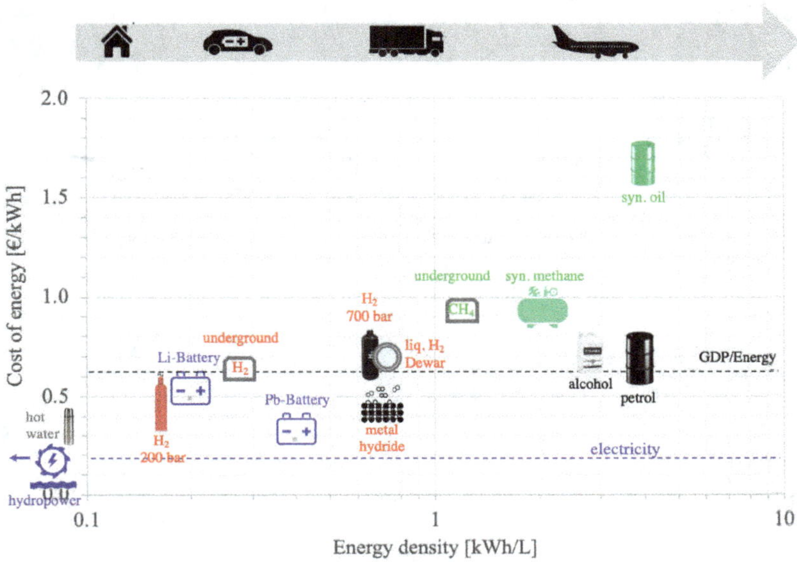

Fig. 5. Cost of energy (electricity) versus volumetric energy density for batteries, hydrogen, and synthetic hydrocarbons.

to produce the energy carriers selectively and efficiently. Furthermore, economically the cost of energy for the coming 5–30 years needs to be invested in order to build up the energy conversion.

The investment and the final cost of the energy unit are of great importance for the future development of the whole economy. In the industrial countries, the energy provided to industry leads to an economic benefit [9] of 0.6€/kWh. Only one billion people live with more than 20,000 €/year and 7 billion have less available and are on the linear relationship between energy demand and Grand Domestic Product (GDP). Only the wealthy countries deviate from this linear relationship due to the increasing import of energy-rich products and an increasing fraction of the economy is not energy related, e.g., fashion, cosmetics, and financial sector. Past developments in Switzerland have demonstrated the feasibility as well as the benefit of long-term investments in technology for a sustainable development, e.g., the Gotthard Tunnel [10], the electrification of the railway [11], and the mountain trains [12]. All three examples stand for an enormous investment leading to less pollution and an economic beneficial development, however, the profit arrived a few generations later.

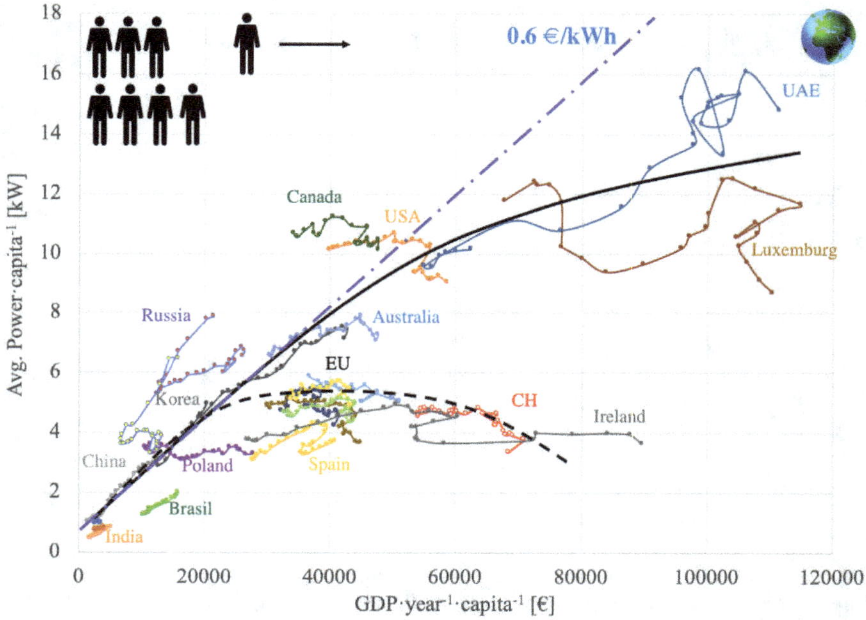

Fig. 6. Average power per capita versus GDP per year and capacity for selected countries.

References

[1] S. Bilgen, *Renewable Sustainable Energy Rev.* **38**, 890 (2014).

[2] P. Friedlingstein, *et al.*, *Earth Syst. Sci. Data* **11**, 1783 (2019).

[3] H. Ritchie, M. Roser, and P. Rosado. Energy (2020). Published online at OurWorldInData.org. https://ourworldindata.org/energy.

[4] A. Valero and A. Valero, *Resour. Conserv. Recycl.* **68**, 126 (2012).

[5] H. Ritchie, *et al.*, CO_2 and greenhouse gas emissions (2020). Published online at OurWorldInData.org. https://ourworldindata.org/co2-and-other-greenhouse-gas-emissions.

[6] Züttel, *et al.*, *Front. Energy Res. Process Energy Syst. Eng.* **9** (2022), https://doi.org/10.3389/fenrg.2021.785908.

[7] N. Gallandat, K. Romanowicz, and A. Züttel, *J. Power Energy Eng.* **5**, 34 (2017).

[8] P. L. Spath and D. C. Dayton. Preliminary screening — Technical and economic assessment of synthesis gas to fuels and chemicals with emphasis on the potential for biomass-derived syngas, NREL/TP510-34929 (2003),

p.95.http://www.fischer-tropsch.org/DOE/DOE_reports/510/510-34929/510-34929.pdf.

[9] https://ourworldindata.org/grapher/energy-use-per-capita-vs-gdp-per-capita.

[10] Neue Zürcher Zeitung 1. März 1880.

[11] Die Elektrifizierung der Schweizerischen Bahnen mit besonderer Berücksichtigung der ehemaligen Gotthardbahn, Sonderbericht der Schweizerischen Studienkommission für elektrischen Bahnbetrieb erstattet an die Generaldirektion der Schweizerischen Bundesbahnen. Mai 1912.

[12] https://www.jungfrau.ch/de-ch/jungfraujoch-top-of-europe/bau-der-jungfraubahn/.

https://doi.org/10.1142/9789811282324_0028

Protonics for Electrochemical Transformations in a Sustainable Future

Sossina M. Haile

Northwestern University, Evanston, IL, USA

Introduction

The fact that solar energy reaches the Earth at a rate which far exceeds that at which human society currently uses global energy resources is well established [1]. Recognition of this reality sparked the vibrant field of "solar fuels" in which the energy of the sun is stored in chemical bonds for use on demand, addressing the inherent intermittency of solar energy [2]. In the last 15 years since the establishment of large central efforts in solar fuels development, the price of solar electricity has plummeted. Accordingly, the idea of using solar electrons to drive chemical reactions, once considered a staggering waste of a precious resource, is gaining serious traction. Today, solar fuels produced through a two-step process using photovoltaics for electricity generation in the first step and an electrochemical conversion device in the second step are much more attractive than a mere decade ago. The electrochemical conversion technologies being pursued in this context, moreover, need not be limited to solar electrons; they are equally well suited to wind, geothermal, and hydrothermal electricity sources. Thus, solar fuels research has morphed, in some quarters, into electrochemical studies of the transformation of benign reagents such as H_2O, N_2, O_2, and even CO_2 (though certainly not benign!) into value-added products. Of the many electrochemical reactions under consideration, we focus here on splitting water into hydrogen and oxygen, splitting ammonia into hydrogen and nitrogen, and producing ammonia from nitrogen and hydrogen.

Hydrogen has been seen as both hero and villain in a sustainable energy future — hero because the product of using its energy content to carry out useful work is water, and villain because creating hydrogen in the first place has historically resulted in carbon emissions. As an energy carrier, hydrogen is typically considered in the context of polymer electrolyte membrane (PEM) fuel cells, which produce electrical power exceptionally well when the hydrogen is extremely pure. Today's average automotive internal combustion engine operates at an efficiency of ≈30% [3], whereas an automotive fuel cell, as reported by Hyundai, for example, for its Nexo fuel cell vehicle, stands at ≈60% system efficiency [4]. Despite this impressive mark, widespread deployment of hydrogen fuel cells for automotive applications has stalled for a variety of reasons including insufficient PEMFC lifetime and excessive precious metals loadings. Perhaps most significant, however, is the staggering cost of a hydrogen delivery infrastructure [5]. Herein lies another aspect of hydrogen as villain. The willingness of policymakers to wait for the completion of such a massive construction project before setting carbon emissions targets has been viewed in some quarters as a stalling tactic. Today, hydrogen as hero is overtaking the image of hydrogen as villain. Use of carbon-free electricity to split water by electrolysis largely eliminates carbon emissions in the production step [6], whereas use of liquid hydrogen carriers stands to circumvent the requirement of deploying costly, single-purpose delivery pipelines [7]. Simultaneously, infrastructure projects for hydrogen delivery have been undertaken in many parts of the world, accelerating the pace at which a transmission solution, of one type or another, will be available.

Electrolytes and the constraints they imply

Electrochemical transformations involving hydrogen, whether they be electrolysis, electric power generation, or ammonia production, have typically employed one of the three types of electrolytes: aqueous acid/base electrolytes with mobile protons/hydroxyl groups, proton exchange polymer electrolytes, or proton conducting oxides. Aqueous electrolytes, as their names imply, are acids, typically H_2SO_4 or HCl, or bases, typically KOH or NaOH, in water at some moderate concentration. The most highly deployed polymer in electrochemical systems is the sulfonated fluoropolymer Nafion [8]. Systems based on aqueous liquid or polymer electrolytes are inherently restricted to operation at near-ambient temperatures. Accordingly, they are not able to benefit from enhanced reaction rates afforded by higher temperature conditions and thus rely heavily on precious group metal (PGM) catalysts. Moreover, their acid/base nature is inherently corrosive to auxiliary components. Proton conducting oxides (PCOs), typically doped derivatives of $BaZrO_3$ or $BaCeO_3$, on the other hand, require operation at temperatures of ~450°C or higher [9], at which the degradation of auxiliary components is thermally accelerated. Furthermore, these electrolytes permit non-negligible electron–hole transport, penalizing the Faradaic efficiency of the transformation of interest [10].

A relatively underexplored class of proton-conducting electrolytes are solid acids which incorporate polyanion groups into their structure. The prototypical material in this class is CsH_2PO_4, caesium dihydrogen phosphate, which adopts a cubic crystal structure with rotational disorder of the H_2PO_4 groups above a polymorphic transition at 228°C (Fig. 1) [11]. While some details of the proton transport, an example of the Grotthuss mechanism, remain unclear, it is well established that the rotational disorder is essential to the high conductivity and low activation energy for proton migration. Several compounds display similarly high conductivity (exceeding $\approx 1 \times 10^{-3}$ S/cm with an activation energy of ≤ 0.5 eV) with similar structural disorder, forming the class of materials known as superprotonic conductors [12]. Of these, only CsH_2PO_4 has been considered, to any significant extent, for electrochemical applications, a consequence of its stability in both oxidizing and reducing environments [13, 14]. The temperature

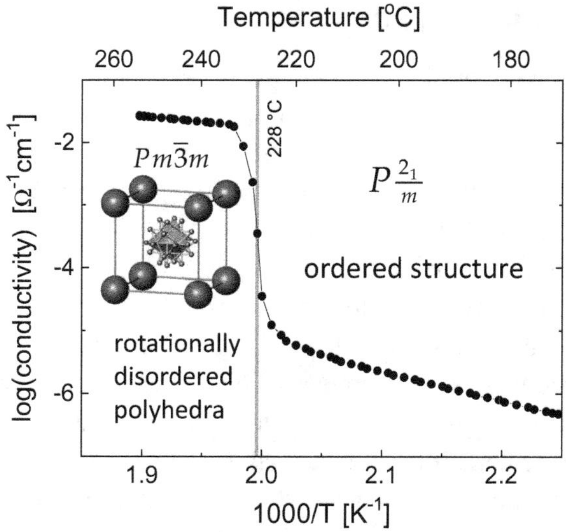

Fig. 1. Conductivity of CsH_2PO_4 (caesium dihydrogen phosphate, CDP) showing the superprotonic transport above the polymorphic transition at 228°C. Inset shows the CsCl-type structure (Cs ions are large red spheres) with the six equivalent orientations of the rotationally disordered $(H_2PO_4)^-$ group. Conductivity measured on heating under a steam partial pressure of 0.4 atm, applied to prevent electrolyte dehydration/decomposition.

of operability of electrochemical devices employing solid acid electrolytes (140–280°C) not only achieves enhanced catalysis rates without accelerating degradation of auxiliary components, but also exposes the device electrodes to gaseous reactants rather than to reactants dissolved in an aqueous (liquid) phase. At first glance, this latter feature might be considered a minor technical difference from the operation of lower temperature systems. However, ensuring reactant delivery to reaction sites and removal of gaseous products has, in fact, emerged as a major obstacle to the effective operation of aqueous and polymer electrolyte systems [15].

The features of the main electrolyte systems are broadly captured in Fig. 2. While challenges remain and the case for investing in intermediate temperature systems is strong, remarkable progress has been made in recent years in electrochemical device operation using all three classes of proton transport electrolytes. In light of publications of several excellent reviews of the high- and low-temperature systems and electrolyte classes,

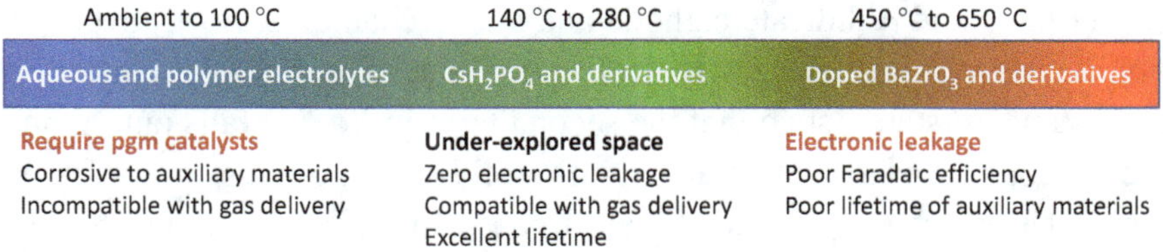

Fig. 2. Comparison of candidate systems for electrolysis based on the nature of the electrolyte (pgm = precious group metal). Aqueous and polymer electrolytes are typically H^+ conductors but may be OH^- conductors. The high-temperature, solid-state electrolytes are H^+ conductors.

the discussion here provides greater coverage of the intermediate temperature system than the other two.

Electric power generation in fuel cell mode

A fuel cell with a proton-conducting electrolyte generates electricity by facilitating hydrogen electrooxidation at the anode and oxygen electroreduction at the cathode:

$$\text{Anode: } 2H_2 \rightarrow 4H^+ + 4e^- \quad (1)$$

$$\text{Cathode: } O_2 + 4H^+ + 4e^- \rightarrow 2H_2O \quad (2)$$

At open circuit, when no electrons flow through the exterior circuit, the cell voltage is ideally given by the Nernst potential, E_N, associated with the sum of the anode and cathode half-reactions:

$$E_N = \frac{\Delta_{rxn}G}{4F} = \frac{\Delta_{rxn}G^0}{4F} + \frac{RT}{4F}\ln\left(\frac{\hat{p}^2_{H_2O,c}}{\hat{p}_{O_2,c}\hat{p}^2_{H_2,a}}\right) \quad (3)$$

where $\Delta_{rxn}G$ is the Gibbs energy of the net reaction (hydrogen and oxygen forming water), $\Delta_{rxn}G^0$ is the standard Gibbs energy of the reaction at the temperature of interest, and $\hat{p}_i$ is the pressure of species i relative to the standard pressure (of 1 bar) and is, for systems operated without pressurization, numerically equal to the partial pressure of species i. The subscripts c and a indicate cathode and anode chambers, respectively. In liquid electrochemistry, it is typical to treat the reactants as occurring in their standard states, such that the second term in Eq. (3) falls out, giving the standard Nernst potential, E_N^0. In solid-state systems, however, in which gases are supplied to the electrode chambers, accounting for the non-standard conditions is the norm. In all cases, the Nernst potential and hence expected open circuit voltage is ≈1 V (or slightly greater).

Upon drawing current from the cell to deliver electrical power, the voltage drops, and the resulting voltage vs. current density relationship defines the polarization curve, which in turn defines the power density curve, Fig. 3. In the low-temperature systems, a nonlinear activation voltage loss is typically evident at small values of current density (resulting from low catalytic activity at the electrodes). This feature is typically absent from high-temperature systems, and the electrodes behave linearly. Nevertheless, they are often analysed in terms of Butler–Volmer kinetics, which inherently imply a nonlinear voltage–current relationship. Deviation

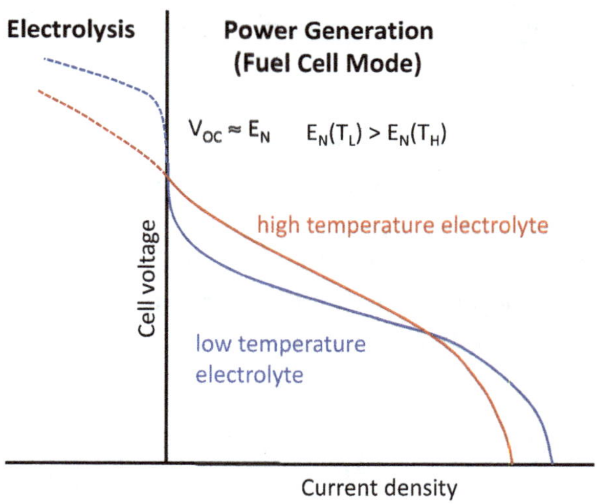

Fig. 3. Schematic polarization curves for electrochemical device operation with a hydrogen-rich gas supplied to the fuel electrode and oxygen-rich gas to the air electrode. In low-temperature (near ambient) systems, large overpotentials associated with the electrochemical reactions are typically encountered, reflected in strongly nonlinear behaviour near open-circuit (zero-current density) conditions. The open-circuit voltage (V_{OC}) is ideally given by the Nernst potential (E_N, Eq. (3)).

Table 1. Characteristics of representative, state-of-the-art fuel cells built around proton conducting electrolytes.

System	T (°C)	Electrolyte	Cathode	Anode	Power density @ 0.6 V (W/cm^2)
PEMFC [16, 17]	70	Nafion	0.30 mg Pt/cm^2 1 atm O_2	0.30 mg Pt/cm^2 1 atm H_2	0.91
SAFC [18]	240	CsH_2PO_4[a]	1.75 mg Pt/cm^2 0.6 atm O_2	0.50 mg Pt/cm^2 0.6 atm H_2	0.38
PCFC [19]	500	BZCYYb4411[b]	PBSCF[c] air	Ni 0.97 atm H_2[d]	0.52

Notes: [a]0.4 atm H_2O supplied to both electrodes to suppress electrolyte decomposition;
[b]$Ba(Zr_{0.4}Ce_{0.4}Y_{0.1}Yb_{0.1})O_3$;
[c]$PrBa_{0.5}Sr_{0.5}Co_{1.5}Fe_{0.5}O_{5+\delta}$;
[d]balance (0.03 atm) H_2O.

of the open-circuit voltage from the Nernst potential is generally attributed to activation losses or fuel cross-over in low-temperature systems. In high-temperature systems, electronic leakage through the electrolyte is more commonly the culprit.

The power densities achieved from representative, state-of-the-art PEMFCs, solid acid fuel cells (SAFCs) and protonic ceramic fuel cells (PCFCs) vary widely, Table 1. The relative advantages and disadvantages of these systems extend beyond this single metric, but the comparison is a useful starting point for assessing those additional factors in future efforts.

Electrolysis

Electrolysis using a proton-conducting electrolyte proceeds by a "simple" reversal of the fuel cell reactions. Subtleties arise because the electrocatalysts useful for the forward reactions (fuel cell mode) may not be suitable in the opposite direction. This is particularly true of solid acid systems, in which the Pt catalyst readily undergoes oxidation at the anode under electrolysis conditions. Another subtlety surrounds the voltage required to drive the electrolysis reaction. It is often stated that 1.23 V $\left(= \frac{\Delta_{rxn}G^0}{4F}\right)$ is the minimum voltage for the water-splitting reaction. In fact, this is only true in systems in which hydrogen and oxygen are produced as gases with a pressure of 1 atm, as occurs in liquid electrolysis. At high temperatures, because the gases can attain arbitrary *partial* pressures, steam electrolysis can proceed at lower voltages. Nevertheless, to enable comparison to lower temperature systems, high-temperature systems are often operated with 1 atm hydrogen supplied to the fuel electrode (now the cathode) and 1 atm oxygen supplied to the air electrode (now the anode), and the applied voltage/current increases the respective hydrogen and oxygen partial pressures beyond the values at which the gases are supplied. Operation in this manner furthermore enables assessment of suitability of the system to generate high-pressure hydrogen. Nevertheless, the possibility of splitting steam at low voltage, well below 1.23 V, should not be discounted as it opens up possibilities of integrating with a broad range of photovoltaic systems with moderate open-circuit values. Returning to the typical condition in which reactant gases are supplied at close to standard pressures, operation at 1.4 V results in waste heat generation that balances the endothermic reaction enthalpy and hence is termed the thermoneutral point. While high-temperature systems provide appreciable current at this voltage, typical low-temperature systems require a larger driving force to achieve the water-splitting reaction. Representative, state-of-the-art electrolysis performance characteristics across the selected systems are summarized in Table 2.

Table 2. Characteristics of representative, state-of-the-art electrolyzers. Here the aqueous system incorporates a hydroxyl (OH^-) ion conductor rather than a proton conductor. In recent years, aqueous alkaline electrolytes have been increasingly replaced by anion exchange membranes [20].

System	T (°C)	Electrolyte	Anode	Cathode	Current density (A/cm^2)
Alkali [21]	25 (20 bar)	1 M KOH, O_2 saturated	Ni	Ni	0.16 @ 1.6 V
PEM [6, 22]	55 (10 bar)	Nafion	IrO_2	Pt@C	0.47 @ 1.8 V
PCFC [23]	500	BZCYYb4411[a]	PBSCF[b] 0.03 atm H_2O[c]	Ni 0.97 atm H_2[c]	0.72 @ 1.35 V

Notes: [a]$Ba(Zr_{0.4}Ce_{0.4}Y_{0.1}Yb_{0.1})O_3$;
[b]$PrBa_{0.5}Sr_{0.5}Co_{1.5}Fe_{0.5}O_{5+\delta}$;
[c]Balance (0.03 atm) H_2O.

Returning to the question of electrocatalyst selection and design, the ideal electrochemical system would support reversible operation between electrolysis and electrical power generation. Such a system would enable chemical energy storage when electricity is available in excess and then provide electricity when demand exceeds the supply available from the wind or solar resource. This strategy has advantages over battery storage because, among other benefits, the size of the conversion device and that of the storage unit are decoupled. However, the round-trip efficiency of reversible electrochemical cells generally lies below that of batteries, underscoring the need to develop more active catalysts. To date, among systems employing proton conductors, only those based on ceramic electrolytes have been operated in reversible mode [23]. In such materials, Faradaic efficiency losses are particularly severe because the electrolyte itself becomes oxidized during electrolysis, increasing its p-type electronic conductivity [10].

Ammonia oxidation for hydrogen production and power generation

Among potential liquid hydrogen carriers, ammonia offers obvious benefits because the product that remains after extracting the hydrogen is benign and abundant N_2. As nitrogen makes up 78% of Earth's atmosphere, there is no need to return the depleted carrier for regeneration. This stands in a stark contrast to alternative candidates such as methylcyclohexane, which is transformed to toluene upon dehydrogenation. Furthermore, the mass fraction of extractable hydrogen is highest for ammonia among candidate carriers, and many parts of the world already have an extensive ammonia delivery infrastructure due to the tremendous importance of this molecule to agriculture.

Electrooxidation of ammonia in an electrochemical system based on a proton-conducting electrolyte involves the following anode reaction:

$$\text{Anode: } 2NH_3 \rightarrow 3N_2 + 6H^+ + 6e^- \quad (4)$$

For direct use of ammonia in power generation, this anode reaction is coupled to the same cathode reaction relevant to a hydrogen-powered fuel cell. For the production of hydrogen, the cathode reaction is the hydrogen evolution reaction, identical to that which occurs in electrolysis. The toxicity of ammonia renders it unlikely to be used in consumer applications, in particular passenger vehicles. A more plausible scenario is one in which ammonia is "electrolyzed" into hydrogen and nitrogen at distributed locations at which passenger vehicles are supplied with hydrogen to operate conventional PEMFCs. Here the challenge is the extreme toxicity of ammonia to the PEMFC anode catalysts. Thus, the electrochemical production must yield high-purity hydrogen, in addition to operating at low overpotential so as to minimize the electrical energy input and thus the cost of the process. Additionally, it is imperative to avoid oxidation of the nitrogen that would generate NO_x, an all-too-common product in low-temperature systems [24]. In this application space, solid acid electrochemical cells have taken the lead. The author's laboratory has demonstrated 100% Faradaic efficiency for extraction of H_2 from NH_3, achieving a hydrogen production rate of 3 ml(H_2) min^{-1} cm^{-2} at a total operating voltage of just 0.39 V using a dual-layer anode, with Cs-promoted Ru serving as an ammonia decomposition catalyst and Pt as a hydrogen electrooxidation catalysts [25]. At the temperature of operation of 240°C, the Pt electrocatalyst is not poisoned by the residual NH_3 remaining after the gas passes the decomposition layer and no NO_x gases are formed. The combination of performance metrics achieved is unrivalled. Nevertheless, opportunities for further decreasing the voltage requirement and increasing the catalyst lifetime remain.

Electrosynthesis of ammonia

Electrosynthesis of value-added chemicals using proton-conducting electrolytes is a nascent area. In addition to N_2 reduction to generate NH_3, the reverse of Eq. (4), CO_2 reduction, has garnered significant recent attention. The former has clear significance in terms of a hydrogen delivery infrastructure; the role of the latter in a sustainable future is unclear. Ammonia today is produced by the famed Haber–Bosch process, in which H_2 and N_2 are reacted under high-temperature, high-pressure conditions. Ammonia synthesis is recognized as a significant source of CO_2

emissions, but this is largely because the hydrogen used in the process is derived from fossil fuel steam reforming [26]. The energy input (for reaching the high-pressure conditions for reacting H_2 and N_2) amounts to just 15–20% of the total [27], thus even today's technologies, if combined with carbon-free hydrogen, have real potential for reducing the carbon footprint of ammonia production. Electrochemical synthesis offers two possible advantages: (1) The voltage in the device generates an effective high pressure, at least of hydrogen, that is much higher than can be reached by mechanical pumping. In turn, this pushes the system thermodynamics to favour NH_3 production over retaining the H_2 and N_2 reactants, opening a possible route to increasing the energy efficiency of the process. (2) The reactor conditions can be achieved in small, modular systems, as compared to the immense plants currently deployed in the Haber–Bosch plants, democratizing access to the technology. In the latter case, large systems are required to render the cost of investing in high-pressure equipment profitable [28].

The chemical challenge of ammonia electrosynthesis arises from the facile evolution of H_2 at the device cathode, which can then outcompete N_2 reduction. To date, no electrocatalyst system meeting the efficiency requirements for cost-competitive ammonia production has been reported [29]. Furthermore, in several cases, background NH_3 has exceeded the amount generated, sending researchers along unproductive paths [30]. This challenge notwithstanding, systems based on proton conducting oxides, typically fabricated with Ag-Pd electrodes, have shown the most promise; ammonia production rates in the range of $1–5 \times 10^{-9}$ mol s^{-1} cm^{-2} have been reported by several laboratories using such cells [31]. Perhaps the most intriguing results have been obtained using a protonic ceramic electrolyte in conjunction with a metal nitride cathode, VN [32]. The hypothesis here is that the cathode reaction proceeds via a Mars–van Krevelen mechanism in which protons react with nitrogen atoms in the lattice, and the resulting vacancies are refilled by dissociated gas-phase nitrogen. The rate of ammonia production was an impressive 8×10^{-9} mol s^{-1} cm^{-2} at 0.3 V overpotential and a temperature of 550°C. Suppression of the hydrogen evolution to increase the Faradaic efficiency beyond the ≈5% observed is surely the next target for this line of investigation.

Concluding remarks

The dramatic decline in the price of wind and solar electricity opens up new possibilities for the use of carbon-free electrical power to generate useful chemicals. At the same time, the intermittency of these sources necessitates renewed focus on energy storage for on-demand use. Hydrogen production by electrolysis, a possible storage strategy, is approaching commercial viability. In fact, even today, about 4% of the world's hydrogen is produced by electrolysis. Shipping hydrogen, however, remains daunting. Ammonia is an intriguing potential solution to this challenge. The decomposition of ammonia to generate high-purity hydrogen has been demonstrated. Achieving technoeconomic success in ammonia electrosynthesis, the other half of the ammonia-as-hydrogen-carrier approach, awaits breakthroughs in catalyst design, with nitrides currently the leading contenders to address this need.

Acknowledgements

The author acknowledges the financial support of the McCormick College of Engineering Catalyst Award.

References

[1] N. S. Lewis and D. G. Nocera, *Proc. Natl. Acad. Sci. USA* **103**, 15729 (2006).
[2] H. B. Gray, *Nat. Chem.* **1**, 7 (2009).
[3] S. Kim, S. Park, S. Kim, and S.-H. Rhi, *J. Electron. Mater.* **40**, 812 (2011).
[4] R. C. Samsun, L. Antoni, M. Rex, and D. Stolten, *Energy Env.* (Julich Forschungszentrum, 2021).
[5] Y. Lee, U. Lee, and K. Kim, *Int. J. Hydrogen Energy* **46**, 14857 (2021).
[6] D. Ferrero, A. Lanzini, M. Santarelli, and P. Leone, *Int. J. Hydrogen Energy* **38**, 3523 (2013).
[7] P. Preuster, C. Papp, and P. Wasserscheid, *Acc. Chem. Res.* **50**, 74 (2017).
[8] K. A. Mauritz and R. B. Moore, *Chem. Rev.* **104**, 4535 (2004).
[9] C. C. Duan, J. K. Huang, N. Sullivan, and R. O'Hayre, *Appl. Phys. Rev.* **7**, 40 (2020).
[10] H. Y. Zhu, S. Ricote, and R. J. Kee, *J. Phys. Energy* **4**, 17 (2022).

[11] S. M. Haile, C. R. I. Chisholm, K. Sasaki, D. A. Boysen, and T. Uda, *Faraday Discuss.* **134**, 17 (2007).

[12] S. M. Haile, D. A. Boysen, C. R. I. Chisholm, and R. B. Merle, *Nature* **410**, 910 (2001).

[13] D. A. Boysen, T. Uda, C. R. I. Chisholm, and S. M. Haile, *Science* **303**, 68 (2004).

[14] R. B. Merle, C. R. I. Chisholm, D. A. Boysen, and S. M. Haile, *Energy Fuels* **17**, 210 (2003).

[15] J. M. Spurgeon and N. S. Lewis, *Energy Environ. Sci.* **4**, 2993 (2011).

[16] A. G. Olabi *et al.*, *Energies* **15**, 35 (2022).

[17] B. G. Pollet, *Electrochem. Commun.* **11**, 1445 (2009).

[18] D. C. Orozco, O. Dyck, A. B. Papandrew, and T. A. Zawodzinski, *J. Power Sour.* **408**, 7 (2018).

[19] S. Choi *et al.*, *Nat. Energy* **3**, 202 (2018).

[20] Immanuel Vincent, Dmitri Bessarabov, "Low cost hydrogen production by anion exchange membrane electrolysis: A review," Renewable and Sustainable Energy Reviews, **81**(2), 1690–1704 (2018). doi: 10.1016/j.rser.2017.05.258

[21] A. Ursua, L. M. Gandia and P. Sanchis, "Hydrogen Production From Water Electrolysis: Current Status and Future Trends," in *Proceedings of the IEEE*, **100**(2) 410–426 (2012). doi: 10.1109/JPROC.2011.2156750.

[22] F. Marangio, M. Santarelli, and M. Calì, *Int. J. Hydrogen Energy* **34**, 1143 (2009).

[23] S. Choi, T. C. Davenport, and S. M. Haile, *Energy Environ. Sci.* **12**, 206 (2019).

[24] S. A. Lee, M. G. Lee, and H. W. Jang, *Sci. China Mater.* (2022).

[25] D. K. Lim *et al.*, *Electrochimica Acta* **288**, 12 (2018).

[26] C. Smith, A. K. Hill, and L. Torrente-Murciano, *Energy Environ. Sci.* **13**, 331 (2020).

[27] I. Rafiqul, C. Weber, B. Lehmann, and A. Voss, *Energy* **30**, 2487 (2005).

[28] G. Hochman, A. Goldman, and F. A. Felder, In *Biomass, Biofuels, Biochemicals,* G. S. Murthy, E. Gnansounou, S. K. Khanal, and A. Pandey, (Eds.) (Elsevier) (2022), pp. 243–258.

[29] N. Lazouski *et al.*, *ACS Energy Lett.* **7**, 2627 (2022).

[30] S. Z. Andersen *et al.*, *Nature* **570**, 504 (2019).

[31] V. Kyriakou, I. Garagounis, E. Vasileiou, A. Vourros, and M. Stoukides, *Catal. Today* **286**, 2 (2017).

[32] V. Kyriakou, I. Garagounis, A. Vourros, E. Vasileiou, and M. Stoukides, *Joule* **4**, 142 (2020).

https://doi.org/10.1142/9789811282324_0029

Finding a Path to CO_2 Utilization on a Scale that Matters

Matthew W. Kanan

Department of Chemistry, Stanford University, Stanford, 94305 CA, USA

Opportunities and current state of research in CO_2 utilization

Anthropogenic greenhouse gas (GHG) emissions have increased the partial pressure of atmospheric CO_2 to its highest level in the past 3 million years [1]. A rapid emissions drawdown is needed to minimize the risk of triggering climate tipping points that would substantially change Earth system components with severe negative impacts on the planet's biomes and human welfare [2]. At odds with this imperative is the enormous and growing demand for energy, 80% of which is supplied by fossil fuels. Eliminating GHG emissions from the global energy system requires massive shifts, foremost being a transition from coal and natural gas power plants to renewable electricity. With a renewable grid, it is possible to decarbonize many energy services including light-duty transportation, heating, and cooling through direct electrification [3]. However, eliminating carbon-based products from other energy services is infeasible for the foreseeable future. Liquid fuels are indispensable for heavy-duty transport, aviation, and marine shipping, which comprise >60% of the ~100 million barrels consumed daily [4]. Demand for these services will increase with global economic growth. Carbon-based chemicals and materials are essential for every industry. Plastics alone account for 8% of oil and gas use and 4% of GHG emissions, with these numbers projected to double by 2050 [5]. The emissions problem cannot be solved without CO_2 utilization — the production of fuels and chemicals from CO_2, H_2O, and renewable energy. Decoupling carbon-based products from fossil resources also provides opportunities to improve global energy security. While these considerations have motivated researchers for many years, the unrivalled scale and time sensitivity of the problem pose an urgent need for actionable advances.

Non-fossil fuels and chemicals are currently made from biomass, thereby letting natural photosynthesis perform CO_2 conversion. With few exceptions, the feedstocks used for bio-based products are glucose or vegetable oils derived from agriculture. This practice competes with food production, which in a global market ultimately leads to the destruction of natural carbon sinks and biodiversity harbours (e.g., rainforests) to create more agricultural land [6]. To avoid this calamitous pitfall, it is critical to utilize inedible biomass (lignocellulose) that is obtained without land use change, ideally residuals from existing processes [7]. Judicious lignocellulose utilization could provide carbon-neutral replacements for petrochemical products. While much attention has focused on fuels, chemicals and materials are better matched to the scale of truly sustainable feedstocks. The research challenge is transforming recalcitrant and heterogeneous lignocellulosic feedstocks into useful products without using large quantities of non-renewable inputs.

Bio-based products are limited by the low solar-to-biomass efficiency of natural photosynthesis, which is typically <1% in plants [8]. Collecting and converting biomass to a useful product takes the overall sunlight-to-product efficiency even lower. Synthetic approaches are therefore essential to producing sustainable carbon-based products on the gigaton scale. Using renewable electricity to power the conversion of CO_2 and H_2O into fuels and chemicals could achieve far greater efficiency than natural photosynthesis. As a simple example, a 10% sunlight-to-product efficiency could be achieved by combining a 20% efficient photovoltaic with a 50% efficient power-to-product process.

Renewable electricity can be used to power CO_2 conversion directly via CO_2 electrolysis. Building on seminal early studies [9], research in this area has grown exponentially in the past decade. Electrolysers that convert CO_2 into CO and O_2 have reached nascent commercialization [10]. Since CO is a common intermediate in chemical and fuel production, these systems could in principle provide renewable routes to high-volume products. Reaching an impactful scale hinges on improving their energy efficiency and manufacturability, which will likely require breakthroughs in materials (e.g., catalysts and separators) and better control of the transport phenomena that govern performance under process-relevant conditions [11]. Beyond CO, electrochemical CO_2 conversion using Cu catalysts produces more valuable products like ethylene, an emissions intensive super-commodity. While laboratory cells have shown promising ethylene selectivity and production rates, their development is currently far behind CO_2-to-CO electrolysis.

An alternative to CO_2 electrolysis is to make H_2 via water electrolysis and then perform CO_2 hydrogenation. Water electrolysis is an established industrial technology, with >80% efficiency for the electrolyser and >65% for the full system (electricity to H_2 lower heating value) [12]. CO_2 can be hydrogenated to produce CH_4, CO, or CH_3OH selectively depending on the catalyst and process conditions. There are commercial processes to transform syngas (CO and H_2) or CH_3OH into multi-carbon fuels and chemicals. Combining CO_2 hydrogenation with these technologies would provide a renewable route to many high-volume products while leveraging existing infrastructure. Innovations in catalysts and process design are needed to improve CO_2 hydrogenation efficiency and its integration with downstream conversions to make these systems competitive with fossil fuels. Another approach is to hydrogenate CO_2 to multi-carbon fuels in a single step, thereby eliminating the need for a separate reactor to make CH_3OH or CO [13]. This approach could be advantageous if the rates and product distributions can be greatly improved.

Carbon dioxide utilization depends on the provision of a concentrated CO_2 supply. Several tens of millions of tons of relatively high purity (95–99%) CO_2 is emitted from ethanol fermentation and ammonia synthesis plants [14]. As of 2022, operational carbon capture and sequestration (CCS) facilities have the capacity to capture >40 Mton/yr and >200 Mton/yr additional capacity is under development [15]. With amine scrubbing technology, capturing CO_2 from coal flue gas requires <250 kWh of equivalent work per ton [16], or ~3% the energy required to convert CO_2 to fuel at 50% energy efficiency. To supply gigaton-scale utilization, CO_2 will eventually need to be captured directly from air. Leading direct air capture (DAC) technologies require ~2 MWh of electricity per ton if a heat pump is used and 0.7 MWh if waste heat is used [17], which could be provided by integrating DAC with CO_2 utilization.

My recent research contributions to CO_2 utilization

Over the past several years, my group has developed carbonate-promoted and carbonate-catalysed reactions to unlock new ways of utilizing CO_2 for chemical and fuel synthesis. Our approach was originally inspired by the Calvin cycle, which fixes CO_2 using the enzyme RuBisCO to perform C–H carboxylation — the conversion of a C–H bond and CO_2 into a carboxylate ($C–CO_2^-$) using a base to remove the proton. RuBisCO only carboxylates ribulose-1,5-bisphosphate, which has an activated C–H bond. Attracted by its conceptual simplicity, we began investigating ways to generalize C–H carboxylation for the synthesis of carboxylic acids with high-volume applications.

The challenge with C–H carboxylation is that most C–H bonds have very high pK_as (>30). Deprotonating such bonds typically requires very strong bases (e.g., organolithiums) that have emissions footprints far outweighing the CO_2 that is consumed in a C–H carboxylation reaction. To avoid indirectly releasing more CO_2 than is consumed, one needs a base that can be regenerated with minimal energy and resource use. We concluded that carbonate (CO_3^{2-}) is the only viable option [18]. In conventional solution-phase chemistry, CO_3^{2-} is a weak base that cannot deprotonate high-pK_a C–H bonds. We therefore began exploring solvent-free systems to see if it is possible to increase CO_3^{2-} basicity in these unusual media. Salts composed of alkali cations and carboxylate anions proved fruitful. Depending on the cation and the carboxylate structure, these salts can form molten phases at temperatures below the onset of decomposition. We discovered that when CO_3^{2-} is dissolved in an alkali carboxylate melt, it can reversibly deprotonate high-pK_a C–H bonds in the presence of CO_2, which leads to C–H carboxylation [18, 19]. Using the alkali carboxylate itself as the substrate, simply heating a mixture of alkali carboxylate and alkali carbonate under CO_2 beyond a melting transition results in CO_3^{2-}-promoted C–H carboxylation.

We have used CO_3^{2-}-promoted C–H carboxylation to develop a new route to furan-2,5-dicarboxylic acid (FDCA) (Fig. 1) [18, 20]. FDCA is a substitute for purified terephthalic acid (PTA), a petrochemical used to make polyethylene terephthalate (PET), which is the world's second largest commodity plastic [21]. In our route, CO_3^{2-}-promoted C–H carboxylation converts alkali furan-2-carboxylate into the dicarboxylate, from which FDCA is isolated by protonation. The salt by-product is converted back to alkali carbonate and acid at a low energy input using electrodialysis technology. Furan-2-carboxylic acid can be made by aerobic oxidation of furfural, which is the only chemical currently produced on large scale from lignocellulose. Thus, CO_3^{2-}-promoted C–H carboxylation enables

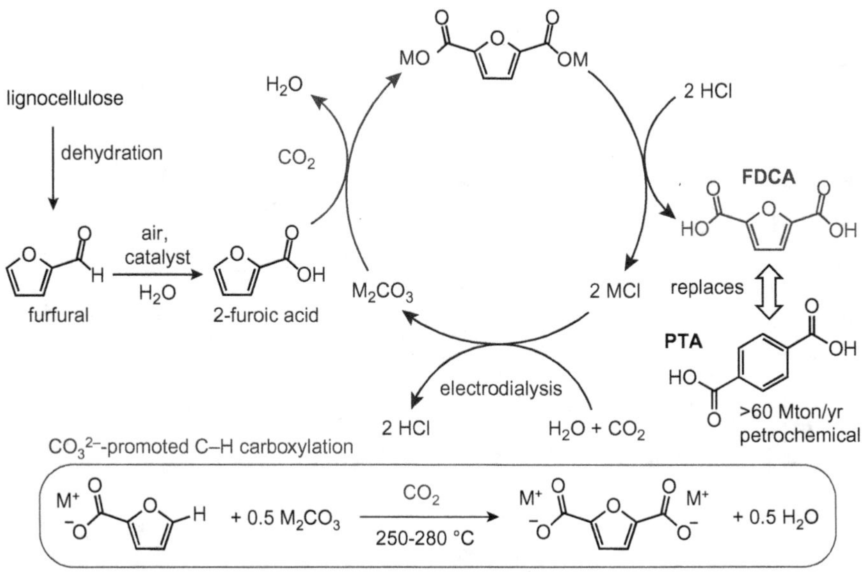

Fig. 1. Carbonate-promoted C–H carboxylation of furan-2-carboxylate and its use to synthesize FDCA from lignocellulose, air, and CO_2. FDCA is a PTA replacement.

the conversion of lignocellulose, air, and CO_2 into a building block for commodity polymers. More recently, we have extended this C–H carboxylation chemistry to make lignocellulose-derived furanic polyamides with favourable thermal properties [22–24].

To expand the generality of this approach, a new strategy was needed to enable CO_3^{2-} to promote C–H carboxylation of gas-phase reactants. We found that dispersing M_2CO_3 into mesoporous supports (oxide or carbon materials) disrupts the M_2CO_3 crystallinity, generating an amorphous, high surface area M_2CO_3 with undercoordinated CO_3^{2-} ions that behave as superbases at elevated temperatures (Fig. 2(a)). We have used these materials ("dispersed carbonates") to perform C–H carboxylation of arenes (e.g., benzene) and heteroarenes (e.g., 1-methyl indole) by simply heating the dispersed carbonates under elevated pressure of CO_2 and substrate at 280–400°C [25, 26]. The carboxylate products can be converted into isolable methyl esters with concomitant regeneration of the dispersed carbonates by treating the product with methanol and CO_2 or with dimethyl carbonate at 160–280°C. Thus, two-step cycles of C–H carboxylation and methylation can be used to convert (hetero)arenes into aromatic esters (Fig. 2(b)).

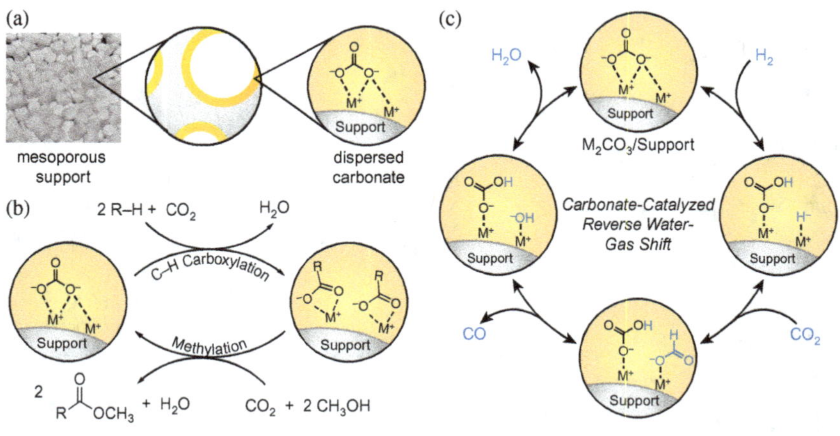

Fig. 2. (a) Schematic depiction of alkali carbonate (M_2CO_3) dispersed in a mesoporous support. (b) Synthesis of (hetero)arene esters from (hetero)arenes using C–H carboxylation and methylation. (c) Carbonate-catalysed reverse water–gas shift catalysis.

Carboxylation in the gas phase provided an entry point to CO_2 hydrogenation catalysis. Interestingly, benzene and H_2 have the same heterolytic bond dissociation enthalpy (400 kcal/mol). Given their ability to promote C–H carboxylation of benzene, we envisioned that dispersed carbonates would catalyse the reverse water–gas shift (RWGS) reaction (CO_2 hydrogenation to CO) via a cycle shown schematically in Fig. 2(c). Analogous to C–H carboxylation, deprotonation of H_2 by CO_3^{2-} forms a hydride (H^-) that attacks CO_2 to form formate (HCO_2^-). Unlike C–CO_2^-, however, HCO_2^- is not a terminal product. Proton transfer from C to O promotes dissociation to CO and OH^-, which deprotonates HCO_3^- to form H_2O and regenerate CO_3^{2-}, thereby closing the catalytic cycle.

RWGS is endothermic ($\Delta H° = 41$ kj/mol). One of the biggest challenges for RWGS catalysis is suppressing CO_2 hydrogenation to CH_4, which is very exothermic ($\Delta H° = -165$ kj/mol). Current RWGS technologies utilize Ni steam reforming catalysts that are unselective. To suppress methane, these catalysts must be operated at very high temperatures (typically >900°C) where the endothermic RWGS reaction becomes thermodynamically favoured [27]. By contrast, we envisioned that CO_3^{2-}-catalysed CO_2 hydrogenation would be highly selective for RWGS because alkali carbonates have very little affinity for CO and therefore no way to reduce CO further to CH_4.

Gratifyingly, dispersed carbonates have proven to be highly active, robust, and extremely selective RWGS catalysts in a temperature regime where CH_4 is strongly thermodynamically favoured [28]. High-performing catalysts can be prepared from very low-cost materials — e.g., K_2CO_3 and mesoporous γ-Al_2O_3 — using procedures commonly employed for high-volume catalyst manufacturing. Because they do not contain transition metals, dispersed carbonates are tolerant to impurities found in CO_2 streams, such as H_2S.

By producing the CO component of syngas, RWGS catalysis provides a link between established water electrolysis and syngas-to-liquids technologies to enable integrated power-to-liquid (PtL) systems for producing sustainable liquid fuels. Water electrolysis is projected to reach >200 GW operational capacity in the next decade [29]. Syngas conversion using Fischer–Tropsch technology is proven on scale in plants that produce >100,000 barrels per day of liquid hydrocarbon fuels and chemicals [30], while recently commercialized gas fermentation technology converts syngas into ethanol [31]. The efficiency and cost of PtL systems that utilize water electrolysis and syngas-to-liquid processes depend critically on the performance of the RWGS unit that links them together. The very high-temperature regime required for conventional Ni catalysts necessitates expensive materials of construction and complicates heat integration with the rest of a PtL process. Ni is also a liability for high-volume catalyst applications given the surge in demand from batteries. Carbonate-catalysed RWGS provides opportunities to build new PtL process designs that lower the barrier to scaling.

Outlook to future developments of research on CO_2 utilization

The next 10 years are critical for demonstrating that CO_2 utilization can be a significant source of carbon-based products and contributor to emissions drawdown. A healthy competition between electrochemical and chemical (e.g., hydrogenation) approaches will increase the likelihood that CO_2 utilization technologies compete with fossil incumbents. Successful technologies will not only optimize efficiency but also creatively integrate with existing infrastructure and industrial emission streams. There are abundant opportunities for the research community to accelerate the trajectory of established pathways or unveil new ones. Dividing effort between fundamental and applied research is counterproductive. Integrating science with device engineering and process design has the greatest likelihood of creating advances that will translate to real technologies. Given the urgency and enormity of the carbon problem, the only meaningful measure of the impact of a research development in CO_2 utilization will be the scale at which it is ultimately utilized and the emissions reductions that result. The "impact factor" for CO_2 utilization research will eventually be measured only in tons.

Acknowledgements

I am grateful for my talented and dedicated co-workers, past and present, who have created our program in CO_2 utilization. The projects described here were supported by the Camille and Henry Dreyfus Foundation, the TomKat Center for Sustainable Energy, the U.S. Department of Energy, and the Stanford Sustainability Accelerator.

References

[1] J. W. B. Rae, Y. G. Zhang, X. Liu, G. L. Foster, and H. M. Stoll. Whiteford, *Annu. Rev. Earth Planet. Sci.* **49**, 609 (2021).
[2] D. I. Armstrong McKay, A. Staal, J. F. Abrams, R. Winkelmann, B. Sakschewski, *et al.*, *Science* **377**, 1171 (2022).
[3] S. J. Davis, N. S. Lewis, M. Shaner, S. Aggarwal, I. L. Azevedo, *et al.*, *Science* **360**, eaas9793 (2018).
[4] ExxonMobil 2022 Outlook for Energy (2022).
[5] M. C. Heller, M. H. Mazor, and G. A. Keoleian, *Environ. Res. Lett.* **15**, 094034 (2020).
[6] T. J. Lark, N. P. Hendricks, A. Smith, N. Pates, S. A. Spawn-Lee, *et al.*, *Proc. Natl. Acad. Sci. USA* **119**, e2101084119 (2022).
[7] R. A. Efroymson, M. H. Langholtz, K. Johnson, B. Stokes, C. C. Brandt, *et al.*, 2016 Billion-ton Report: Advancing Domestic Resources for a Thriving Bioeconomy, Volume 2: Environmental Sustainability Effects of Select Scenarios from Volume 1. No. ORNL/TM-2016/727. Oak Ridge National Lab, Oak Ridge, TN, USA (2017).
[8] J. Barber, *Chem. Soc. Rev.* **38**, 185 (2009).
[9] Y. Hori, In *Solar to Chemical Energy Conversion: Theory and Application.* Lecture Notes in Energy, M. Sugiyama, K. Fujii, and S. Nakamura (eds.), Vol. 32, pp. 191–211 (2016).
[10] R. Küngas, *J. Electrochem. Soc.* **167**, 044508 (2020).
[11] J. A. Rabinowitz and M. W. Kanan, *Nat. Commun.* **11**, 5231 (2020).
[12] IRENA, Green Hydrogen Cost Reduction: Scaling Up Electrolyzers to Meet the 1.5°C Climate Goal, International Renewable Energy Agency, Abu Dhabi (2020).
[13] C. Chen, M. Garedew, and S. W. Sheehan, *ACS Energy Lett.* **7**, 988 (2022).
[14] Y. Xu, L. Isom, and M. A. Hanna, Bioresour. Technol. **101**, 3311 (2010).
[15] Global CCS Institute, Global Status of CCS 2022 (2022).

[16] M. E. Boot-Handford, J. C. Abanades, E. J. Anthony, M. J. Blunt, S. Brandani, *et al.*, *Energ. Environ. Sci.* **7**, 130 (2014).

[17] S. Deutz and A. Bardow, *Nat. Energy* **6**, 203 (2021).

[18] A. Banerjee, G. R. Dick, T. Yoshino, and M. W. Kanan, *Nature* **531**, 215 (2016).

[19] A. D. Frankhouser and M. W. Kanan, *J. Phys. Chem. Lett.* **11**, 7544 (2020).

[20] G. R. Dick, A. D. Frankhouser, A. Banerjee, and M. W. Kanan, *Green Chem.* **19**, 2966 (2017).

[21] S. R. Nicholson, N. A. Rorrer, A. C. Carpenter, and G. T. Beckham, *Joule* **5**, 673 (2021).

[22] A. W. Lankenau and M. W. Kanan, *Chem. Sci.* **11**, 248 (2020).

[23] C. P. Woroch, A. W. Lankenau, and M. W. Kanan, *Macromolecules* **54**, 9978 (2021).

[24] C. P. Woroch, I. W. Cox, and M. W. Kanan, *J. Am. Chem. Soc.* **145**, 697 (2023).

[25] D. J. Xiao, E. D. Chant, A. D. Frankhouser, Y. Chen., A. Yau *et al.*, *Nat. Chem.* **11**, 940 (2019).

[26] T. M. Porter and M. W. Kanan, *Chem. Sci.* **11**, 11936 (2020).

[27] S. Thor Wismann, K. E. Larsen, and P. Mølgaard Mortensen, *Angew. Chem. Int. Ed.* **134**, e202109696 (2022).

[28] C. S. Li, A. D. Frankhouser, and M. W. Kanan, *Cell Rep. Phys. Sci.* **3**, 101021 (2022).

[29] Hydrogen Council, McKinsey & Company, Hydrogen Insights Report (May 2023).

[30] V. Dieterich, A. Buttler, A. Hanel, H. Spliethoff, and S. Fendt, *Energ. Environ. Sci.* **13**, 3207 (2020).

[31] S. Gehrmann and N. Tenhumberg, *Chem. Ing. Tech.* **92**, 1444 (2020).

https://doi.org/10.1142/9789811282324_0030

Design and Evolution for New Metabolism with Speed and Scale

James C. Liao
Academia Sinica, 128 Academia Road,
Section 2, Taipei 115, Taiwan

Biology plays a major role in the carbon cycle on Earth. It was estimated that terrestrial photosynthesis draws down 120 giga tons of carbon per year, half of which is released back into the atmosphere by respiration. The remaining carbon is slowly converted to CO_2 and CH_4 due to microbial metabolism, with about 10% stored in soil. The carbon cycle reached an equilibrium to establish a relatively stable level of atmospheric CO_2, until the industrial revolution. To date, burning of fossil fuel emits more than 10 giga tons of carbon annually, which upsets the natural balance of the carbon cycle and caused a rapid increase of atmospheric greenhouse gas (GHG) in the past 100 years or so. The rise in GHG in the air has caused global warming and more frequent occurrence of extreme weather events.

The 2018 IPCC report "Global Warming of 1.5°C" recommends that CO_2 emission must be reduced by 50% by 2030 and reach net zero by 2050 in order to limit global warming to 1.5°C. In addition, methane emission also needs to be reduced by 50% by 2030. To meet this target, the current fossil fuel-based economy must be changed to renewable-based economy to avoid carbon emission. Thus, conversion of C1 compounds (CO_2, methanol, and methane) to useful chemicals is essential to achieve "net zero" by 2050. In this regard, biology becomes an important player because of the specificity and diversity of biological chemistry. However, the biological chemistry nature evolves does not address the problems that we face today which must be solved with speed and scale. Thus, new metabolism is needed to meet our needs.

Redesign and re-evolution of metabolism

The initial stage of metabolic redesign is focusing on demonstrating the plasticity of metabolic systems by converting a carbon substrate to a specific chemical in an organism that does not natively do so. At this stage, the common strategy is the identification of enzymes from various organisms and expresses them properly in a desirable organism that is amenable to genetic manipulation or possesses a specific phenotype of interest. An example is the isobutanol production using *E. coli* [1], which is a non-native yet toxic product for the organism. The redesign of a metabolic pathway may be straightforward, but the rewiring of regulatory systems for optimal production often presents a major challenge.

Another example is the production of 1-butanol in *E. coli* [2]. The pathway requires only a handful of heterologous genes, which should be relatively easy to transfer from various organisms to *E. coli* to fit in the existing biotechnology and industrial platform. However, despite many trials by multiple groups, *E. coli* was unable to match the level of production demonstrated by the native butanol producer — *Clostridium*. We then developed an evolution strategy to couple production with growth and showed that by manipulating the driving force of metabolism, *E. coli* can produce to a level matching the *Clostridium* performance. This example demonstrates that rational design coupled with *in vivo* evolution is a powerful way to develop biological chemistry with speed and efficiency. To date, numerous chemicals of various complexity have been engineered for multiple platform organisms, demonstrating the plasticity of downstream metabolism after substrate assimilation. However, few are focusing on the speed and scale required to combat the climate problem.

To do so, the following problems must be addressed: (1) the redesign of the metabolic infrastructure, including CO_2 fixation, photorespiration, and sugar metabolism, and (2) the strategy for re-evolution *in vivo* to increase the speed and scale (efficiency).

Preventing carbon loss in photosynthesis

Rubisco is responsible for the majority of CO_2 fixation on Earth. Rubisco itself is not sensitive to oxygen. However, Rubisco cannot distinguish CO_2 from O_2, causing oxygenase activity as a side reaction. The result of this activity is the oxygenation of the Rubisco substrate ribulose 1,5-bisphosphate to 3-phosphoglycerate and 2-phosphoglycolate. The latter is then oxidized to CO_2 in peroxisome and mitochondria, resulting in the loss of carbon. This phenomenon is called "photorespiration" and is particularly severe for C3 plants which account for the majority of the plants on Earth. It is thought that Rubisco in nature has evolved to a limit that is balanced between specificity and activity.

To alleviate this problem, several authors have engineered plants to recycle 2-phosphoglycolate to productive pathways. The first approach [3] is to construct an oxidation pathway to break down 2-phosphoglycolate to two CO_2 in chloroplasts, instead of mitochondria. This would allow Rubisco to re-assimilate CO_2 and partially reduce the carbon loss. The second approach [4] is to convert two molecules of 2-phosphoglycolate to glycerate with a loss of CO_2. The former can then be assimilated into the CBB cycle. The third approach [5] is to convert 2-phosphoglycolate to a common biosynthetic precursor, acetyl-CoA without carbon loss. This would allow the plant to synthesize compounds such as fatty acids that are useful as fuel and chemical. These approaches utilize enzymes in different organisms and express them in the chloroplasts of plant cells. The last approach [6] evolves a new-to-nature enzyme, glycolyl-CoA carboxylase (GCC), to fix another CO_2 to glycolate and eventually convert it to glycerate for re-assimilation in the CBB cycle.

These strategies highlight the different approaches for establishing "new metabolism" in cells. Taking advantage of enzymes nature already evolved would accelerate the progress, but creating new enzymes may also have advantages in the future. In either case, the construction of new metabolism by rational design must be followed by *in vivo* evolution to adapt to the cellular environment.

Rubisco-independent CO_2 fixation

Another approach to solve the carbon loss problem in photorespiration is to avoid using Rubisco in carbon fixation. Nature has evolved seven pathways for CO_2 fixation, of which only the CBB pathway uses Rubisco. Since it is oxygen-tolerant, it is widely used by phototrophic organisms that use light to split water. The other CO_2 fixation pathways are all oxygen-sensitive and involve enzymes that depend heavily on oxygen-sensitive iron–sulfur clusters. However, there are some oxygen-insensitive carboxylases that are used for carboxylation reactions in the cell and can be repurposed to form CO_2 fixation pathways. Many of the theoretical CO_2 fixation pathways have been designed based on kinetics and thermodynamics considerations [7]. However, most of them involve oxygen-sensitive enzymes and are difficult to implement. So far, two CO_2 fixation pathways have been constructed using cell-free systems to demonstrate the feasibility of the pathway *in vitro*. The first one (CETCH cycle) [8] uses only an enoyl-CoA carboxylases/reductases (ECR) to fix a carbon on crotonyl-CoA (C4) and acrylyl-CoA (C3), respectively, and form a synthetic cycle. The cycle produces glyoxylate, a C2 compound. The second synthetic cycle (rPS-MCG) [9] (Fig. 1(a)) uses a PEP carboxylase (PPC) and an ECR to fix two CO_2 and produces acetyl-CoA, which is also an intermediate in the cycle. The cycle can also output any C2, C3, and C4 intermediate, forming a self-replenishing cycle. The rPS-MCG cycle also features a possibility to recover some energy from NADPH in the form of FADH2, analogous to the TCA cycle. These cycles are demonstrated in "one-pot" *in vitro* systems to prove the chemical principle and demonstrate the efficiency. The *in vitro* systems tested protein–protein interactions, stabilities, and cofactor recycling controls, which must be carefully balanced, even in *in vitro* systems. These results demonstrate the potential challenges in implementing a fundamental pathway in cells and highlight the importance of *in vivo* evolution in implementing new metabolism, which we address shortly.

The carbon loss in glycolysis

Once CO_2 is assimilated, it is converted to sugar and enters the glycolytic *Embden–Meyerhof–Parnas (EMP) pathway*, the most conserved pathway present in almost all organisms. This pathway partially oxidizes glucose to make the 2-carbon unit acetyl-CoA for the synthesis of fatty acids, amino acids, and most of the bioproducts of interest to industry. However, during this process, one-third of the carbon is lost as CO_2, resulting in a great reduction in carbon yield. This was the reason for the relatively low carbon yield of all bioproducts and the significantly reduced competitiveness relative to petroleum-derived products, and the problem is independent of the carbon substrates, as long as the metabolism uses the EMP pathway to produce acetyl-CoA. To address this problem, we designed a synthetic non-oxidative glycolysis (NOG) pathway [10] (Fig. 1(b)) that avoids losing carbon during the breakdown of sugars. The proof of concept was established by demonstrating NOG's function *in vitro* and *in vivo*. We further developed an *E. coli* strain that uses NOG as the sole pathway for sugar catabolism [11] by rational design followed by stepwise *in vivo* evolution. The *in vivo* evolution process turns out to be time-consuming and suffers from escape strains that utilize alternative pathways for growth. This result highlights the need for a more systematic *in vivo* evolution process and a strategy to align the "design goal" with the organism's goal for efficient growth. Nevertheless, the development of a strain that uses NOG to replace EMP demonstrates that even a fundamental pathway can be altered by a combination of design and evolution strategies.

The methanol utilization problem

Similar efforts in redesigning and re-evolving metabolism for methanol utilization in microorganisms have also met with comparable problems. For example, a methanol condensation cycle (MCC) [12] to ethanol or butanol without carbon loss has been designed and shown *in vitro*. An *E. coli* that can utilize methanol as the sole carbon source for growth has been designed and evolved [13]. In short, while designing novel pathways has seen

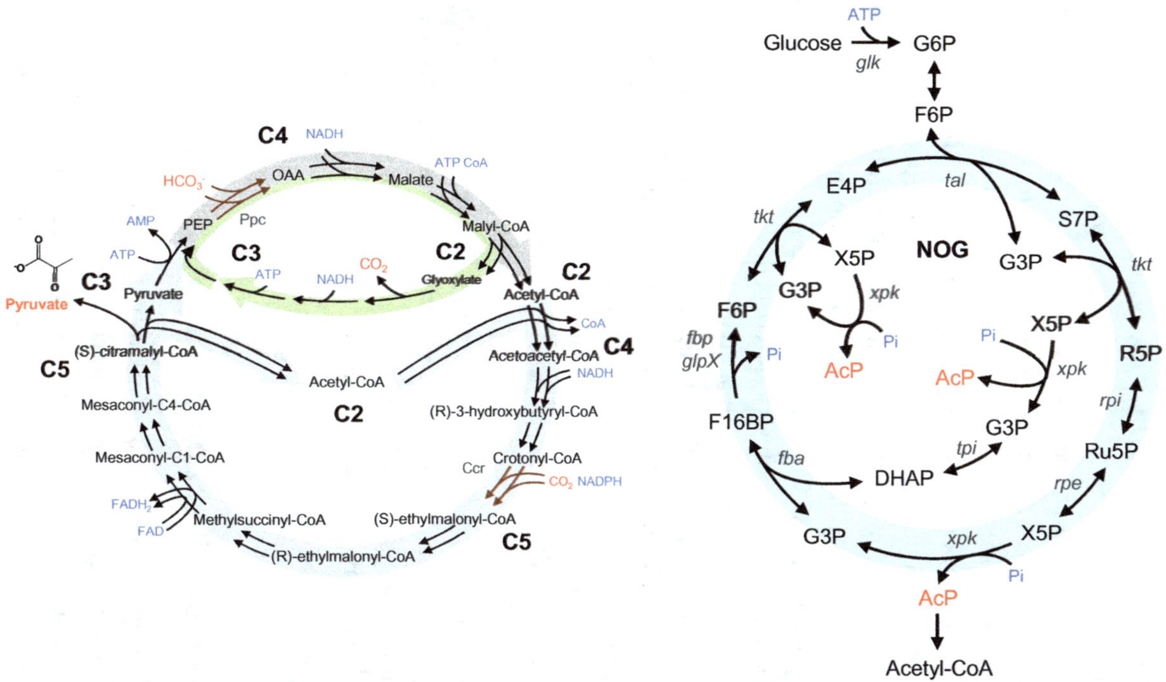

Fig. 1. (a) The synthetic rPS-MCG cycle [9] for CO_2 fixation without Rubisco. (b) The non-oxidative glycolysis (NOG) pathway [10].

repeated success, implementing them in microorganisms to a level that can perform new metabolism with speed and at scale is still challenging.

Challenges and opportunities in *in vivo* evolution

Previous examples have shown that purely relying on rational design could not accomplish the goal of "establishing new metabolism for speed and scale". A few strategies have been developed to move towards this goal. The most common one is to artificially create a library of variants in a promoter, ribosome-binding site, gene knockout, or protein sequence, followed by screening. The process has also been aided by machine learning and automation. However, the constraint on the library size significantly limits the success. Hence, additional tools are required to establish new metabolism with speed and scale that may impact the climate solution requires additional tools.

In conjunction with rational design, we explored natural evolution and searched for properties that can be exploited in creating the *E. coli* strain that can grow with methanol as a sole carbon source. *E. coli* has all but three enzymes that are required to grow with methanol as a sole carbon source [13]. However, despite years of efforts by many groups worldwide, the task has proved to be more difficult than initially expected. We first used a metabolic model to identify and then removed two potential kinetic traps that may cause a metabolic dead-end. The remaining steps were relying on *in vivo* laboratory evolution. The final result showed that the cell used a combination of single nucleotide polymorphism (SNP) and copy number variation (CNV) to create diversity and achieved a delicate balance between detoxification and assimilation. Through evolution, the metabolic pathway for methanol assimilation was established and optimized by balancing the production and utilization of the toxic intermediate, formaldehyde. Through this example, we discovered that *E. coli* can dynamically tune the copy number of a specific fragment of DNA on the chromosome. The organism utilizes two mechanisms to adjust the copy number of a DNA fragment. One depends on IS elements and the other does not. We have identified the essential elements that are required for the dynamic CNV. Through these mechanisms, the cell can amplify the expression of a large number of genes. Upon selection, strains with appropriately amplified gene sets are enriched. By effectively combining SNPs with CNV, the *in vivo* evolution can be exploited.

In another example, we knocked out an essential gene, *panD*, coding for aspartate 1-decarboxylase which is essential for β-alanine synthesis in *E. coli* for growth in glucose minimal medium [14]. We performed an *in vivo* evolution experiment and select cells that can grow without any nutritional supplement. The organism was able to "repair" this metabolic block by rerouting metabolism in different ways. The first pathway is to repurpose the uracil degradation pathway, coupled with necessary auxiliary functions for detoxification. The second is to mutate ornithine decarboxylase (SpeC) to gain a simultaneous decarboxylation and oxidative deamination function. This novel activity coupled with other transamination reactions and dehydrogenase reactions also "repaired" the *panD* mutation. This example demonstrated that *in vivo* evolution may have the ability to evolve novel activities that are not present in nature.

Outlook to future developments

Currently, using biological approaches to solve climate problems depends on a few fundamental metabolic pathways that were not evolved for such a purpose. To make a practical impact, the speed and scale of biological reactions must be improved. Both rational design and evolution approaches are needed in this endeavour. In particular, *in vivo* evolution that is responsible for establishing natural biological reactions must be further explored. While the basic principles for DNA replication, repair, and recombination are reasonably established, new mechanistic details are required to explore the full capacity of *in vivo* evolution. Together with rational designs, these approaches are expected to play a major role in building new metabolism that can achieve the speed and scale required to convert single carbon greenhouse gases to chemicals to alleviate the dependence on fossil sources.

References

[1] S. Atsumi, T. Hanai, and J. C. Liao., *Nature* **451**, 86 (2008).
[2] C. R. Shen, E. I. Lan, Y. Dekishima, A. Baez, and K. M. Cho *et al.*, *Appl. Environ. Microbiol.* **77**, 2905 (2011).

[3] P. F. South, A. P. Cavanagh, J. W. Liu, and D. R. Ort, *Science* **363**, 6422 (2019).

[4] R. Kebeish, M. Niessen, K. Thiruveedhi, R. Bari, H.-J.Hirsch, *et al.*, *Nat. Biotechnol.* **5**, 593 (2007).

[5] H. Yu, X. Li, F. Duchoud, D. S. Chuang, and J. C. Liao, *Nat. Commun.* **9**, 2008 (2018).

[6] M. Scheffen, D. G. Marchal, T. Beneyton, S. K. Schuller, M. Klose *et al.*, *Nat. Catal.* **4**, 105 (2021).

[7] A. Bar-Even, E. Noor, N. E. Lewis, and R. Milo, *Proc. Natl. Acad. Sci. USA* **107**, 8889 (2010).

[8] T. Schwander, L. S. von Borzyskowski, S. Burgener, S., N. S. Cortina, and T.J. Erb. *Science* **354**, 900 (2016).

[9] S. Luo, P. P. Lin, L. Y. Nieh, G. B. Liao, B. W. Tang, *et al.*, *Nat. Catal.* **5**, 154 (2022).

[10] I. W. Bogorad, T. S. Lin, and J. C. Liao, *Nature* **502**, 693 (2013).

[11] P.P. Lin, A. J. Jaeger, T. Y. Wu, S. C. Xu, A. S. Lee, *et al.*, *Proc. Natl. Acad. Sci. USA* **115**, 3538 (2018).

[12] I. W. Bogorad, C. T. Chen, M. K. Theisen, T. Y. Wu, A. R. Schlenz, *et al.*, *Proc. Natl. Acad. Sci. USA*. **111**, 15928 (2014).

[13] F. Y-H. Chen, H.-W. Jung, C. Y. Tsuei, and J. C. Liao, *Cell* **182**, 933 (2020).

[14] S. Pontrelli, R. Fricke, S. T. Teoh, S. T. W. A. Laviña, S.P. Putri, *et al.*, *Nat Chem Biol* **14**, 1005 (2018).

https://doi.org/10.1142/9789811282324_0031

Discussions of Session 4 — New Chemistry for Renewable Energy

Chair: Daniel Nocera
Auditors: E. Debroye*, B. Elias†

**KU Leuven, Department of Chemistry, Celestijnenlaan 200F, box 2404, B-3001 Leuven, Belgium*
†UCLouvain, Institute of Condensed Matter and Nanosciences (IMCN), Molecular Chemistry, Materials and Catalysis Division (MOST), Place Louis Pasteur 1, bte L4.01.02, 1348 Louvain-la-Neuve, Belgium

Daniel Nocera:

You might be afraid, seeing the challenge of the problem. I wanted to tell you just one little quick thing before we dive. I used to do a lot of public speaking for energy, and I was with Kurt Vonnegut on a stage. I used to say we need to save the planet and he was a little bit of a different thinker. He pulled me aside and said: "Dan, you're so worried about the planet". He put his hand on my shoulder and said: "Don't be, she is an organism and like all organisms she has a beautiful immunological system. And like any organism when the invader compromises the organism, her immune system will kick in and eliminate us. So, don't worry as we suffocate her, and we try. She'll just have her immune system kick in and eliminate humans. She's going to be just fine." So actually, that makes me happy in the energy problem. As you move on, the world is going to be just fine. So, it's all about you and not about the planet and you should realize that. So, that should make you happy, that's a happy thought.

I think you saw the complexity of the problem, that's number one, I wanted to present how wide in scope it is. So, you saw first materials and the importance of materials. You saw chemical synthesis at scale which is important for the CO_2 problem. Roles of biology that can be played, and then again, a lot of material science in terms of the batteries and PV. So, I did that because I knew there was a broad audience here. First, I'll give it to the panel, they can make one statement of what they think is something that maybe they needed to say and didn't. But then all I want to do since you have these six people covering such a broad area of science, is take more questions from you. The energy problem is hard, it's large, and the misconception even among scientists, forget the public, is unbelievable. And so, here's a chance where you can try to at least put some of the questions you have on the table to this panel of experts. I'll start with Clare, we're only going to do this once. We're going to speak only once and then only get questions from you. If you don't, then I will fill it in with questions, but I believe I would like to hear from all of you. So, Clare, if you have one thing that you would like to say that you didn't.

Clare Grey:

Well, I thought I covered most things, but I think there is a challenge in this field where we do need to innovate out of the current system and you have a tension between what's feasible in terms of the economics which then limits you to certain chemistries. But actually, if we're going to really change the way we do things we have to, sometimes, not be limited by funders who say: "he's never going to meet the DOE target, it's not going to be that". So, we have to keep an eye on what the target is, but we also have to be innovative, and that sometimes means ignoring the short-term targets, but thinking about the longer-term targets.

Sossina Haile:

I'd say that there are a lot of ways in which electrochemistry can help to address the energy problem. So, it would be hard to say what is one specific thing that I overlooked. That's just to say there are many electrochemical energy transformations that are critical here and so hopefully we can have a larger discussion about those.

Matthew Kanan:

What I was hoping to convey at the end, but I think I kind of rushed through it, is that our job, at least the way I see it, is to create viable options for technology developers to try to implement and it can be daunting sort of thinking about this scale. But there are many existing pathways with clearly identified problems that need to be addressed to help those become real technologies and then there are of course opportunities for new pathways. But you have to be mindful of material availability, so material choice in terms of the space that you are going to explore, and manufacturability if you want to do something on scale so thinking about what type of process could be used to make what you are trying to develop. Those are factors that shouldn't impede research of course, but I think it's worth exploring those, and being mindful of those as the outset is to decide problem selection and how to approach something.

Sossina Haile:

I guess one thing that perhaps I could and should have said is that when we think about these large-scale solutions, if we think about extrapolating any single solution, it becomes impossible. So, I just say that we should probably keep that in mind as well: that it's ok to go after part of the problem. We don't have to think that 'this particular solution' is going to address all of energy.

Henry Snaith:

So, with respect to solar energy, it's already working pretty well. That shouldn't make us think that we shouldn't work on it or don't need to work on it. Firstly, it's really important that we keep pushing the efficiency up, that's going to be necessary to keep this deployment curve up. Every percentage increase in performance is more power produced from renewables linearly. And then the opportunities: I've talked a lot about perovskites, and tandems, and multi-junctions. That's one way to make higher efficiency. There are a whole host of other opportunities and other methods, some of them more practical than others and there's a whole range of potential materials that could also be the next perovskite or well, probably if you are in the industry, you might say the next silicon because perovskite isn't quite the next silicon yet. But there are lots of opportunities there and lots of opportunities for chemistry to contribute massively to it.

James Liao:

So, if we look at the sustainability problem, you can roughly divide it into two parts. One is 'how do we generate electricity without giving out carbon?'. The other one is 'how do we reabsorb carbon that we had to give up by making various chemicals?'. So, at all sides I think the chemistry and biology approaches are very useful and there are many interesting and practical problems that remain to be solved. I think this is a field that needs a lot of people to invest their careers in.

Andreas Züttel:

I think we should see the transition to renewable energy as a unique chance. It will not only be cost, there will also be a lot of benefits in the future. We should not forget that 150 years ago when the steam engine allowed industrialization, this was not seen positively in the beginning. They were fighting against the steam engine. And so, that's why today we shouldn't see only problems in the transition but also see the chance for the future.

Daniel Nocera:

Ok, then I just want to tell you how much fun this field is. I know it sounded very applied, but I'll just tell you one little quick thing. So, if you wanted to have the poor use water, you can't have clean water. Actually, you need to have water hydrogen from urine and that just made us work backwards and say: 'what do we need?'. It turned out we made a catalyst that does self-assembly, it went to all the other talks or sessions so far. Most catalysts need to work in concentrated base, and we made a catalyst that self-assembled. The energy to do catalysis was greater than the energy for self-assembly so when it breaks down it reforms again, and it was very much like the session yesterday morning. So, once you start working backwards there's an unbelievable number of just fundamental interesting problems. But in this field, you have to really keep your eye on the target and basically work backwards to the problem if you're going to do it quickly. Also, I think in your questions you can ask any of us where the important fundamental science breakthrough came. So, with that, I will begin taking questions.

David MacMillan:

First of all, thanks to the whole panel, I thought that was a fantastic session. My question is really for James, and I thought he nicely summarized these two components that are going on. Coming back to Dan's point: all of this is incredibly important and it's important to work on all of it. But in terms of what's the most emergent problem, is it getting the carbon out? Because at the rate we get to renewables, it could be too late before we trigger the immune system.

James Liao:

It depends on the country; it is very local. I'm more familiar with the situation in Taiwan: we don't have any domestically produced energy, so production of electricity is the most important thing for us. Once you have electricity you can get the economy going. I don't know if this community is aware of RE100: that's the industrial organization. I think it was started in the United States by big corporations like Apple, Google, Microsoft, and so on..., asking their suppliers to use renewable energy by year, I forgot, 2050, or 2030, or 2035, otherwise they're not going to continue with them as a supplier. So, this has a huge impact on the supply chain across the world. Consequently, I think most of the companies are pushing the government to provide green electricity. So that drives up the green electricity price. In a way that's good because then that makes the development of solar panels a profitable business so people can afford to invest in the electricity panel. On the other hand, in terms of carbon, I think the EU has just begun to talk about the carbon tax, the border carbon tax, carbon credit and so on and so forth but it is still in the talking. The United States is refusing to introduce the carbon tax, but most of the companies are developing their technologies on the assumption that one ton of carbon costs 1000 dollars, basically one dollar per kilo. If you don't meet that criterion, companies are not going to invest. So, both are emerging simultaneously, it is just that it doesn't transpire to the general public and the academic community is not generally familiar with the complications of all these politics. To get back to your point, both are important but electricity is more urgent and it is a real problem, it is the energy problem that you cannot live without. Carbon can be a political issue and is a price issue. A hundred dollars per ton is the goal, if you cannot meet that then you don't have a process. So eventually, that will be extremely important.

Daniel Nocera:
And Dave, just for point of clarification, did you mean CO_2 in terms of the biological problem or specifically or more generally the societal problem?

David MacMillan:
The rate that we're going towards renewables: in that timeframe we're going to get past the tipping point of carbon in the atmosphere so that the more pressing question is how you get carbon out of the atmosphere. Even though all these solutions, all these things that are incredibly important to work on, is that the more pressing of all the issues? That's my question.

Matthew Kanan:
I think what you're saying is: should we be focusing on trying to get control technologically over the CO_2 content in the atmosphere or just focus on decarbonizing as quickly as possible? So, the latest I've seen on the trajectories is that we're on a 2.6 C-trajectory among the different IPCC scenarios. All these tipping points, it's just a matter of risk assessment and so there's actually a recent paper in Science that walked through kind of the status of understanding there. So, what you're talking about, getting it out is a removal, that's carbon removal. I view that as actually a tremendous opportunity and it's kind of the ultimate CO_2 conversion. The reason it's an opportunity is because there are natural resources, there are silicates, minerals, magnesium-rich silicates called ultramafics, that have the capacity to remove hundreds of thousands of gigatons of CO_2. They naturally remove CO_2 on the order of up to a gigaton per year, that's just a

kinetic problem. So thermodynamically it's downhill, magnesium silicate is like a "mega-oxide" trapped in a silicate. So, if you can crack the kinetic problem of how to accelerate that with minimal energy, then you have the opportunity to spend renewable energy and get a lot more carbon benefit than you would be spending energy to replace an existing fossil product for example. So, I view that as a very healthy competition, it's a very controversial area which is why I didn't want to mention it in my talk. If you just think in dollars per carbon benefit, I can either replace a product with a lower carbon product and someone's going to pay a premium because it's going to cost more to make that product. Or I can use that same renewable input to remove carbon and if I can remove more carbon than I would say that that should be done. So, I think that that's an emerging technological option.

David MacMillan:
Great answer. I was just thinking along the lines of renewable fuels, like thermodynamically the rate where ultimately we should be. But in the interim, if there was a tipping point through carbon content, do we need to be working or worrying as much for capture? So thanks.

Daniel Nocera:
You will always have to cut through the scams unfortunately in our field. So, there are people trying to just get Science and Nature papers and they're writing one page paragraph intro and they have nothing to do with the real problem. The other is, there's money to be made. So right now, unfortunately, mechanically taking it out you'll hear people… It's around 800 dollars to 1000 dollars per metric ton and some of the best groups in the world; some at MIT, Stanford,… If you talk to the ones who aren't scamming they're going to say it's going to be very hard mechanically to get lower cost. So, what Matthew's saying it's a real role for chemistry versus a bunch of Emmy's.

Claire Grey:
The argument against it is it does allow the fossil fuel industries to continue. So, I think one does need to be aware of that and there's this very strong lobbying force for carbon capture because of that.

Daniel Nocera:

There's a New York Times editorial from the persons who started carbon capture and started a company, they wrote an editorial. I think three weeks ago, they've given up on it for that reason. There's a whole political piece to this that becomes even more overbearing.

Sossina Haile:

If I could just add on that political piece. One thing that I find challenging also, after you do the CO_2 capture. I like the magnesium silicate solution because you're not tempted to then burn that carbon again. Because if you're capturing it and burning it again, you're keeping the approximate level the same in the atmosphere. So, unless you're pulling it out and doing something else that you're not tempted to burn it again, it's just an exercise in futility.

Thomas Ebbesen:

I have one comment about Europe. Europe does have a carbon fee but it's just too low for obvious reasons, and also because no other countries are joining this. So, it's very hard to impose this and we'd have to calculate the carbon content of products that are imported and that's what some people are trying to suggest. And this is a very large market that might help the rest of the world align. I want to ask, we're talking politics and this is very fascinating this whole issue obviously and everyone is concerned with this. So, would it improve if you had higher mobility or higher conductivity in your material, would that make a difference?

Sossina Haile:

For the most part, the electrolytes are quite good. It's the electrochemical reaction step which is rate limiting. For both fuel cell mode and electrolysis mode.

Thomas Ebbesen:

Thank you.

Joachim Sauer:

So, it's nice to talk politics but we can contribute to chemistry and I just wondered, of course a CO_2 tax which is high is an international problem and as long as we cannot agree in the world about the CO_2 tax, it is not very effective. The other point is, as a scientist we should look at all possible solutions. And Claire, I agree with this one, that in the US there is the wrong incentive for CO_2 capture, because they would use it to put it back to the earth where it was coming from, to get more methane out. But we have heard good examples in the cement industry if you are capturing at the source, and maybe not forever but for now, and also other things we have heard that are useful for some period. So we should not say because it is misused in a certain political or economic situation that we should not understand how it works and should not improve on it. Of course, when it comes to funding from states and organizations then there will be a political direction where to put the money first.

Daniel Nocera:

Just one thing, I mean that's the imperative on Science. I started years ago hoping for a carbon tax, I've given up and it just makes the science harder. For instance, we're just saving now metric tons of CO_2 in agriculture. It's because the ammonia equivalents being made is now commercially on par with what's being sold in a bag. So, I think you can always hope for a carbon tax, but the scientists assume it's not there, it just raises the level of the Science you need to do.

Gunnar Von Heijne:

So, you kind of skipped over one issue about batteries and that's recycling. So, what's the status on that and perhaps more generally, we heard yesterday about windmill wings becoming a waste problem. So, are there other waste recycling problems lurking inside these technologies that we've been hearing about? Assuming they're scaled up to the global scale?

Clare Grey:

So yes, batteries are going to have to be recycled and I think one of the interesting research areas is how you rethink the design of a battery so that it's more easy to recycle, so that you don't have to bring the components back to the individual parts. One of the difficulties there's been, is that the technology is being changing. So, that to lock on to a specific battery type, as the ones going into the cars are changing, has made it more difficult.

The other challenge is that the batteries, when they're recycled, often contain a lot of lithium metal, so they're dangerous, they explode. They contain PF_6^- so when in contact with water, they form toxic gasses. And sort of rethinking how you put the components in so that they're easy to recycle is a very important research area. And on the anodes for example, the binders used to be PVDF. Now, they're cellulose based binders so that's an easier recycling. On the cathode, it's very difficult to find a cellulose based binder that is stable at those sorts of voltages. So, at some point, there may come a trade-off where you might decide actually you're not going for the one ton battery that has the 400 mile/500 km range that's sort of the current trajectory for electric vehicles. We may have to accept compromises in terms of range and I would argue that, if you do that with fast charging, that it results in a paradigm shift. So, I think the recycling and the sustainability issues may result in compromises in battery packs which ultimately will make for more sustainable ones. But yes, recycling is an interesting question. And you can do things by robotics and sort of make things more efficient and ultimately, you've got to put the recycling into the whole lifecycle analysis and the costing of the whole thing.

Daniel Nocera:
Before I take another question. I think it's a big problem and also a big issue for PV cells I would like to hear Henry on the PV side.

Henry Snaith:
Photovoltaic solar energy is recycled or there's an intention to recycle it, I should say. So, there's one company that does a very good job for solar. They actually don't make silicon modules, they use cadmium telluride and they recycle as part of their whole process to get the cadmium out because it can't go into landfill. And that works very effectively, it adds a little bit of cost to the module, but it works. The main manufactured silicon: the companies don't presently recycle their own modules at the end of the life and it's left to wherever they're deployed but this is something that certainly has to change going forward. Certainly, within the EU, there's a big sort-off push towards a cyclic economy where we look at the whole

life-cycle analysis from cradle to cradle and looking at re-using the materials even. But of course, they may not be need to be reused directly in the same product. But if we look at the amount of glass that we are going to need, and the amount of plastic lamination foils, regardless of what the core solar technology is, then that needs recycling, absolutely. For the perovskites, we've been working with other collaborators to look at recycling these materials. And actually, you can remove the lead with unity effectiveness just using hot water, and then allow the water to cool and precipitate it out and that water can be re-used and re-used. So, this is a very simple recycling process that doesn't involve harsh chemicals or acidic environments. But there's a lot of drive, all the companies working on lamination foils are thinking about recycling as well and not just 'how does a module last 25 years', but how do you get it apart at the end and then recuperate the materials.

Clare Grey:

Can I just add one more thing? One of the interesting things in the battery space, you've got the old technologies like lithium-ion phosphate and carbons coming in from China with the blade technologies, and there you've got a much more sustainable cathode, but you've lost the financial incentive to recycle, because you know you got iron and phosphors. You've got to rethink how you then re-improve or go back and not recycle back to its constituents because there's no money to be made out of it. So that's the tension actually, as you go to these more environmentally friendly materials: there's less money to be made in recycling.

Thomas Ebbesen:

Can I just make a comment on this? Inside the EU again, the companies are required to take back the products. So, this is a super incentive in recycling because they don't want to have to take apart a compound that isn't easily recyclable. That was the intent of that law and still that applies. You can take a fridge back, so a store they have to take it and they have to do something with it. So, force them to face this from the beginning when they make the product. So this, I think, is a huge incentive.

Daniel Nocera:
Chad, did you have a question?

Chad Mirkin:
So, you gave us a quantitative google-based analysis of the problem which was good and very compelling. A big part of it is to understand the problem, but what's equally important, maybe even more important, is: this is a chemistry-based conference. What is the chemistry that we should be betting on as a group? So, we talked a lot about regulations and all sorts of things that we don't control as chemists but as you guys sit here looking at all the different solutions (and we've heard a variety of different types), does the field look at this and create real-time analysis of where the best bets are? What's the CRISPR of this field right now? Where is the most excitement as opposed to just an activity that may or may not contribute down the road?

Daniel Nocera:
I'm in lots of different Science fields, this is the one that does that best. Anybody who's doing something important will look to see the problem like you said and then literally you can work backwards and there's always an important Science problem. The one thing I don't advocate for, I don't even think will happen, is that there will be a doctrine of the ten commandments of what Science should be for energy, because it's going to have lots of different solutions. But in the battery field, I bet you Clare could go on for hours about what the key Science things are, as Sossina could for fuel cells. Without each one of those, again, if you keep your eye on the target, the people making the impact do work backwards to find that questions. And I think each of us will have a laundry list of important things, but I don't think you are going to see the ten commandments of energy Science.

Matthew Kanan:
I think I can maybe provide a more technical answer just specific to my area, for CO_2 conversion, the way I see it. I think the challenges are finding the simplest processes, because there's an efficiency loss with every step in the process, so I like hydrogenation certainly as a route to liquids

and commodity chemicals. Because liquid fuels it's not one chemical, it's a mixture of hundreds or thousands of chemicals. It's defined by the property of the liquid as a fuel. We know how to convert certain intermediates to those targets. I think in any one of those pathways that I laid out, some of them I didn't even get a chance to talk about. I can go over methanol and then methanol to olefin and then olefin to liquid. There are efficiency losses along the way, so you can identify specific efficiency losses and say "If I have a lower temperature methanol synthesis catalyst, I don't have to run at high temperature and multiple passes through my methanol reactor with immediate saving, and that route looks a whole lot more attractive" for example. So, a lot of people in the catalysis community are trying to tackle that problem, it's a really tough problem. I like this idea, I talked about going to syngas, and then going syngas to taking the existing infrastructure, go syngas to liquids. But again, that will compete with if somebody comes along with a really compelling direct CO_2 hydrogen conversion in single pass reactor to valuable liquids out. That process is going to be a lot more scalable, if the product distribution is amenable to conversion to the liquid fuel with the right species, with relatively straightforward separation. I think, at least for my perspective, for the big targets there are a lot of pathways already, but they all have either too many steps or there's a problem with some of those individual steps that need to be addressed. But there's potential because we already have established how to convert certain intermediates into basically all the hydrocarbons that we use on scale. There's potential to leverage that and really make something that essentially can integrate seamlessly with existing assets.

Sossina Haile:

Can I just add to that? Maybe this goes to the question about 'Are the electrolytes conductive enough?'. I would say there's a lot of ways in which we can integrate the electrolytes for electrochemical systems with new electrocatalysts. And so, there are aspects of being able to work at just warm temperatures that allow electrocatalysis to proceed in a way that's much faster than at room temperature but, as I said, not as degradation prone to the very high temperature systems. I think that is a space that's really not been much explored. I think there are a lot of oxide electrochemical compounds that can help do these various transformations.

Again, by operating at this intermediate temperature space, I think we have a lot of opportunities there.

Daniel Nocera:

And then in biology, I think as you heard from James, it's rate of throughput. One way is to metabolically engineer a pathway. What's happening with this ammonia thing which is now getting commercialized. We did a lot of genetics but found out how the bacteria grew and the other way to do throughput is just to grow at high, what's called, colony forming units. Those bacteria for the microbiologists, we're now growing those things at 10^{11} colony forming unit, most bacteria grow at 10^{8}. You can actually just be scale or density of organism start getting to high throughput. That's, I think, James' main line of research: throughput in biology.

James Liao:

I think Chad asked a very interesting question: 'What would be the CRISPR of the energy sector?' Of course, from my standpoint, we're working on something that we hope becomes the CRISPR. This is the Synec DNA which I talked about, where we're developing a fast evolution technology inside the cell so that we can very quickly evolve organisms to take on C1-compounds including CO_2, formic acid, methane, methanol. And once you have these C1-simulationed organisms worked out, then you can start thinking about recycling. Once an area in my mind is that probably we would not need to replace the current fossil fuel based chemical industry, because that's too huge. But if you can replace most of the energy through non-carbon green energy then you save about 50% of the carbon footprint in fossil fuel in the chemical industry. And then for the remaining part, you try to use the CO_2 recycling type of technology but not the traditional CCUS, not this traditional capture and storage, that type of thing. We should think about recycling and re-utilization, either via the chemical or biological way. I'm focusing on the biological ways, I think other people are focusing on the chemical ways. I think those will be a better picture for the future.

Arne Thomas:

Besides CO_2, hydrogen is the other molecule we should of course be concerned about. And that would be a question mainly to probably Sossina and Andreas Züttel. So, we inferred that one of the biggest issues or bottlenecks for a hydrogen economy is the storage and transport of hydrogen. I would like to hear your personal opinion, and probably also of the other panel members. What would be the future molecule? Sossina will probably say ammonia now, but there are other molecules of course where you can think of storing hydrogen, or there are materials that I have seen on one of the slides from Andreas Züttel, I think metal hydrides for example. So, what's your personal opinion, where will we go with hydrogen storage? Or would you say we will stay with the electron economy in the end and there will be no hydrogen economy?

Andreas Züttel:

Thank you for the question. You know the big difficulty is the seasonal storage. There are two main challenges: seasonal storage and aviation. Aviation will be done with the synthetic fuels, there is no way to get around it. As for the seasonal storage, I believe that hydrogen will play an important role there. The different storage options we have, or the cost of these different storage options, depends very much on how often we cycle the hydrogen through the storage. In seasonal storage the difficulty is, it's just done once per year and therefore you need a large volume for the lowest possible cost. That's why for hydrogen basically only the huge underground storages are a feasible solution for the future. Because, if we built such a storage it's going to be cycled only once a year. That means 50 times in 50 years and that can only be economically feasible if it's cheap and cheap because of the huge volume. But then it becomes a technical challenge to build these huge volumes. Of course, there is the technical solution that's what we try to realize in Switzerland and then there's also the natural solution. In gas fields for example, if they are tight enough you can fill them up with hydrogen and then store the hydrogen in these natural cavities underground and then of course the cost is much lower than then you have to build such a storage. There is an ongoing project in Austria where they finally discovered that when they pump the hydrogen down, it becomes methane spontaneously. There are some archeo-organisms down there and apparently, they convert the hydrogen into methane. Now, they changed the project and they are also pumping down CO_2 and try to see whether they can convert all the hydrogen into methane. You know, in methane storage at the same pressure, you have about 4 times more energy than in pure hydrogen storage, so that's an advantage. If the CO_2 comes out of the atmosphere, then it is also CO_2 neutral. These are approaches which look promising for the future.

Sossina Haile:

To follow up on that, I think for local storage it absolutely makes sense to just go to hydrogen. That we can do with electrolysis. I didn't talk about the electrolysis that we've been able to achieve. Again, using proton conducting electrolytes and the electrochemistry associated with that. The round-trip efficiency is not as much as batteries but as was said, you're not cycling this back and forth as many times. If you want to now do the transport as well, that's where I would say ammonia, or other organic molecules that have hydrogen that can be removed, are valuable. The benefit of ammonia of course, is that you don't have to return the carrier. If you're doing an organic molecule you have to return the carrier and put hydrogen on it. The challenge with ammonia is pretty evident, we heard about it from Karthish yesterday. It's how do you do the synthesis then, how do you go from nitrogen to making ammonia? And electrochemically the challenge is that if you drive the proton across the electrolyte to try to react with the nitrogen, you're probably just going to evolve hydrogen rather than doing the reduction of nitrogen. So, that's an area with the real challenge where the chemical contributions could be. The other potential idea is to start developing some nitrogen-ion conducting electrolytes. So again, to the question of 'do we need new electrolytes?': the conductivities are good enough, but we need ones with new functionalities. So, if we can have a nitrogen-ion conductor that would be great.

Matthew Kanan:

I'm agreeing with Sossina. I don't think it makes sense to transport hydrogen over significant distances. If you look at what's happening in these big water electrolyser projects that are being built now: Shell has one in the Netherlands, I think it's a 120 megawatt electrolyser, that's feeding right into their refinery. So, I think hydrogen should be produced on scale and used right where it's produced or stored locally if it's going to be in a reversible fuel cell electrolyser energy storage system. Transporting it can be difficult from a leak perspective. There's sort of new Science emerging.

Hydrogen obviously doesn't have any IR absorption, but it's a so-called indirect greenhouse gas because it interacts through atmospheric chemistry with other gasses in the atmosphere to increase their GHG potential. So, hydrogen is not innocent, you actually don't want to leak it from a climate perspective. So, I think it should be used where it's made.

Daniel Nocera:
Ben, did you have a question?

Ben Feringa:
Yes, maybe I misunderstood. I'm a little bit surprised because as you know the chemistry is built on the petrochemistry at the moment, the whole chemistry tree as well. I got the impression that if we focus on the energy, which is great of course, this is the major problem. That if we solve this 90% that is used for fuel issue, that the 10% which is left is fine. We can continue with using this for chemistry, no problem. Isn't that also the real challenge? That we as a chemical community try to build the new chemistries of the future using electrochemical methods, hydrogen, and CO_2 to give us our building blocks. We cannot continue by saying 'Oh, we will do the same petrochemistry with this 10% because there will be enough for the next 400 years'. Shouldn't we connect that? Can we see that separately? I don't think so.

Henry Snaith:
I mean, this is not my area obviously but at the end of the day you're not necessarily burning the product you make. So, as long as the energy input is coming from clean carbon-free sources, does it matter that you're locking the fossil fuels up in products to then be recycled and go back into the system at the end? So, do we care if we're using fossil fuels as a primary source?

Ben Feringa:
I don't deny that recycling is a key point that we have to work on, I mentioned this also in my talk. Instead of making bonds, also breaking bonds and go through the cycle and the process. But I think there must be tremendous opportunities now also to rethink how we are making our building blocks, how we are doing our chemistries. Why not directly use

electrochemistry? Of course there are several groups in the work that focus on that now, but why do all these steps in between?

Matthew Kanan:
One thing to add about that. You can't take a barrel of oil and convert it just into the feedstocks for chemicals. I think it's 10% of a barrel goes to jet now, and it hasn't necessarily been optimized but it's hard to push it much beyond that. So you can't just say: 'Okay, we are going to eliminate gasoline and then all of a sudden we have all these extra barrels that we can funnel into whatever we want'. It's not the way refinery works. I'm agreeing with you that yes, new ways of making feedstocks are needed. I think ethylene is a big one. Making aromatics is a challenge by the way, aromatics are necessary in aviation and other places. But yes, I agree with you that that is an important consideration.

Daniel Nocera:
Ben, that was Dave's session yesterday. Catalysis is going to be really important. I think there is a huge research opportunity for the community, to start rethinking chemical processes, like Dave said, to hexanes, to C12, to NSA, that would be great.

James Liao:
I'd like to address Ben's question. I totally agree with what you just said. Eventually, we should build everything from C1 type of raw materials so that we completely recycle. But it's a kinetic step, getting there is not going to make money. Once you don't have the economic driver it's very difficult to get there. For example, if you want to make a perfume using CO_2, it's a lot of work for very little, even though your margin is very high.

Ben Feringa:
Now my Science heart starts to beat. Because I'm not thinking in terms of dollars, but I'm thinking in terms of scientific challenges, and I think there are tremendous opportunities here to change our perspective of how we should do chemistry. And I think Matthew already gave us some examples, how can we do syngas chemistry, away from the traditional feedstocks, and linking CO_2 and electrons and hydrogen for instance?

James Liao:

I totally agree with that. So, there are two ways to do that: one is using the electrochemistry or directly using hydrogen, that means indirectly using electrochemistry, or using biochemistry. In biochemistry you still need energy to come from sunlight, or come from electricity, or whatever. So that's one of our major focuses, trying to develop the industrial workhorse of microorganisms that can rely on C1 as a feedstock. So, once you get into the bacteria, then the rest is channeling the carbon flow into whatever you want to make the compound you want.

Karen Goldberg:

Well, a couple of things: I want to reiterate the point that Matthew made. We currently have petrochemicals because we're refining gasoline, it's a byproduct and it's not going to be that in the future. I wanted to go back to something early on where Dan gave up on a carbon tax, I was holding out that you were going to accomplish that, Dan. But I want to say that there's not nothing we can do about that, we're in a room full of scientists who all have students, and they all have to raise their voices for this. We have to get Science out to the general public. This is such an important issue and so, I would encourage you all to help your students get out and speak up! If they have the opportunity to get involved in any kind of policy, help them! Because we need science voices in policy. So, I want people to speak up like you Dan, and I'm sorry that you gave up, because you speak so eloquently on the issue and when you were up there talking, I thought this is what our politicians need to be hearing.

Daniel Nocera:

But let me just tell you there are good things happening. There's a big part of investment that, because of voices, is decarbonizing the investment for hedge funds. So, it's called ESG and that actually is having a profound drive. ESG is Environmental Social Government, so all your pension funds, etc because, I think a lot of young people are concerned. ESG is really driving investment and discovery in this field right now, so it's happening but in a different way. Tackling the politicians is great except they're only there for a day or two and then there's another one that you're saying the same thing to. So, I always like it when market is driving it and that political voice is having a profound impact on investment.

Karthish Manthiram:

For the specific that you're working on, what are the gaps in observation and understanding? What are the things that you wished you could see but for which there aren't tools to be able to observe these things just yet? What are perhaps ways in which the theories that are deployed are explanatory but not yet predictive? So, if you could just highlight what these technical or scientific gaps are?

Daniel Nocera:

Theoretically, so if we get down to brass tacks. Computationally we're really far away from what's happening. Just take something simple: at an electrode surface. It's a really hard computational problem. While computing is moving on, it's slowly getting there, but computation has a long way to be predictive in this field. What tends to happen is that the computation isolates problems but doesn't treat the full problem. And a lot of these systems are messy. But with machine learning techniques, and people are starting to use machine learning and computational power, we're slowly getting there. There are people concentrating on the entire system and that means solvent, ions, electric fields, putting everything in. I'm just choosing electrochemistry since it's your field. And I think that it's coming along and that it will be a really great tool for the field and community.

Matthew Kanan:

I think, sort of two areas. One is Operando Science, it still has a long way to go to provide insight under real operating conditions that enable insight into what may be going wrong, or what does the catalyst really look like when it's operating. I think on the synthesis side the paradigm, at least as far as I understand it, we're going to leverage AI and other computational tools to discover the best material and then we'll figure out how we're going to make it and so on and so forth. But at least from my perspective and for heterogeneous catalysis, knowing what material may or may not be a great catalyst is potentially enabling, but there's still huge question marks as to whether or not that's actually going to be a viable catalyst to synthesize, a catalyst that stays in that state, whatever it is, on the support, and one that has some sort of durability. So, I view it more as, if we want to do sort of high throughput science, a recipe to make a catalyst, and the recipes, and the ways in which you can use those recipes are quite limited in terms of the types of synthesis techniques that are really amenable to scale. But would be good if we could close the loop between: "Okay here's the recipe, here's the function of that catalyst, and then how do I feedback and tweak the recipe". So, we sort of iterate based on function and recipe because those are the things. Once I know a recipe that gives me a good function, you could spend a career then figuring out what exactly is the active structure, or ensemble of structures more likely, that are responsible for the catalysis. But I think trying to figure out how to close the throughput loop that way, would have a better chance of having a more near-term impact.

Sossina Haile:

In terms of understanding of the electrochemical pathways, I'm sure that you would agree this is heterogeneous catalysis, so the surfaces are essentiel. And then the question is always: is the reaction proceeding via some defect sites, as came up yesterday? Is it some dislocation that's terminating at the surface? Is it some grain boundary that's terminating at the surface? Is it an impurity that was there? Things that don't get included in the computation because we didn't realize that they were important. So that's the challenge then in terms of exploiting the computational advances to be able to make real catalysts that work.

Henry Snaith:

For solar materials, the big challenge is really understanding the defect chemistry in the materials and how the materials change chemically over time when ageing and under use. At the moment, the sort of typical approaches to understanding defect chemistry is to make some changes to the material. Whether it's how you crystallize it, or some additives that you add in, and then combine that with some first principle calculations to estimate what happens when you make these changes. And also postulating, trying to calculate the energy levels for certain defects that you can imagine could occur, and then trying to fit some measured parameter to those calculated defect energies, or might be a diffusion coefficient of an interstitial, or something like that. The problem is that you're applying the theory to fit a sort of not great data set, if I'm being honest. Very often you can always fit the data with the theory, and it doesn't prove that that's happening. It's just something that's self-consistent until some results show up that are not consistent with that, then you maybe have to rethink it. Or often the data sets are ignored, or it's a dead-end and it doesn't make sense and it's put down. But understanding defect chemistry, I think, is really central. There's more advanced electron microscopy already trying to image static defects, it doesn't tell you anything about the electronic properties. But then again, you couple those defects that you now can see are present from, say, very high-resolution TEM where you haven't damaged your sample, cryo-TEM for instance, but then they need to be coupled with the model. We've got to a level where we can't just look, we don't have a microscope to probe all the properties at the atomic scale of the systems. So, we have to couple it with theory and check whether that theory is just giving us an answer that is consistent, at what point can we advance that theory and those calculations to actually definitively give us much higher level of confidence on the actual property of the defect or the system.

Clare Grey:

Going back to the point about Operando techniques. I think one of the challenges is that there are multiple techniques and in the battery space we're doing quite a good job at it. The difficulty is there are so many different competing reactions, and working out that you've really got that reaction that's going to impact the degradation of a battery over multiple years. And because you've got such a highly complex, interrelated process, understanding how the different components react and interact with each other is challenging. So, I think overused complexity and understanding, that is sort of one of the big issues. And then the other one is, as you move into next-generation technologies where there is lithium-iron or lithium-sulfur, you start getting into chemistries where there are 3 phase boundaries whether it's with the oxygen, a solid, and a liquid, or it might be a boundary with a redox mediator in the liquid that's doing some of the redox chemistry, but then you've also got tunneling from an electrode through a solid. So, it's not a simple one electron transport process, it's a multiple rearrangement of solvents and these things even are important in normal lithium-ion batteries because they control degradation, they control the formation of the solid electrolyte interphase (SEI), they control the degradation of the oxides. So, if the oxides are oxidized too high, oxygen is released, you get reconstructions. So, it's sort of understanding these complex surface reconstructions that are often involving multiple phases and multiple competing reactions.

Andrew Tuberfield:
To an outsider, the materials science that underlies batteries and PV is a slightly disconcerting mixture of very rational design and completely random discovery. You know, where did perovskites come from? So, the question is: is the key to transatlantic electric powered flight or a 45% solar cell something you don't know you're looking for yet, a completely different material system? And if it is, how do you find it?

Henry Snaith:
I wish we knew the answer. So, the 45% efficiency solar cell is possibly feasible with a 4-junction perovskite solar cell, and so that may be the CRISPR for solar, but of course it might be another material that is discovered. I think in this community, collectively, there are lots of different approaches to developing materials, discovering materials. This is something that has been mentioned in the session this morning or yesterday, I can't remember, "the lab of the future". I think we need to think about how we do chemistry and discovery differently. I run a physics lab and I very much run it like an old school chemistry lab, where one student spends about most of the two-year period basically trying to find one new material. And it's like, how do we speed that up? Because there is the whole periodic table and more, because we can make organic compounds as well to discover new materials. And I really think that we should be spending time and effort on working out 'How do we do this whole process quicker?' and get the right answers to screen through materials, combining it with computation. Although I think that a brute force experimental approach will probably win, but that's my bias. I think that will be important.

Clare Grey:

Can I mention that, in the battery space, I think there's an area where you know the voltage that typical elements operate at, and you've got materials' genome projects that have gone through and computed everything, and we sort of know what the fundamental limits are. And so, you can put lithium plus oxygen as the ultimate energy density. I think what the field needs is to increase efficiency and that efficiency might be the overpotential of a reaction but also the degradation. I think we can have lots of discovery projects for new materials, but it's not going to be game changing, it's really guessing the types of compounds we know to operate better and to look and think about stability and efficiency. But maybe I'm a pessimist and others will disagree.

Daniel Nocera:

The high throughput screening works for composition and that's good for some pieces, but for the electrocatalysis in chemistry it's kinetics and that's not in high screening. In our field, this field of electrocatalysis, there's been huge projects and huge amounts of money put into screening materials. They get new compositions and not one really great catalyst has come out of it. Because you have to then overlay on the composition kinetics and reaction chemistry. So, you got to use that sort of screening at great place and then not fool yourself when you have reactivity that you have to deal with. We don't have really great screens for reactivity actually for electrocatalysis.

Bert Weckhuysen:

I have two comments and two questions. The first comment is that Dan, I really liked your introduction. The second is that the operando methods which we have been discussing I think that there would be a benefit that the fields of solid catalysts, fuel cells, batteries are much more working together to exchange their methodologies because we can really learn from each other there. I think there's a lot of benefit there, especially when it comes to synchrotron type of experimentation, where a lot of the methodologies can be just transferred from one material to another. My two questions are: Refinery of the future, I call it, what can we keep and what do we have to replace? If you take the current refinery, we have a lot of steel in the ground and certain methodologies we can maybe keep but what can we then replace and to what cost? Because, we have a lot of green electrons, but do we then use our green electrons to the best of our capabilities? That's my first question maybe for Dan or Matthew. The second is, when we talk about going for new feedstocks then we have a lot of impurities. If it's water electrolysis, is it salt water? Then we have chlorine, so stability will become essential, stability of catalysts of all these materials. The same for cement and metallurgy where we take point sources and CO_2 comes from there, how do we then deal with these impurities? So, all the selectivity and activity have a lot of attention. I think stability at the end maybe will prevail to make it flexible, dynamic towards changing feedstock. What we currently have with the oil refineries, they are robust, therefore they can handle all the feedstocks. Those are my two questions.

Matthew Kanan:

With respect to your first question, I think you start by replacing or at least ramping down steam reforming as you bring online hydrogen. That's what Shell's doing as far as I understand it. You first take the hydrogen use for just hydrogenating in the late stages of the refining process. But then eventually, you're going to want to replace the whole front-end, I think, with a CO_2 to CO infrastructure be that reverse water-gas shift, which I talked about, or electrochemical, or otherwise. But then the downstream stuff actually I think it's better, it's cleaner as opposed to cracking something you pull out of the ground and then putting the pieces back together. There's typically a lot more impurities in that. The liquid fuel you get of the Fischer-Tropsch plants, at least the one in Qatar, is some of the cleanest diesel in the world, as someone mentioned the other day. So, for water electrolysis the analysis I've seen, I'm not an expert in this — maybe Karthish or someone else knows better. But as far as I've seen, the purification cost to get it to the purity grade you would need, certainly for alkaline and possible also for PEM, is a very minor component of the overall cost to hydrogen production. If you're using conventional reverse osmosis, if you're even going to start with seawater, if you're going to start with another source that's pure, it's even easier than that. So, you need pure water for electrolysers to last a long time, certainly the PEM ones less so for the alkaline. But that seems to be a quite manageable prospect. The bigger challenge I think is the impurity on the CO_2, the sulfur ones are just going to get scrubbed if there's any transition metal catalyst anywhere in that process, they're going to get scrubbed. Biology gives you options there actually, some of the syngas conversion biotech. The huge benefit is that it can tolerate just about any impurity, the bugs just kind of let them go. That's been a big advantage in the initial commercialization of some of that technology. For any sort of electrochemical CO_2 conversion process, scrubbing the impurities out of the CO_2 is a big issue.

Daniel Nocera:
Biology gives a big advantage in that the bug replaces itself, so there's actually very tiny energy penalty for reproduction and you can hit a steady state and then do things so that's a really big advantage of biological systems.

Martyn Poliakoff:
Several minor comments I want to make. First of all, to respond to James about fragrances, to point out that most fragrances or natural fragrances are made from CO_2 by plants. And that fragrances are an enormously good example of where we could reduce the amount of chemicals that are needed. If we could make a stronger smelling chemical which smells more or less the same, we can reduce the amount of chemicals that people need. Because they don't put on perfume or aftershave because they want to buy a particular concentration of fragrance and ethanol, they will use whatever smells nice. But the other thing which we haven't really talked about in terms of all these biological processes, is the need for phosphorus. And phosphorus is an element which is getting into lower quantities, it is also politically quite sensitive, because a lot of phosphate fertilizers come from Russia and Ukraine, therefore I think that we need to think quite carefully about recycling phosphorus. You may remember that, in the 1930 novel Brave New World, everybody who died was sent to the phosphorus reclamation plant to recover the phosphorus before they were cremated. But I think that this is likely to become a limiting element in the future, it's not just platinum and palladium that we need to worry about but phosphorus might well become important as well.

Daniel Nocera:

Martyn, you may have missed the fragrance part of James' talk, but I saw putrescine on one of your slides, which is a very powerful odorant. Just on the phosphorus comment, so first those bugs. I just ran over it; they accumulate phosphorous so we grow them off of wastewater and they accumulate as polyphosphate, but Kit Cummins at MIT has written some variety of articles on the need for chemistry to recycle phosphorus, it's going to become a huge problem.

Peter Palese:

Hoping that our planet does not induce an immune response against us, my question is: is there a red line in terms of percentage of CO_2 which we have to reach so that our life as we know it is sustainable? Is there a number which we should try to reach so that we don't have our planet develop an immune response against us?

Matthew Kanan:

I don't think you can quote a hard number and say this is the limit. What is done is, there are risk assessments at given CO_2 levels, and an assessment of 'What is the temperature rise going to be associated with that CO_2 level', and then what are the sort of consequences with respect to various tipping points. The only thing you can say with certainty is that risk goes up and up and up the more the concentration goes up. So, it just matters what the certainty in the risk analysis is, which varies widely depending on the tipping point you're talking about, and then your risk tolerance relative to the cost or difficulty of trying to do some sort of forced accelerated decarbonization, which tends to impact poor people first and then rich people later. So, I think there's not a hard number.

Peter Palese:

So, you don't want to give a number?

Matthew Kanan:

No, there are excellent discussions of this in the literature that have a whole range of numbers. Again, there is a huge variety: there's risk level at this number, there's elevated risk level at this number, etc. It all just depends then on how much you trust those risk assessments and then what your risk tolerance is.

Daniel Nocera:

I've always said, if you made CO_2 purple, everybody would be scared, and we wouldn't be having this discussion.

Sossina Haile:

Just a follow-up on a couple of comments that came through. My view on carbon is that the more we leave in the ground the better, because even if we think about CO_2 reduction then we're adding to that cycle. It's always expensive to put CO_2 back into the ground. So, I just think we have to consider the overall lifecycle of the carbon that we take out of the ground. Just a very small comment on impurity tolerance and the question of how our catalysts sustain impurities. When we go to those moderate temperatures, impurity tolerance is just not an issue. So, if you noticed, when we were doing ammonia decomposition, we're not decomposing all the ammonia in the reforming catalyst prior to the hydrogen oxidation catalyst. So, our platinum at 250°C can handle lots and lots of ammonia and we're thinking about 'could you make a direct ammonia fuel cell?'. But it's less useful because as I said you don't really want people pumping up ammonia into their systems. But the point is that when you have a little bit higher temperature you have all of those poisons just desorb off of those catalysts. So, we've run our fuel cells even with 5% CO and it's really no problem because, again, at high temperatures those pollutants just desorb from the platinum.

Karen Goldberg:

I appreciate that we should keep as much carbon in the ground, but if we go back to Matthew's plot where he showed what the projections were out to 2050, the petroleum and the natural gas are going up. Renewables is a steep swap, but that's accounting for the increase in energy, we're not changing the amount that we're currently taking out of the ground.

Sossina Haile:

I appreciate your comment, Karen. I was just saying to Clare that I have those plots from 10–20 years ago on my computer and you can see how bad the projections were. We thought coal was going to go down 10 years ago and then it went up. Wait a minute, we did not expect renewables to be doing anywhere as well as they're doing now. So, even if you look at the projections from 2 years ago, they are very pessimistic for renewables as compared to where they are today.

Karen Goldberg:
I hope you're right.

Sossina Haile:
That was just a statement of those data points, right? It doesn't mean that the new projections are equally bad, but it's just simply to say that if you look at the projections where they were 2 years ago, they are very pessimistic compared to the projections today.

Henry Snaith:
Just on the note of projections. Most scenarios have solar at about 600 gigawatts production capacity in 2040, whereas the industry plans to be at 600 gigawatts by 2030. Quite arguably with competition they may be higher than that. So, today's projections of solar and wind deployment are just so far off. And if you look at all the 2050 scenarios, they only have renewables contributing to 50% of the power generation. So, it's certainly going to make a dent in it. That has to not happen. We have to have close to 100% certainly by 2050 and arguably by 2040 or before. So, they're wrong basically.

Sossina Haile:
Well, they're not wrong. We have to get back to this issue of policy and governments getting involved, in order to bring those numbers down. Because as long as those are cheaper, those are going to continue to be used.

Clare Grey:
But they're not cheaper now, solar energy is now cheaper.

Henry Snaith:
This is a point, especially with renewables on the generation capacity. Even though there's a very clear trajectory, none of the models put in that sort of

linear progression in the trajectory. They all sort of plateau out. And again, if you plot the projections over time historically, they all plateau out and they just plateau out a little bit later. Perception is that cost will go down for about 5 more years and that will be it and that has just not been reality. Industry knows that's not reality. Economics I think will win it. Doesn't mean we don't need to do anything, doesn't mean we don't need to worry about policy, because things can slow that down or there's incentives that can speed it up. But just simply the economics will drive towards renewable generation. The real requirement is that the storage comes online so we don't have issues from too much intermittent power generation.

Matthew Kanan:

The point about the liquid projections, and I agree there's uncertainty in those. The most recent ones take into account massive electrification of vehicles, and I think the question you could raise is that projection, is that growth going to be flat or maybe even negative? That would depend on some sort of major change scaling really fast in the fuel for commercial transportation, aviation, shipping, trucking. There's the beginning of tech demos in that space, but if you think of what's required to scale that in the next 20 years to the scale of transportation to sustain commerce, putting on top of that the expected growth in the economy and population growth, it's very hard to imagine that demand coming down. And on the material side, on the plastic side, everyone loves to hate plastic but they're pervasive and they're essential to every single industry. Things like polyester fiber that's half the polyester market. You think clothing is going to decline in the next 20–30 years? I don't think so. So, I think carbon is here to stay for certain sectors.

Clare Grey:

Just to quickly follow up on policy. We haven't really talked about the regulation, which of course is very coupled to policy. The grid is the thing that needs to be regulated and changed. At the moment, there are very few incentives to sell back to the grid in most countries, and that has to dramatically change in order to use, for example, electric vehicles and change the balance. And then we incentivize: for example, in the UK, there's a power line being built from Morocco to the UK for solar and they're also providing solar in Morocco to the local community. So there just has to be a rethink of the grid and that probably is a policy.

Daniel Nocera:
I just want to make a note that we have 10 minutes before this session ends.

Donald Hilvert:
I'd like to change the question a little bit. Specifically, about the biotechnological approaches that were presented. Questions I guess to James or to Dan: where do you see the challenges in scaling up these processes and to what extend can one scale them up?

James Liao:
Biological scaling up has a long history dating back to probably the 1960s, so all these issues have been pretty well spit out. Now it's the chemistry part that's not there yet. So, right now we're focusing more on the chemistry part; getting better enzymes, better bugs... Once that's done, then plug it into the scaling up technologies that dated back, I don't know how many years.

Donald Hilvert:
So, you just see banks of fermenters to grow the microorganisms?

James Liao:
Yes, depending on what kind of micro-organisms you are talking about, in the gas or liquid phase, there's a whole bunch of technologies people developed in the 1980's and even earlier. So, the traditional problems will still be there, we still need to worry about that, but I don't think that's the problem that will be the showstopper. I think that's a difficulty that we will have to overcome but it will not be a showstopper.

Gerald Joyce:
I just want to press on this one and follow-up on Dan's questions. Gigaton per atom CO_2 removal from the atmosphere: how many liters of fermenters are we talking about? And then when the carbon is brought into E. coli or yeast or whatever it is, how recalcitrant is that carbon?

James Liao:
We have to rely on not only one approach, we have to rely on multiple approaches. It's not possible to solve the gigaton problem using a single approach, it's just not there. So far, even the biggest product that we have is about a gigaton, which is probably the building material, cement or steel, nothing else comes even close to that scale. We will have to develop many different approaches. If we say that this is not possible, you can destroy any approach just by that statement. For example, eventually most of the gigaton problem will be solved by using green energy. In a world where 75% of the CO_2 emission comes out of the energy sector, that's the world average. In developed countries, in Japan its 93%, in Taiwan its 90%. Imagine that the energy sector is replaced by PV or anything that's greener, at least 60–70% will be gone. Then we deal with the parts that are not possible to replace without carbon, for example materials. I don't think anyone could imagine that we live in a world without carbon. Carbon is a building block for just about everything, so we still need to use that. But in a process using that, some of the material stays there, it doesn't cause problems. It's the energy we put in to make that material that causes a problem. So ultimately, it's an energy problem, so if we deal with the energy problem, the gigaton problem will gradually reduce and become smaller and smaller and smaller.

Gerald Joyce:
Because coming back to what Dave said. Can we get there just by the energy problem as you call it, or do we need to scrub? And to scrub we've got to do maybe 2 gigatons a year?

James Liao:
Yes, I think that's called the legacy CO_2 problem. So, quite a few people argue that even if we stop emitting CO_2 today, we'll still have to take care of the legacy CO_2 in the air. Otherwise, it stays there according to calculation for a thousand years and will continue to cause problems.

Andreas Züttel:
Now I'm surprised. You showed on one of your slides that about half of the CO_2 is naturally absorbed again. So why do you now claim that it's going to stay for a thousand years? Now, there are 5 gigatons of carbon in the

form of CO_2 absorbed every year by the planet, so it will go down if we don't emit. In 200 years, it will be back down to 280 ppm if we don't emit.

James Liao:

It's a little bit more complicated than that. It's a balance and even people argue that the anthropogenic CO_2 may or may not contribute to that much, but of course then we enter into another game of debate which I don't think we should talk about right now. Anyway, people are arguing, which I don't know if it's true or not, that the CO_2 half-life in the air by some method that is longer than what we can sustain. So, people are arguing that we will have to take care of that. So, what that net zero means is that we will still have to use some carbon emissions processes to sustain the economy. At the same time, we have to absorb that much out of the air to achieve net zero.

Daniel Nocera:

I think Matthew will have a quick follow-up, then we're going to end with a question from a panel member.

Matthew Kanan:

Just a quick follow-up on this. First of all, if you look at the IPCC scenarios for 2°C, 95% of them assumes 600 gigatons of carbon removal by 2100, which is about 8 gigatons per year starting now and there's essentially no tech for that. Ok yes, the planet absorbs CO_2 but at least half of it, probably even more, goes to the ocean which is a problem in the short term because of acidification, short-term being time-scale of human civilization. Geology regulates CO_2 over longer time scales by silicate weathering, so the only viable sink on a 100+ gigaton, or whatever large scale you want, is to take the natural base equivalents that the earth has and carbonate those. There's a natural carbon cycle on million-year timescales where those go down and they eventually get blow out again through a volcano. But mineralization, in my opinion, is the only resource for that scale removal.

Sossina Haile:

Just one small comment. We actually had CO_2 emissions go down during the pandemic, but atmospheric CO_2 continued to go up.

Daniel Nocera:
Okay, we're going to end this session. Henry has a question for all of us.

Henry Snaith:
You've put some pressure on me, Dan. It's a rather specific question. Well, it's actually the problem of aviation fuel because with all the things we've discussed, aviation fuel is a certain fraction of our power demand and usage, and it seems very difficult to think about how that's going to be 'greened'. So, I just wanted the comments, not only of those on the panel, but anyone from the audience as well about how that's going to be done if we can't mechanically capture the carbon to then use that to turn it into a fuel. If the sort of main carbon capture processes are only really going to work well at the source of generation ultimately, we need to stop burning the fossil fuels so we shouldn't have those high concentration sources of carbon. So, how are we going to create aviation fuels in the future, in a sustainable manner?

Clare Grey:
I can take this very quickly from a battery perspective. Well, despite Matthew's comment there aren't going to be any battery powered planes. There will be in the microlite industry, so as long as you're an executive you can still get to fly planes. But let's democratize the conversation.

James Liao:
I want to get back to the gigaton problem and biological approaches. Although people are not thinking about this very much, but I think eventually someone has to think about the geo-engineering approaches. What I talked about today in the biological approaches is not just inside the bioreactor. It could be in the oceans. Once you are talking about the ocean, then the CO_2 fixation could be very large. I think Gates' Foundation is doing some research in that area. It's very complicated because you disturb the ecological system dramatically so that needs to be studied very carefully and some people are doing that.

Matthew Kanan:
I can answer Henry's question specifically, at least the way I see it. You're going to start with the large, pure CO_2 emission streams. In the States,

that's coming off of the corn ethanol fermenters. Its 20+ million tons a year, 98+% CO_2 coming off. There are other chemical processes that have greater than 50% CO_2 streams coming off. So, you're going to go after the low hanging fruit as you build up processes. As I said, I like reverse water-gas shift Fischer-Tropsch. You could do a methanol route, you could do an ethanol route. All those are going to compete, the best process is going to win. And then, as and if that scales, then you start incentivizing larger scale capture. In the capture technology, there's a lot of noise in that space but some of it is getting better. Then you're going to start capturing from coal flue with 10%, then gas flue with 4%, and then hopefully work your way down. I see it evolving, starting first with relatively accessible emission sources that are high purity.

Ben Feringa:
As far as I know, earlier this year Shell made 5 tons of kerosine via the Fischer-Tropsch from their CO_2 concentrate at the refinery, of course this is from a refinery source. They made this and put it in a plane that flew to Barcelona back and forth. So, as far as I know, that is at least the state of affairs with a company like Shell.

Daniel Nocera:
Thank you for the questions and have a nice night.

Ben Feringa:
Let's thank Dan and his team once again for the great presentations and a wonderful discussion.

Thioester recycling - the key to unlocking the door to amide synthesis in water
Acyl-Coenzyme A and related thioesters are key intermediates in enzyme-catalysed acylations which enable amide formation and beyond in water. Generic formation and recycling of non-native thioesters is presented opening the door to biocatalytic amide synthesis and bioorthogonal peptide labelling.

Figure by M. Hayes and C. Schnepel (used with permission from AstraZeneca).

Session 5

Directed Protein Evolution for Green Chemistry

https://doi.org/10.1142/9789811282324_0032

Application of Computational Tools for the Design of Enzyme Cascades

Sabine L. Flitsch

School of Chemistry and MIB, The University of Manchester, Manchester M1 7DN, UK

Our view on the present state of research on biocatalysis

The combination of sequential biocatalytic reactions in non-natural synthetic cascades is a rapidly developing field and has led to the generation of complex valuable chemicals from simple precursors [1–5]. These enzyme cascades can often be telescoped into one single reactor, either using cell-free enzymes, whole cells, or a mixture thereof, because many enzymatic reactions use similar reaction conditions. With the toolbox of natural and engineered biocatalysts increasing dramatically through metagenomics data and protein engineering, so do the options for biocatalytic retrosynthesis of a target molecule, leading to new routes employing enzymatic transformations [6]. Until recently, the retrosynthetic analysis and the design of enzyme cascades were performed manually and limited highly skilled and trained specialists who have intricate knowledge of the field of biocatalysis. As the field expands dramatically in its coverage of chemical reactions and processes and becomes increasingly data-rich, computational tools are emerging and helping capture information from the literature and design enzyme cascades. These computational tools have become useful for the expert but are also designed to provide for a wider chemical community to find biocatalytic synthetic strategies for developing more efficient and green synthetic processes.

Our recent research contributions to biocatalysis

The planning of a synthetic strategy starts with considering the broad types of reactions (reaction rules) that could be used in a stepwise fashion towards the target from accessible starting materials. In organic chemistry, the strategy of "retrosynthesis", i.e., planning backwards from the target, has proven very useful. Based on the chemical retrosynthesis concept, a collection of tools for automated biocatalytic cascade design ("RetroBioCat"; https://retrobiocat.com) was developed [7] (Fig. 1). A database of reaction rules was established, compiled of chemical reactions that have been used for biocatalytic transformations. In the first instance, RetroBioCat applies these reaction rules iteratively for retrosynthesis towards a suitable starting material. For example, for the piperidine target in Fig. 1, RetroBioCat would suggest three strategies involving either three steps (carboxylic acid reduction by CAR, transamination by a TA, and imine reduction by IRED) or two steps (alcohol oxidation by AlOx and reductive amination by RedAM or amine oxidation AmOx followed by imine reduction by IRED) from the different respective starting materials.

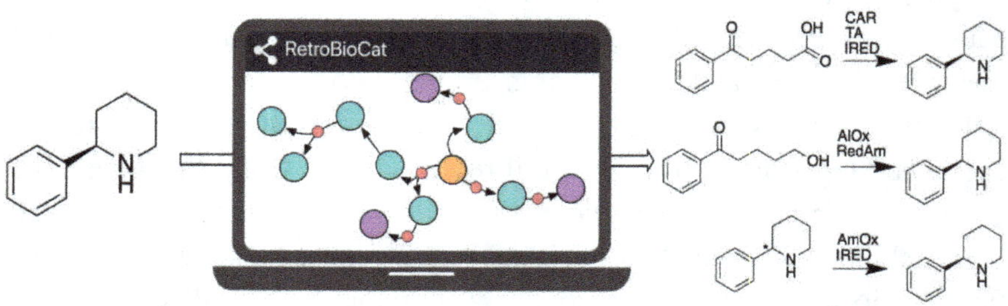

Fig. 1. RetroBioCat provides a collection of tools for automated biocatalytic cascade design. For example, three pathways towards the chiral piperidine target are suggested.

A number of pathways suggested by RetroBioCat have already been implemented successfully. Two recent examples from our laboratory are the chiral amino polyols **1** and **2** in Fig. 2, both highly polar targets with dense functionality and stereochemistry that are challenging to access using synthetic chemistry. Target **1** and analogues were prepared from biorenewable and easily accessible amino polyol starting materials through an oxidation–cyclization–reduction sequence that could all be performed in one pot [8]. Key to the success is finding suitable specific enzymes through protein engineering, directed evolution, or metagenomics database analysis. Here, the first step was catalysed by a mutant of galactose oxidase (F2) that had been engineered through directed

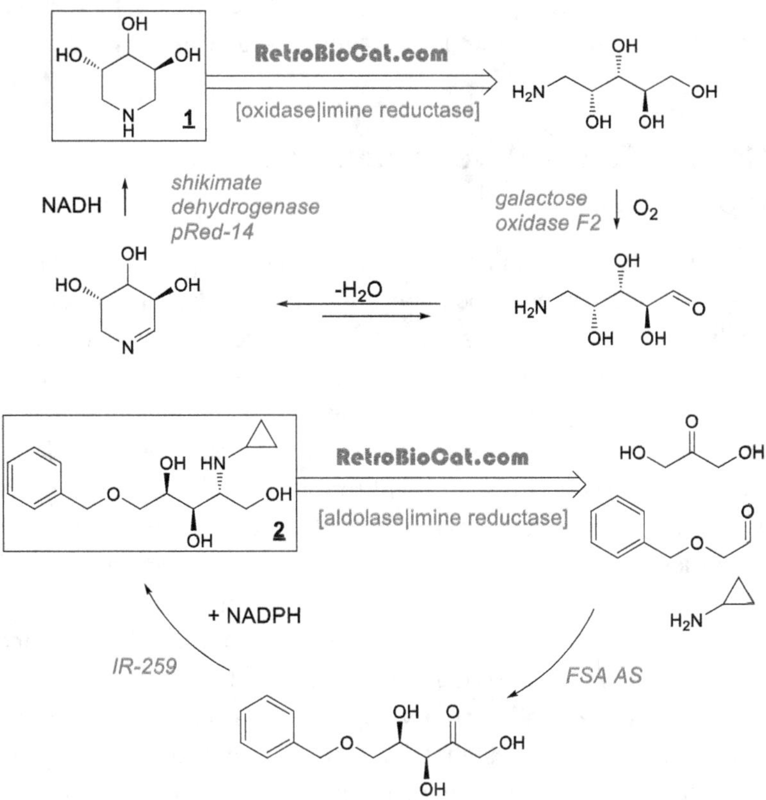

Fig. 2. Two recent examples of enzyme cascades designed using RetroBioCat.

evolution to accept a broad range of alcohol substrates [9, 10]. The second enzyme (pRed14) was identified from metagenomic database analysis. Interestingly, pRed14 had been annotated in protein databases as an alcohol dehydrogenase of the shikimic acid pathway, demonstrating how non-natural promiscuous activity can be used successfully in biocatalysis.

The sequence leading to target amino diol **2** (Fig. 2) is an example of carbon–carbon bond-forming reactions (such as aldol reactions) alongside functional group interconversions in biocatalytic cascades. Overall, the reaction represents a highly stereoselective three-component synthesis from easily available prochiral starting materials in two steps [11] using biocatalysis mediated by aldolase FSA AS followed by imine reductase IR-259. The reversibility of the first aldol reaction and cross-reactivity of the carbonyl substrates prohibited a one-pot reaction. Such issues of cross-reactivity can be overcome by using flow biocatalysis [12].

Outlook to future developments of research on biocatalysis

Many chemicals are synthesized through multi-step processes using a diversity of reactions and biocatalysis will need to be equally diverse to make a broader impact as a green and sustainable alternative to chemical processes. There have been numerous examples of successful enzyme cascades, but the challenge remains to increase the “reaction rules”, the classes of transformations that enzymes can catalyse. For example, more biocatalysts are needed for C–C bond formations, C–H activations, halogenations, and even catalytic amide bond formation, and finding these new activities and incorporating them into enzyme cascade are very active and exciting areas of research. There are several approaches that are being pursued at the moment, including studying the biosynthesis of secondary metabolites to find new enzymatic reactions, looking for promiscuity to find non-natural enzymatic activity, protein engineering, directed evolution, and *de novo* protein design. Given that most biocatalysts are proteins with common design, production methods, and reaction conditions, biocatalysis lends itself to the use of computational and automated tools that are increasingly able to deal with the numbers and complexity of protein sequences and structures. Protein structure prediction has been a bottleneck in biology for a long time and has recently made a large step forward through machine learning, although the prediction of function at a precise molecular level remains challenging. The scientist interested in finding new enzyme activity has very impressive computational and experimental toolkits available that can generate and deal with very large datasets. However, what makes the subject particularly interesting intellectually is the need for creativity in developing original mechanistic hypotheses that can then be explored by using these toolkits.

Acknowledgements

We kindly acknowledge financial support from the European Research Council 788231-ProgrES-ERC-2017-ADG.

References

[1] E. L. Bell, W. Finnigan, S. P. France, A. P. Green, M. A. Hayes *et al.*, *Nat. Rev. Methods Primers* **1**, 46 (2021).

[2] J. H. Schrittweiser, S. Velikogne, M. Hall, and W. Kroutil, *Chem. Rev.* **118**, 270 (2018).

[3] A. I. Benitez-Mateos, D. Roura Padrosa, and F. Paradisi, *Nat. Chem.* **14**, 489 (2022).

[4] S. P. France, L. J. Hepworth, N. J. Turner, and S. L. Flitsch, *ACS Catal.* **7**, 710 (2017).

[5] E. Romero, B. S. Jones, B. N. Hogg, A. R. Casamajo, M. A. Hayes *et al.*, *Angew. Chem. Int. Ed.* **60**, 16824 (2021).

[6] F. H. Arnold, *Angew. Chem. Int. Ed.* **58**, 14420 (2019).

[7] W. Finnigan, L. P. Hepworth, S. L. Flitsch, and N. J. Turner, *Nat. Catal.* **4**, 98 (2021).

[8] C. Swanson, G. J. Ford, A. P. Mattey, L. Gourbeyre *et al.*, *ChemRxiv.* (2022).

[9] L. Sun, I. P. Petrounia, M. Yagasaki, G. Bandara, and F. H. Arnold, *Protein Eng.* **14**, 699 (2001).

[10] J. B. Rannes, A. Ioannou, S. C. Willies, G. Grogan, and C. Behrens *et al.*, *J. Am. Chem. Soc.* **133**, 8436 (2011).

[11] G. J. Ford, C. R. Baldwin, R. T. Bradshaw Allen, J. R. Marshall, A. P. Mattey, *et al. J. Am. Chem. Soc. AU* (2022).
[12] A. P. Mattey, G. J. Ford, J. Citoler, C. Baldwin, J. R. Marshall *et al.*, *Angew. Chem. Int. Ed.* **60**, 18660 (2021).

https://doi.org/10.1142/9789811282324_0033

De Novo Enzyme Design

Donald Hilvert

Laboratory of Organic Chemistry, ETH Zürich,
CH-8093 Zürich, Switzerland

In his 1902 Nobel lecture, the great German chemist Emil Fischer foresaw an era in which "chemistry will not only make extensive use of the natural enzymes as catalytic agents, but will also prepare synthetic ferments for its own purposes" [1]. Today, more than a century later, enzymes have become readily available for the production of everything from pharmaceuticals and agrochemicals to biofuels. Indeed, biocatalysis is increasingly viewed as an enabling technology for a greener and more efficient chemical industry [2, 3].

The advent of powerful engineering tools to tailor the properties of enzymes from nature for new reactions of chemical interest has enabled their broad application [4]. Thanks to directed evolution, for example, altering the substrate and stereochemical preferences of natural enzymes is almost routine [5–7]. Nevertheless, the success of such endeavours generally requires some starting activity. If none is detectable, engineering can often supply it. In favourable cases, natural proteins can be modified rationally to access interesting abiological reactivity. Sometimes a few mutations suffice to alter functions dramatically [8]. Alternatively, noncanonical amino acids, metal ions, or other cofactors can be exploited as sources of novel chemistry [9–11].

Creating enzymes from scratch is far more challenging than tailoring the properties of an existing catalyst. In one approach, the mammalian immune system has been harnessed to create antibodies possessing catalytic activity [12]. The properties of these catalysts are programmed by the structure of a stable transition state analogue that is used to elicit an immune response. Although more than 100 different chemical transformations have been successfully catalysed in this way, including normally disfavoured processes and reactions lacking biological counterparts, even the best antibody catalysts are orders of magnitude less efficient than their natural counterparts [13].

A more generally productive pathway to *de novo* protein catalysts that exhibit true enzyme-like rates and selectivities combines state-of-the-art computational methods and high-throughput evolutionary optimization [14–16]. Conceptually, computational enzyme design is like catalytic antibody technology, but rather than utilizing an imperfect transition-state analogue to provide chemical instruction, the rate-limiting transition state of the target reaction, including potentially stabilizing functional groups, is modelled computationally and docked *in silico* into structurally characterized protein scaffolds. After optimization of active site packing, the designs are ranked according to their calculated energies, and the top scorers are tested experimentally.

In collaboration with several computational groups, we have used the latter approach to repurpose natural protein scaffolds for the catalysis of mechanistically distinct chemical transformations. These include simple proton transfer [17–19], a stereoselective Diels–Alder cycloaddition [20, 21], and a multi-step aldol reaction [22–24] (Fig. 1). Although the starting designs typically exhibit only modest efficiencies, these have proven ideal starting points for laboratory evolution, and iterative rounds of mutagenesis and screening have yielded artificial enzymes that match the speed and stereoselectivity of their natural counterparts. Optimization frequently entails dramatic active site remodelling to create more complex arrays of functional groups and/or minimize unproductive states by modulating protein conformation landscapes. The best resulting catalysts not only achieve billionfold rate accelerations but, on a preparative scale, also produce their target products as single stereoisomers.

Natural proteins offer a wide range of architectures for enzyme engineering. Because their complex sequence–structure relationships reflect

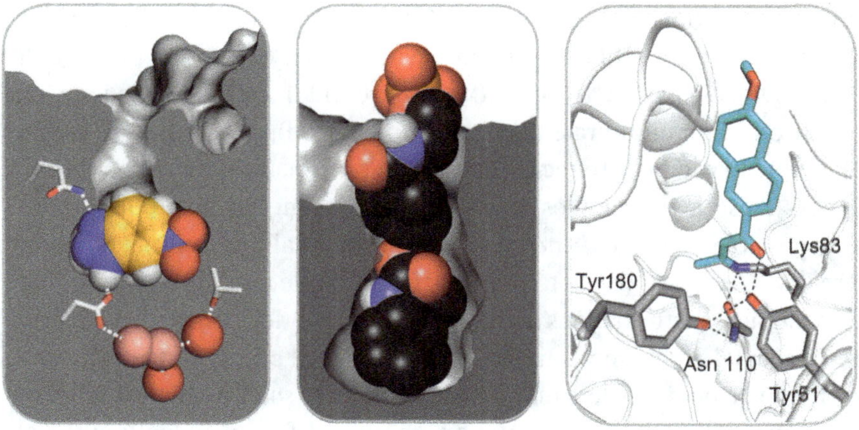

(a) (b) (c)

Fig. 1. *De novo* enzymes generated by directed evolution of modestly active computational designs. (a) A Kemp eliminase effectively utilizes acid–base chemistry in a shape complementary pocket to accelerate an elementary proton transfer 6×10^8-fold [18]. (b) A Diels–Alderase produces a single product diastereomer by employing hydrogen bond donors and acceptors to preorganize the diene and dienophile substrates and stabilize the cycloaddition transition state electronically [21]. (c) The $>10^9$ rate enhancement achieved by an aldolase is ascribed to a catalytic tetrad that arose residue by residue during evolutionary optimization [24].

unique evolutionary histories, they may respond to sequence modification in unexpected ways. *De novo*-designed proteins, which are often hyperstable and possess well-understood sequence–structure relationships, represent potentially more robust starting points for enzyme design. Artificial retro-aldolases have been produced by computationally customizing the backbone and sequence of a *de novo* eight-stranded β-barrel protein, illustrating the potential of this approach [25]. Alternatively, the intrinsic reactivity of inorganic or organic co-factors can supply sufficient starting activity for subsequent evolutionary optimization [11]. For example, metal ions and metalloporphyrin cofactors have been introduced into designed α-helical bundles to produce protein catalysts for hydrolytic reactions, redox processes, carbene transfers, and other activities [26–28]. This approach was used to transform a computationally designed zinc-binding peptide into an enantio-specific metalloesterase having catalytic efficiency only two orders of magnitude below the diffusion limit [29]

(Fig. 2). Functional diversification of this scaffold by divergent evolution has also yielded efficient, stereoselective catalysts for a bimolecular hetero-Diels–Alder reaction [30], a retro-aldol cleavage, and an ene reduction of an unsaturated ketone (unpublished), attesting to the utility of metal ion catalysis for accessing diverse non-natural functions (Fig. 2).

As these few examples attest, enzyme design has come of age. It is now possible to create *de novo* enzymes fully rivalling their natural counterparts. The task today is to progress from simple model systems to more demanding transformations and complex, real-world challenges. Emerging experimental and computational innovations will be key to the success of such endeavours. Faster, more robust methods such as high-throughput screening and continuous evolution, improved forced fields, multi-state design, and machine learning have much to offer in this context. Their successful implementation promises to bring Emil Fischer's dream of being able to prepare enzymes on demand, for our own purposes, to full realization.

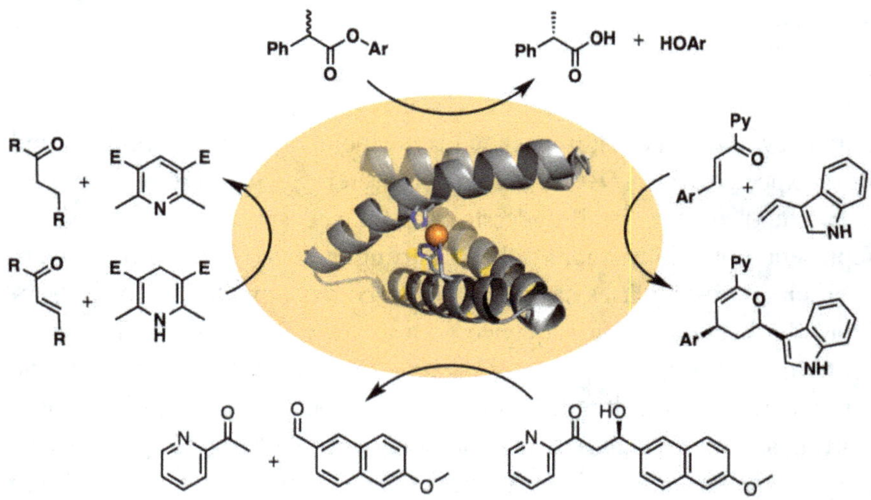

Fig. 2. Divergent evolution of a *de novo* zinc-binding helical bundle. Promiscuous esterase, hetero-Diels–Alderase, retro-aldolase, and ene reductase activities were optimized by iterative rounds of mutagenesis and screening to yield highly efficient and stereoselective metalloenzymes.

Acknowledgements

The author is indebted to the ETH Zurich and the Swiss National Science Foundation for generous support.

References

[1] Emil Fischer — Nobel Lecture. NobelPrize.org. Nobel Prize Outreach AB 2022. https://www.nobelprize.org/prizes/chemistry/1902/fischer/lecture/.
[2] E. L. Bell, W. Finnigan, S. P. France, A. P. Green, M. A. Hayes, L. J. Hepworth, S. L. Lovelock, H. Niikura, S. Osuna, E. Romero, K. S. Ryan, N. J. Turner, and S. L. Flitsch, *Nat. Rev. Methods Primers* **1**, 46 (2021).
[3] R. A. Sheldon and D. Brady, *ChemSusChem.* **15**, e202102628 (2022).
[4] U. T. Bornscheuer, G. W. Huisman, R. J. Kazlauskas, S. Lutz, J. C. Moore, and K. Robins, *Nature* **485**, 185 (2012).
[5] P. Romero and F. Arnold, *Nat. Rev. Mol. Cell Biol.* **10**, 866 (2009).
[6] M. T. Reetz, *Angew. Chem. Int. Ed.* **50**, 138 (2011).
[7] C. Zeymer and D. Hilvert, *Annu. Rev. Biochem.* **87**, 131 (2018).
[8] M. D. Toscano, K. J. Woycechowsky, and D. Hilvert, *Angew. Chem. Int. Ed.* **46**, 3212 (2007).
[9] F. Schwizer, Y. Okamoto, T. Heinisch, Y. Gu, M. M. Pellizzoni, V. Lebrun, R. Reuter, V. Köhler, J. C. Lewis, and T. R. Ward, *Chem. Rev.* **118**, 142 (2018).
[10] I. Drienovská and G. Roelfes, *Nat. Catal.* **3**, 193 (2020).
[11] K. Chen and F. H. Arnold, *Nat. Catal.* **3**, 203 (2020).
[12] R. A. Lerner, S. J. Benkovic, and P. G. Schultz, *Science* **252**, 659 (1991).
[13] D. Hilvert, *Annu. Rev. Biochem.* **69**, 751 (2000).
[14] G. Kiss, N. Çelebi-Ölçüm, R. Moretti, D. Baker, and K. N. Houk, *Angew. Chem. Int. Ed.* **52**, 5700 (2013).
[15] H. Kries, R. Blomberg, and D. Hilvert, *Curr. Opin. Chem. Biol.* **17**, 221 (2013).
[16] S. L. Lovelock, R. Crawshaw, S. Basler, C. Levy, D. Baker, D. Hilvert, and A. P. Green, *Nature* **606**, 49 (2022).
[17] H. K. Privett, G. Kiss, T. M. Lee, R. Blomberg, R. A. Chica, L. M. Thomas, D. Hilvert, K. N. Houk, and S. L. Mayo, *Proc. Natl. Acad. Sci. USA* **109**, 3790 (2012).
[18] R. Blomberg, H. Kries, D. M. Pinkas, P. R. E. Mittl, M. G. Grütter, H. K. Privett, S. L. Mayo, and D. Hilvert, *Nature* **503**, 418 (2013).

[19] R. Otten, R. A. P. Pádua, H. A. Bunzel, V. Nguyen, W. Pitsawong, M. Patterson, S. Sui, S. L. Perry, A. E. Cohen, D. Hilvert, and D. Kern, *Science* **370**, 1442 (2020).

[20] J. B. Siegel, A. Zanghellini, H. M. Lovick, G. Kiss, A. R. Lambert, J. L. St. Clair, J. L. Gallaher, D. Hilvert, M. H. Gelb, B. L. Stoddard, K. N. Houk, F. E. Michael, and D. Baker, *Science* **329**, 309 (2010).

[21] N. Preiswerk, T. Beck, J. D. Schulz, P. Milovníc, C. Mayer, J. B. Siegel, D. Baker, and D. Hilvert, *Proc. Natl. Acad. Sci. USA* **111**, 8013 (2014).

[22] L. Jiang, E. A. Althoff, F. R. Clemente, L. Doyle, D. Röthlisberger, A. Zanghellini, J. L. Gallaher, J. L. Betker, F. Tanaka, C. F. Barbas, D. Hilvert, K. N. Houk, B. L. Stoddard, and D. Baker, *Science* **319**, 1387 (2008).

[23] L. Giger, S. Caner, R. Obexer, P. Kast, D. Baker, N. Ban, and D. Hilvert, *Nat. Chem. Biol.* **9**, 494 (2013).

[24] R. Obexer, A. Godina, X. Garrabou, P. R. E. Mittl, D. Baker, A. D. Griffiths, and D. Hilvert, *Nat. Chem.* **9**, 50 (2017).

[25] Y. Kipnis, A. Ouald Chaib, A. A. Vorobieva, G. Cai, G. Reggiano, B. Basanta, E. Kumar, D. Hilvert, and D. Baker, *Protein Sci.* (in press).

[26] M. L. Zastrow, A. F. A. Peacock, J. A. Stuckey, and V. L. Pecoraro, *Nat. Chem.* **4**, 118 (2012).

[27] W. J. Song and F. A. Tezcan, *Science* **346**, 1525 (2014).

[28] A. Lombardi, F. Pirro, O. Maglio, M. Chino, and W. F. DeGrado, *Acc. Chem. Res.* **52**, 1148 (2019).

[29] S. Studer, D. A. Hansen, Z. L. Pianowski, P. R. E. Mittl, A. Debon, S. L. Guffy, B. S. Der, B. Kuhlman, and D. Hilvert, *Science* **362**, 1285 (2018).

[30] S. Basler, S. Studer, Y. Zou, T. Mori, Y. Ota, A. Camus, H. A. Bunzel, R. C. Helgeson, K. N. Houk, G. Jiménez-Osés, and D. Hilvert, *Nat. Chem.* **13**, 231 (2021).

https://doi.org/10.1142/9789811282324_0034

Computational Enzyme Design

Sílvia Osuna

Universitat de Girona, Institut de Química Computacional i Catàlisi (IQCC) I Departament de Química, carrer Maria Aurèlia Capmany 69, 17003 Girona, Spain
ICREA, Pg. Lluís Companys 23, 08010 Barcelona, Spain

My view of the present state of research on computational enzyme design

Enzymes are typically engineered for catalytic activity, enantioselectivity, thermodynamic stability, substrate specificity, and stability in non-aqueous solvents and co-solvents. Available enzyme design approaches can be classified into rational design and Directed Evolution (DE) [1]. DE is able to provide highly active tailor-made enzymes at the expense of experimentally generating and screening tens of thousands of variants. However, the high economic cost associated with DE limits the broad application of enzyme-catalysed processes for chemical manufacture. Most importantly, it is also unknown how the introduced mutations contribute to enzyme proficiency. Different rational design approaches exist that range from multiple sequence alignments (MSA), structural evaluation of the active site pocket and available tunnels, to the application of sophisticated computational tools, such as Quantum Mechanics (QM), hybrid QM and Molecular

Mechanics (QM/MM), Empirical Valence Bond (EVB), Molecular Dynamics (MD), and Monte Carlo simulations [2]. One of the most popular approaches is the *inside-out* strategy based on modelling the transition state(s) (TS) of the desired transformation (defined as *theozyme*) with Quantum Mechanics (QM) and grafting this ideal arrangement into an existing protein scaffold with Rosetta [3]. These rational approaches hold the promise of providing a comprehensive understanding of the relationship between mutations and their impact on enzymatic activity, yet none of the existing computational approaches is able to generate highly proficient enzymes rivalling natural ones and those generated with DE. In my view, the low activity of rationally designed enzyme variants can be attributed to the following limitations: (1) the high complexity of enzymatic catalysis and the lack of a computational approach able to accurately consider the multiple chemical steps and associated conformational changes taking place along the catalytic itinerary [2], (2) the need for reducing the sequence space, which is often solved by introducing mutations only in the active site pocket or entry/exit channel (as opposed to DE that introduces mutations throughout the structure) [2], and (3) the lack of fast yet accurate computational screening protocols for estimating the catalytic activity.

My recent research contributions to computational enzyme design

Understanding enzymatic function requires the evaluation of the chemical steps along the mechanism and also the exploration of the ensemble of thermally accessible conformations that enzymes adopt in solution. The ensemble of both reactive and unreactive conformations presenting different relative stabilities can be represented in the so-called free energy landscape (FEL, see Fig. 1(a)). We computationally reconstructed the FEL of some natural and laboratory evolution (DE) pathways using extensive MD simulations, Markov state modelling (MSM), and enhanced sampling techniques [4–6]. These studies demonstrated that increased enzymatic activity is often achieved by introducing mutations that alter the enzyme conformational ensemble. The introduced mutations located at the active site and often at distal positions induce a long-range effect that impacts the enzyme active site pocket and thus catalysis. This is achieved by

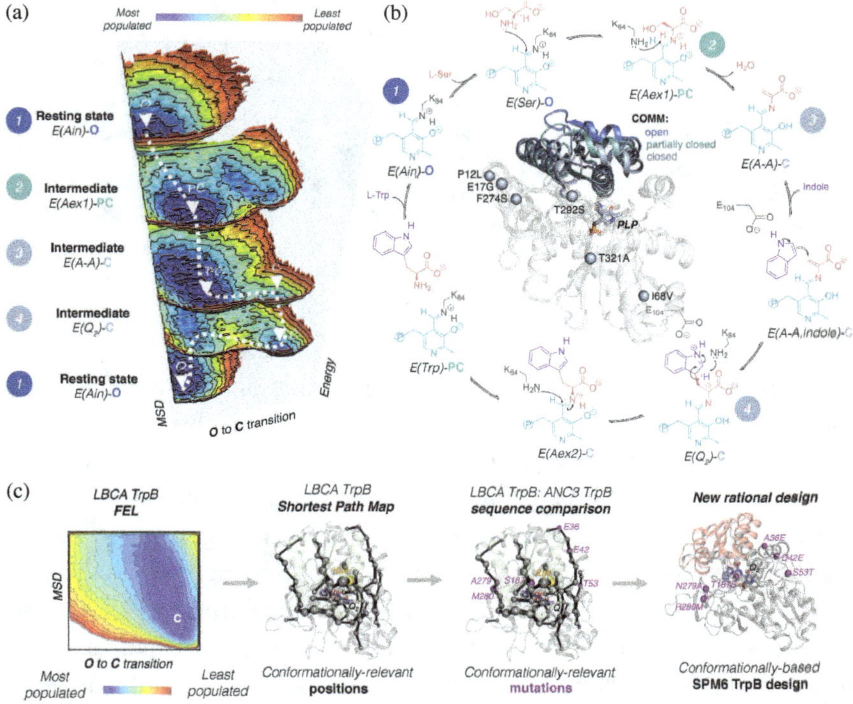

Fig. 1. (a) Computationally reconstructed Free Energy Landscape (FEL) of the laboratory-evolved 0B2-*pf*TrpB tryptophan synthase B that displays stand-alone activity (data from Ref. [5]) at several reaction intermediates along the catalytic itinerary (shown in panel (b)). TrpB adopts a different conformation of the catalytically relevant COMM domain along the process: open (**O**, dark blue) states are adopted in the resting state E(Ain), partially closed (**PC**, teal) at the reaction intermediates E(Aex1) and E(A-A), and closed (**C**, light blue) at E(Q_2) states. The most stable conformations are represented in blue, whereas the least stable ones are in red. (b) Reaction mechanism of TrpB and detail of the COMM domain conformation along the cycle [11]. Overlay of the COMM domain conformation as shown by X-ray data: **O** highlighted in dark blue, **PC** in teal, and **C** in light blue. (c) The mutations introduced with DE to generate 0B2-*pf*TrpB are marked with blue spheres. Computational pipeline developed for rationally designing new stand-alone enzyme variants based on the combination of shortest path map (SPM) and ancestral sequence reconstruction. (d) Development of an X-ray template-based AF2 approach for estimating the conformational heterogeneity of TrpB systems [12]. AF2 predictions are represented on the 2D-FEL representation using vertical lines coloured from orange to dark blue depending on the number of sequences provided in the MSA. From these AF2 structures, short nanosecond timescale MD simulations were run for FEL reconstruction (shown on top of the computationally expensive FELs (in grey) obtained by means of extensive metadynamics simulations).

(d)

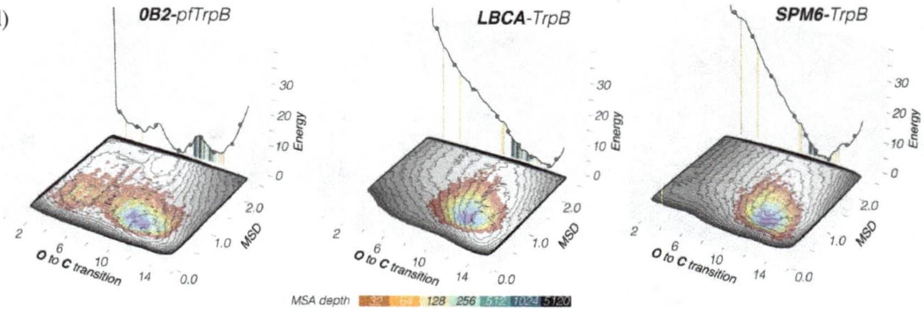

Fig. 1. (*Continued*)

favouring the catalytically productive conformational states and disfavouring the non-productive ones for the novel functionality, thus converting computational enzyme design into a population shift problem [2]. However, computational enzyme design seen as a population shift problem requires the reconstruction of an FEL for each generated variant, which is computationally too expensive for allowing the fast routine design of enzymes [7]. Most importantly, the reconstructed FELs do not provide any clue on which positions either located at the active site or distal might be responsible for stabilizing a desired conformational change. We hypothesize that by using graph theory coupled to the extensive MD simulations for FEL reconstruction, the existing long-range allosteric network of interactions can be revealed and used for predicting distal and active site mutations (Fig. 1(c)) [2, 8]. To that end, we developed the Shortest Path Map (SPM) tool that relies on the construction of a graph based on the computed mean distances and correlation values obtained along MD simulations [2, 6]. SPM decreases the sequence space to a smaller number of conformationally relevant positions and has the potential of identifying the challenging distal activity-enhancing positions. Indeed, we successfully applied SPM to identify DE mutations in retro-aldolase, monoamine oxidase, and tryptophan synthase enzymes [2].

In a recent publication, we combined SPM and ancestral sequence reconstruction to rationally design new stand-alone tryptophan synthase B (TrpB) variants (see Fig. 1(a–c)) [8]. Tryptophan synthase (TrpS) is a heterodimeric enzyme complex composed of two subunits: TrpA and TrpB that are allosterically connected. The tight allosteric communication between subunits involves, in the case of TrpB, open-to-closed transitions of the rigid COMM domain that forms a lid covering the active site (see Fig. 1(b)). The existing allosteric communication between subunits makes both TrpA and TrpB much less efficient when isolated, i.e., their stand-alone activity is low [5, 8]. However, the ancestral reconstruction of TrpB enzymes (LBCA TrpB) revealed a high stand-alone activity for the ancestral variants, which was lost along evolution [9]. The Arnold lab applied DE on *pf*TrpB and generated a new enzyme 0B2-*pf*TrpB that presented higher catalytic activity when isolated [10].

We computationally reconstructed the FEL of the ancestrally reconstructed LBCA TrpB, as well as the wild-type *pf*TrpS complex, isolated *pf*TrpB, and laboratory-evolved stand-alone 0B2-*pf*TrpB enzyme [5, 8]. These works elucidated the conformational ensemble that a stand-alone catalyst has to display for being efficient. We developed a rational computational protocol for achieving stand-alone activity of TrpB subunit based on the following steps (summarized in Fig. 1(c)): (1) reconstruction of the FEL of the ancestral LBCA TrpB displaying stand-alone activity, (2) application of the SPM methodology to detect the conformationally-relevant positions, (3) sequence comparison at the conformationally relevant SPM positions between the reference ancestral scaffold and the target ANC3 TrpB variant that had no stand-alone activity, and (4) transfer of the six non-conserved SPM mutations to the target ANC3 TrpB scaffold for generating the new SPM6-TrpB variant [8]. Interestingly, the experimental validation of the SPM6 TrpB design indicated a 7-fold increase (in terms of k_{cat}) of stand-alone activity. Although we did not reach the isolated activity of the reference LBCA TrpB, it is worth highlighting that by testing only one single variant, the fold increase in k_{cat} was similar to the 9-fold obtained by DE that required the generation and screening of more than 3000 variants [10]. This study therefore provides evidence for the potential of our SPM methodology for computational enzyme design.

The recent success of the Alphafold2 neural network (AF2) in predicting the folded structure from the primary sequence with high levels of precision has revolutionized the field of protein design [13]. Despite AF2's impressive performance, the application of AF2 for understanding and engineering functions directly from the obtained single *static* picture is not straightforward. However, in this direction, we recently tested the applicability of AF2 for elucidating the conformational heterogeneity of several TrpB enzymes [12]. We developed a template-based AF2 approach for estimating TrpB ability to adopt multiple conformations of the catalytically relevant COMM domain, which is required for enhanced standalone activity. Our results revealed the potential of AF2, especially if combined with short nanosecond timescale MD simulations, for estimating the changes induced by mutation in the FEL at a rather reduced computational cost.

Outlook to future developments of research on computational enzyme design

Inspired by the AF2 approach, some deep learning techniques have also recently been developed for protein design that can potentially mitigate some of the limitations mentioned above. The combination of the convolutional neural network trRosetta [14] and Rosetta was shown to be successful for the design of new stable proteins [15]. The AlphaDesign based on AF2 was also developed to predict novel proteins [16]. These examples show the potential of deep learning techniques to generate new functional variants within the allowed biological constraints. The application of AF2 (or other deep learning strategies) to computational enzyme design for any target reaction and substrate still remains largely underdeveloped. In the near future, I anticipate that many hybrid biophysical and deep learning strategies will be developed to solve the mentioned limitations and allow the fast routine rational design of efficient enzymes.

Acknowledgements

S.O. thanks the Generalitat de Catalunya for the consolidated group TCBioSys (SGR 2021 00487), Spanish MICIN for grant projects PID2021-129034NB-I00 and PDC2022-133950-I00. S.O. is grateful to the funding from the European Research Council (ERC) under the European Union's Horizon 2020 research and innovation program

(ERC-2015-StG-679001, ERC-2022-POC-101112805, and ERC-2022-CoG-101088032), and the Human Frontier Science Program (HFSP) for project grant RGP0054/2020.

References

[1] E. L. Bell, W. Finnigan, S. P. France, A. P. Green, M. A. Hayes, L. J. Hepworth, *et al.*, *Nat. Rev. Dis. Primers* **1**, 46 (2021).
[2] S. Osuna, *Wiley Interdiscip. Rev. Comput. Mol. Sci.* **11**, e1502 (2020).
[3] G. Kiss, N. Çelebi-Ölçüm, R. Moretti, D. Baker, and K. N. Houk, *Angew. Chem. Int. Ed.* **52**, 5700 (2013).
[4] C. Curado-Carballada, F. Feixas, J. Iglesias-Fernández, and S. Osuna, *Angew. Chem. Int. Ed.* **58**, 3097 (2019).
[5] M. A. Maria-Solano, J. Iglesias-Fernández, and S. Osuna, *J. Am. Chem. Soc.* **141**, 13049 (2019).
[6] A. Romero-Rivera, M. Garcia-Borràs, and S. Osuna, *ACS Catal.* **7**, 8524 (2017).
[7] M. A. Maria-Solano, E. Serrano-Hervás, A. Romero-Rivera, J. Iglesias-Fernández, and S. Osuna, *Chem. Commun.* **54**, 6622 (2018).
[8] M. A. Maria-Solano, T. Kinateder, J. Iglesias-Fernández, R. Sterner, and S. Osuna, *ACS Catal.* **11**, 13733 (2021).
[9] M. Schupfner, K. Straub, F. Busch, R. Merkl, and R. Sterner, *Proc. Nat. Acad. Sci. USA* **117**, 346 (2020).
[10] A. R. Buller, S. Brinkmann-Chen, D. K. Romney, M. Herger, J. Murciano-Calles, and F. H. Arnold, *Proc. Natl. Acad. Sci. USA* **112**, 14599 (2015).
[11] M. F. Dunn, *Arch. Biochem. Biophys.* **519**, 154 (2012).
[12] G. Casadevall, C. Duran, M. Estévez-Gay, and S. Osuna, *Prot. Sci.* (2022).
[13] J. Jumper, R. Evans, A. Pritzel, T. Green, M. Figurnov, O. Ronneberger, *et al.*, *Nature* **596**, 583 (2021).
[14] I. Anishchenko, S. J. Pellock, T. M. Chidyausiku, T. A. Ramelot, S. Ovchinnikov, J. Hao *et al.*, *Nature* **600**, 547 (2021).
[15] C. Norn, B. I. M. Wicky, D. Juergens, S. Liu, D. Kim, D. Tischer, *et al.*, *Proc. Natl. Acad. Sci. USA* **118**, e2017228118 (2021).
[16] M. Jendrusch, J. O. Korbel, and S. K. Sadiq, *Biorxiv* 2021.2010.2011. 463937 (2021).

https://doi.org/10.1142/9789811282324_0035

Development and Application of Biocatalytic Reactions that Enable the Synthesis of Complex Molecules

Alison R. H. Narayan

Department of Chemistry and Life Sciences Institute, University of Michigan, Ann Arbor, 48109 MI, USA
Narayan Group, 210 Washtenaw Ave., Ann Arbor, 48109 MI, USA

Biocatalysis as an enabling technology in organic synthesis

Small molecules make a disproportional impact on human health as drugs that prevent or treat disease, tools for analysing biological systems, and probes to manipulate whole pathways and individual biomolecules [1]. This limitless potential can be curbed by the practicality and accessibility of target molecules through modern chemical synthesis [2]. The trend away from natural products in drug discovery [3, 4] illustrates this conundrum, as the synthetic challenge associated with accessing these complex structures outweighs their potent activity and therapeutic potential [5]. Chemical methods that facilitate a desired transformation with precise chemo-, site-, or stereoselectivity can allow for more efficient synthetic routes free of protecting or directing groups and unnecessary redox manipulations, thus expanding the practical-to-target molecules [6]. Biocatalytic methods present the opportunity to develop exquisite catalyst-controlled selectivity of enzymes, enabling highly streamlined synthetic routes [7–9]. This is exemplified by nature's ability to make intricate secondary metabolites with potent biological activity, such as taxol [10, 11] and vancomycin [12, 13]. It is also possible to expedite access to synthetic molecules through biocatalytic strategies, as illustrated by Merck's five-enzyme, one-pot sequence to access the HIV drug, islatravir, dramatically reducing the step count and increasing the overall yield in the production of this nucleoside drug [14]. Although biocatalysis has been embraced by industrial chemists for the commercial production of pharmaceutical agents, several factors have prevented the broad adoption and implementation of biocatalysis in mainstream organic synthesis, including limitations in the breadth of well-developed reactions, the unknown substrate scope of functionally characterized enzymes, and the perceived incompatibility with multi-step, preparative-scale sequences [15].

My recent research contributions to directed protein evolution for green chemistry

My research group seeks to provide synthetic chemists with highly efficient, selective, scalable, sustainable, and well-characterized biocatalytic methods that can be smoothly implemented into synthetic approaches towards target molecules. Using enzymes from natural product biosynthetic pathways and targeted protein families as a starting point, we elucidate the natural chemical function and mechanism of a given biocatalytic transformation. From this initial benchmark, we use bioinformatic tools, structural analysis, computational modelling, and evolutionary approaches to assemble refined panels of complementary biocatalysts of utility to the synthetic community.

At this meeting, two approaches will be discussed for the construction of complex molecules: (a) using biocatalysts to generate reactive intermediates that can be intercepted by small molecule reagents *in situ* and (b) employing biocatalysts that execute convergent reactions, whereby various monomers can be cross-coupled on demand. This work to be discussed builds on previous efforts from my research group focused on the development of biocatalytic oxygenation reactions [16, 17] and biocatalytic oxidative carbon–carbon bond formation [18].

Outlook to future developments of research on directed protein evolution for green chemistry

With major advances in the fields of DNA synthesis, directed evolution, DNA sequencing, bioinformatics, and computational modelling of proteins, the potential utility of biocatalysis lies in the hands of chemists. Whether biocatalysis remains underutilized in chemical synthesis will depend not only on the boldness of chemists willing to use large-molecule catalysts but also on the diligence and creativity of chemists characterizing the function of enzymes and developing novel enzymatic transformations.

Acknowledgements

This research was supported by funds from the University of Michigan's Life Sciences Institute and the Chemistry Department, the Alfred P. Sloan Foundation, the Research Corporation Cottrell Scholars program, and the National Institutes of Health (NIH; R35 GM124880).

References

[1] C. J. Gerry and S. L. Schreiber, *Nat. Rev. Drug Discov.* **17**, 333 (2018).

[2] D. E. Fitzpatrick, C. Battilocchio, and S. V. Ley, *ACS Cent. Sci.* **2**, 131 (2016).

[3] F. E. Koehn and G. T. Carter, *Nat. Rev. Drug Discov*. **4**, 206 (2005).

[4] E. Patridge, P. Gareiss, M. S. Kinch, and D. Hoyer, *Drug Discov. Today* **21**, 204 (2016).

[5] A. L. Harvey, R. Edrada-Ebel, and R. J. Quinn, *Nat. Rev. Drug Discov.* **14**, 111 (2015).

[6] P. A. Wender, *Chem. Rev.* **96**, 1 (1996).

[7] S. Chakrabarty, Y. Wang, J. C. Perkins, and A. R. H. Narayan, *Chem. Soc. Rev.* **49**, 8137 (2020).
[8] N. J. Turner and E. O'Reilly, *Nat. Chem. Biol.* **9**, 285 (2013).
[9] R. de Souza, L. S. M. Miranda, and U. T. Bornscheuer, *Chem. Eur. J.* **23**, 12040 (2017).
[10] E. K. Rowinsky and R. C. Donehower, *N. Engl. J. Med.* **332**, 1004 (1995).
[11] M. Hezari and R. Croteau, *Planta Med.* **63**, 291 (1997).
[12] D. Bischoff, S. Pelzer, B. Bister, G. J. Nicholson, S. Stockert, *et al.*, *Angew. Chem. Int. Ed.* **40**, 4688 (2001).
[13] D. H. Williams and B. Bardsley, *Angew. Chem. Int. Ed.* **38**, 1173 (1999).
[14] M. A. Huffman, A. Fryszkowska, O. Alvizo, M. Borra-Garske, K. R. Campos, *et al.*, *Science* **366**, 1255 (2019).
[15] S. K. Wu, R. Snajdrova, J. C. Moore, K. Baldenius, and U. T. Bornscheuer, *Angew. Chem. Int. Ed.* **60**, 88 (2021).
[16] T. J. Doyon, J. C. Perkins, S. A. B. Dockrey, E. O. Romero, and K. C. Skinner, *et al.*, *J. Am. Chem. Soc.* **141**, 20269 (2019).
[17] S. A. B. Dockrey, A. L. Lukowski, M. R. Becker, and A. R. H. Narayan, *Nat. Chem.* **10**, 119 (2018).
[18] L. E. Zetzsche, J. A. Yazarians, S. Chakrabarty, M. E. Hinze, and L. A. M. Murray *et al.*, *Nature* **603**, 79 (2022).

https://doi.org/10.1142/9789811282324_0036

Multi-Functional Biocatalysts in Organic Synthesis

Thomas W. Thorpe, James R. Marshall and Nicholas J. Turner

Department of Chemistry, University of Manchester, Manchester Institute of Biotechnology, 131 Princess Street, Manchester M1 &DN, UK

Introduction

Enzymes display remarkable catalytic activity, evolvability, and reaction promiscuity, features which together have resulted in their increased application in chemical synthesis [1]. Reaction promiscuity is manifested in several different ways and includes both the tolerance of broad substrate scope as well as the ability of enzymes to catalyse mechanistically different reactions. The latter is exemplified by the repurposing of P450 monooxygenases such that their chemistry is altered from oxygenation reactions to carbene/nitrene generation, allowing for the generation of new synthetic pathways that exploit these reactive intermediates [2]. This "new-to-nature" chemistry of enzymes in the laboratory serves to expand the toolbox of biocatalysts available for synthetic purposes and provides new starting points for further engineering and evolution.

The ability of enzymes to catalyse two or more mechanistically distinct reactions also raises the question as to whether these different activities can simultaneously exist within a single active site. Multi-functional biocatalysts of this type would have great value in synthesis and would complement examples developed in the field of chemo-catalysis [3]. In addition, the discovery and development of a new range of multi-functional biocatalysts would have a major impact on synthetic applications, particularly in the context of cascade processes. Currently, there is growing interest in the design and application of processes in which multiple-bond forming processes and functional group interconversions are achieved by combining several biocatalysts in a single vessel. For these processes, each individual biocatalyst is typically engineered separately, followed by addition to the cascade and then subsequent process optimization. Multi-functional biocatalysts could make these approaches more efficient by reducing the enzyme count and catalyst loading, as well as streamlining enzyme engineering by minimizing the sequence space that is required to be explored. The availability of multi-functional biocatalysts presents different challenges that need to be considered when engineering these catalysts in order to co-evolve the different activities encoded at the active site.

The ability of a single enzyme active site to catalyse two or more mechanistically distinct reactions requires a number of conditions to be simultaneously met, namely (i) the ability to bind structurally related substrates and products since the product from one reaction becomes the substrate for the next reaction, etc., (ii) the availability of catalytic active site residues (e.g., proton donors, proton acceptors, and nucleophiles) that can participate in different non-bonding interactions, (iii) in some cases, the availability of cofactors (e.g., NAD(P)H and $FADH_2$) that can mediate different redox processes e.g., reduction of C = O as well as reduction of C = N bonds. However, these challenges are far outweighed by many of the advantages accompanied with multi-functional biocatalysts. Such enzymes essentially generate intra-enzymatic cascades, reducing overall enzyme count and loading. There are also implications for streamlining enzyme immobilization and downstream processing.

This discussion paper is aimed to stimulate discussion around the discovery, engineering, and synthetic application of multi-functional biocatalysts.

Out of the scope, although also of growing interest, are examples of multi-functional biocatalysts generated by creating an additional active site within the protein (Plurizymes) [4] and fusion proteins [5].

To frame the discussions, examples are given from the authors' own work as set out in Fig. 1 in which methods for the conversion of alcohols to chiral amines have progressed and been simplified, by the discovery of enzymes with multiple catalytic activities.

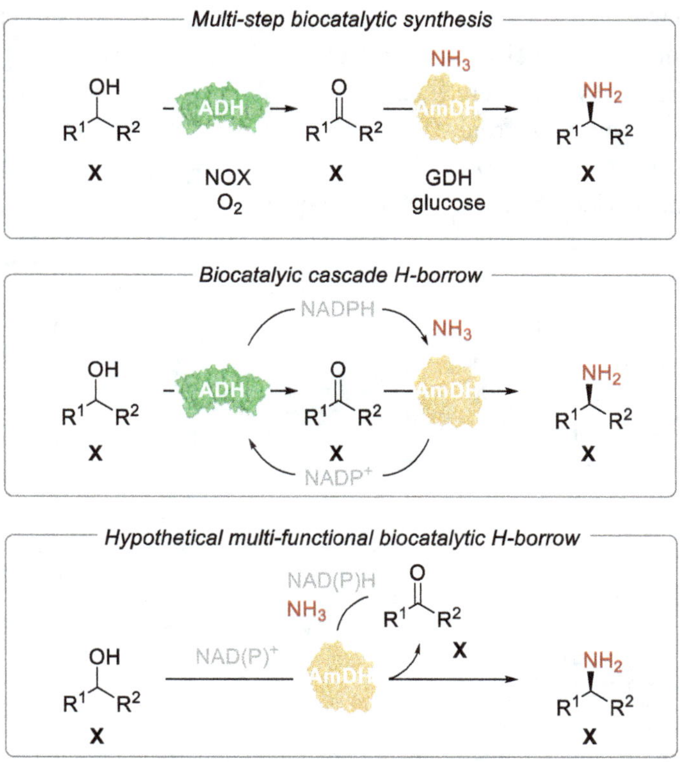

Fig. 1. Advances in biocatalytic systems developed for the conversion of alcohols to amines: (i) multi-enzyme conversion of an alcohol to an amine using alcohol dehydrogenase (ADH) and amine dehydrogenase (AmDH) with co-factor recycling; (ii) replacement of the NADH/NADPH co-factor recycling modules with a shared "hydrogen-borrowing" system; (iii) a "hypothetical" multi-functional biocatalyst enabling a single enzyme/single co-factor system.

EneIRED: A multi-functional biocatalyst that exploits iminium catalysis

Many high-value amine-containing compounds have multiple stereogenic centres. Accessing these molecules and controlling their centres is a challenge as highlighted by complex biomimetic tandem catalysis and multi-enzyme systems required for their synthesis. The discovery of an enzyme (EneIRED) that combines conjugate reduction and reductive amination into a single enzyme allows exquisite control of up to 3 stereocentres generating chiral amines [6]. EneIRED was identified from a collection of metagenomic IREDs [7] which were generated to explore the sequence space around the reductive aminase from *Aspergillus oryzae* [8]. EneIRED was initially identified by its ability to catalyse the coupling of cyclohex-1-enone with allylamine to yield the fully reduced product in high conversion (Fig. 2).

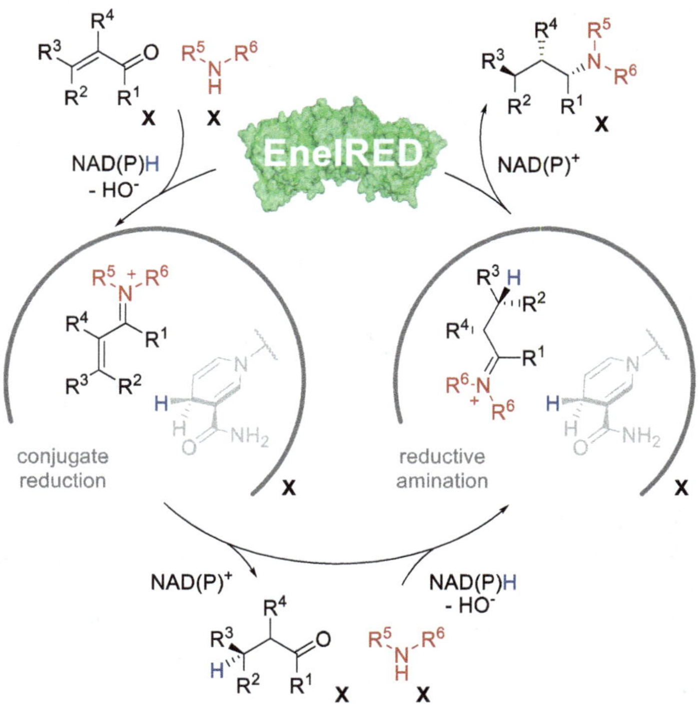

Fig. 2. Catalytic cycle of the multi-functional biocatalyst EneIRED which catalyses conjugate reduction-reductive amination via iminium-type catalysis.

A comprehensive assessment of the substrate scope of EneIRED was undertaken covering enals, linear and cyclic enones, including 3-substituted cyclohexenones with largely excellent conversions to the fully saturated product. A variety of primary amines and as well as pyrrolidines could be coupled with these unsaturated carbonyls. Generally, products contained 2 stereocentres, but when provided with the correct substrates, EneIRED could control 3 stereocentres with excellent selectivity, including the generation of a fluorinated tertiary amine. Furthermore, EneIRED was also shown to catalyse a 6-electron conjugate reduction-reductive amination reaction starting from the di-enyl ketone.

Mechanistic studies from isotopic labelling experiments confirmed that the conjugate reduction preceded the reductive amination step (Fig. 2). Structural differences in the active site were observed in comparison to other IREDs, i.e., EneIRED contains an additional tyrosine residue important for the steric control of the conjugate reduction-reductive amination reactions. EneIRED employs two catalytic cycles that together generate the fully reduced amine *via* two different iminium intermediates, reminiscent of autotandem organocatalytic systems (Fig. 2). EneIRED exemplifies the ability of multi-functional biocatalysts to combine different chemistries into a simple protein biomolecule while generating compounds of high value.

Summary and future perspectives

Although multi-functional biocatalysts have the potential to transform organic synthesis, the field is clearly at an early stage of development. Such enzymes meet the demand for the development of more streamlined and smarter biocatalysts [9, 10]. Coupled with advances in other areas (e.g., metagenomics and photobiocatalysis), enzymes possessing even more complex chemistries will emerge, thereby enabling new retrosynthetic pathways to be developed [11].

For a more sustainable future and in the current climate of the world, companies and institutions are scrutinizing every aspect of their practices to reduce their carbon footprint [12]. Multi-functional enzymes can contribute to this challenge by reducing the overall enzyme loading, with a concomitant reduction in engineering of multiple central and auxiliary enzymes. Substrate localization is another element which is improved, in turn leading to greater reaction efficiency.

Acknowledgements

We acknowledge the ERC for award of an Advanced Grant (BIO-H-BORROW).

References

[1] J. B. Pyser, S. Chakrabarty, E. O. Romero, and A. R. H. Narayan, *ACS Cent. Sci.* **7**, 1105–1116 (2021).

[2] D. C. Miller, S. V. Athavale, and F. H. Arnold, *Nat. Synth.* **1**, 18–23 (2022).

[3] N. Liu, Y.-F. Xie, C. Wang, S.-J. Li, D. Wei, M. Li, and B. Dai, *ACS Catal.* **11**, 9945–9957 (2018).

[4] D. Carballares, R. Morellon-Sterling, and R. Fernandez-Lafuente, *Int. J. Mol. Sci.* **23**, 5304 (2022).

[5] F. S. Aalbers and M. W. Fraaije, *ChemBioChem.* **20**, 20–28 (2019).

[6] T. W. Thorpe, J. R. Marshall, *et al.*, *Nat.* **604**, 86–91 (2022).

[7] J. R. Marshall, P. Yao, *et al.*, *Nat. Chem.* **13**, 140–148 (2021).

[8] G. A. Aleku, S. P. France, *et al.*, *Nat. Chem.* **9**, 961–969 (2017).

[9] K. F. Biegasiewicz, S. J. Cooper, X. Gao, D. G. Oblinsky, J. H. Kim, S. E Garfinkle, L. A. Joyce, B. A. Sandoval, G. D. Scholes, and T. K. Hyster, *Science* **364**, 1166–1169 (2019).

[10] R. A. Sheldon and D. Brady, *ACS Sustain. Chem. Eng.* **9**, 8032–8052 (2021).

[11] N. J. Turner and E. O'Reilly, *Nat. Chem. Biol.* **9**, 285–288 (2013); W. Finnigan, L. J. Hepworth, S. L. Flitsch, and N. J. Turner, *Nat. Catal.* **4**, 98–104 (2021).

[12] M. Okereke, *J. Clim. Change Health* **4**, 100049 (2021).

https://doi.org/10.1142/9789811282324_0037

Engineered Industrial Biocatalysts: Delivering on the Promise of Directed Enzyme Evolution

Stefan Lutz

*Codexis Inc., 400 Penobscot Drive,
Redwood City, 94063 CA, USA*

Biocatalysis and directed enzyme evolution offer a vision and a path to greener, more sustainable processes and products. Building on three decades of exploration, innovation, and development in protein engineering, this promise has been realized today as reflected in the growing number of industrial applications involving engineered high-performance biocatalysts.

Beyond traditional applications of enzymes in commodity products such as proteases in laundry detergents, early developments of engineered high-performance biocatalysts have been driven by the pharmaceutical industry. Here, directed evolution delivered highly active and selective enzymes at scale, replacing expensive and potentially toxic chemical catalysts and reagents, reducing the number of process steps, and enabling operation under environmentally benign conditions for the manufacturing of life-saving drugs including antivirals against COVID and HIV [1–3]. These successes have undoubtedly contributed to recent broader utilization of engineered biocatalysts in other markets including food and beverage ingredients, as well as advanced diagnostic and genomics applications [4, 5]. Finally, protein engineering is delivering promising solutions to global environmental challenges, offering novel industrial biocatalysts for plastic recycling and carbon dioxide capture [6, 7].

Despite significant progress and success, the integration of biocatalysis throughout industry is still in its infancy. Directed enzyme evolution presents a powerful approach to harness and exploit the tremendous versatility and functional diversity of proteins, a potential that we have barely tapped today. Its solutions will enable society to not only benefit from greener, more sustainable manufacturing processes but also successfully tackle growing political, economic, and environmental challenges. Critical to the continuation of this success story are investments in education of a highly skilled and innovative workforce and technological advances to accelerate protein engineering and also engineering solutions to biomanufacturing.

References

[1] U. T. Bornscheuer, G. W. Huisman, R. J. Kazlauskas, S. Lutz, J. C. Moore, *et al.*, *Nature* **485**, 185 (2012).

[2] C. K. Savile, J. M. Janey, E. C. Mundorff, J. C. Moore, S. Tam, *et al.*, *Science* **329**, 305 (2010).

[3] M. A. Huffman, A. Fryszkowska, O. Alvizo, M. Borra-Garske, K. R. Campos, *et al.*, *Science* **366**, 1255 (2019).

[4] S. Wu, R. Snajdrova, J. C. Moore, K. Baldenius, and U. T. Bornscheuer, *Angew. Chem Intl. Ed.* **60**, 88 (2021).

[5] Z. Ouaray, S. A. Benner, M. M. Georgiadis, and N. G. J. Richards, *J. Biol. Chem.* **295**, 17046 (2020).

[6] O. Alvizo, L. J. Nguyen, C. K. Savile, J. A. Bresson, and S. L. Lakhapatri, *et al.*, *Proc. Natl. Acad. Sci.* **111**, 16436 (2014).

[7] V. Tournier, C. M. Topham, A. Gilles, B. David, and C. Folgoas, *et al.*, *Nature* **580**, 216 (2020).

https://doi.org/10.1142/9789811282324_0038

Beyond Nature: Engineering Cyclase Enzymes for Challenging Applications

Bernhard Hauer, Andreas Schneider,
Julian Ludwig and Kristina Schell

Department of Technical Biochemistry,
University of Stuttgart, Stuttgart D-70569, Germany

Biotechnology promised to be a major driver to the challenging problems we face in the 21st century. Under the term of industrial biotechnology, significant contributions are expected in the quest for new system solutions in chemistry. The field of enzyme engineering developed into a mature technology that impacts various areas, such as industrial biocatalysis, biomedicine, and synthetic biology. Traditionally, new enzyme activities have been discovered by screening microorganisms and mutant libraries. However, it became obvious that these methods have their limitations to identify the much-needed novel enzyme functions. However, we have to expand our portfolio of biocatalysts towards new, non-physiological reactions to boost the impact of biotechnology. It is a restriction to biocatalysis that for many important reactions catalysed corresponding enzymes are missing [1, 2].

In 1976, Jensen published a landmark review putting forward the hypotheses that ancient enzymes were characterized by broad substrate and reaction scope. Over the years, many contributed to our current view that enzymes are poly-reactive (promiscuous), and this is vital to the evolution of new enzymes. This concept is also of fundamental importance to biocatalysis. Very early on, scientists used enzymes for non-natural reactions. A striking example is pyruvate decarboxylase, which is technically used for the synthesis of phenylcarbinol. Instead of decarboxylating pyruvate to acetaldehyde, the enzyme is used for CC linkage.

How to obtain new enzyme activity thus remains a central question in this field. In recent years, promiscuity-based as well as other strategies were successfully expanded by chemistry- and date-based approaches. To identify initial activity and to address the catalytic flexibility, small and functionally rich enzyme libraries have been used. Once a starting point has been identified, the starting enzyme can be optimized using the entire repertoire of methods.

This presentation will highlight hidden as well as uninvestigated enzymatic activities and illustrate strategies to unmask them. In my talk, I discuss what opportunities are associated with an enzyme like squalene hopene cyclase (SHC). SHC is a class II cyclase and driven by a Brønsted acid catalysis mechanism. The wild-type enzyme shows a few activities beyond the physiological reaction of cyclization squalene to the hopene and hopanol. Interestingly, it turned out that the active site of the SHC enzyme is highly evolvable.

To initiate our studies, we used a reaction mechanism-guided approach to generate a platform for exploiting the SHCs protonation machinery [3]. Different functional groups can be activated and enable the synthesis of various cyclohexanoids and so uncouple the enzyme from its polycyclization chemistry (Fig. 1(a)) [4]. Tailoring the enzymes beyond the active site and including the entrance tunnel enabled the development of a most efficient biocatalyst for the cyclization of homofarnesol to (–)-ambroxide. The overall catalytic performance was increased by a factor of almost 400 ensuring the stereocontrol of the reaction (Fig. 1(b)) [5]. We then expanded our strategy by introducing anchoring sites for very dynamic substrates. Based on designed hydrogen bonding, it became possible to induce

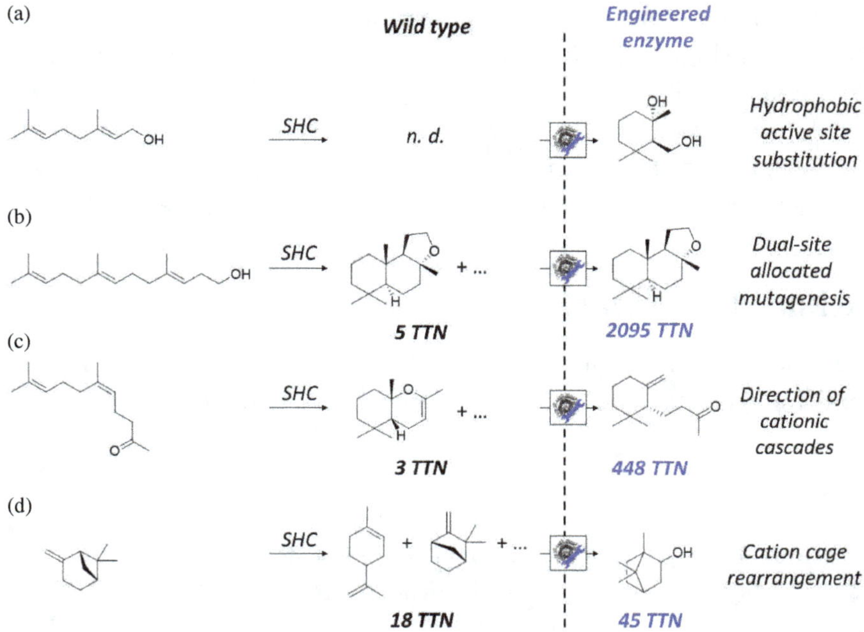

Fig. 1. Evaluability of the SHC as demonstrated in (a) the hydrophobic substitution of the active site enabling the cyclization of small terpene analogue geraniol, (b) the dual-site allocated mutagenesis of the active site and entrance tunnel enabling high-performing diastereo- and enantiopure cyclization of *E,E*-homofarnesol, (c) the structure-guided engineering of the active site enabling the directed monocyclization of neryl acetone, and (d) the cation-cage tuning enabling the highly selective and enantiopure cationic rearrangement of (+)-β-pinene.

directed cyclizations and predestinate termination in otherwise barely accessible intermediates. We could demonstrate the synthesis of apocarotenoids in a single step with high selectivities (>99.5% ee) and yields (up to 89%) (Fig. 1(c)) [6]. In the third example, the precise carbocation control over an unactivated monoterpenes by biocatalytic Brønsted acid catalysis is presented. SHC showed initial activity for the promiscuous isomerization of (+)-β-pinene producing a wide mixture of different olefinic and hydroxylated monoterpenoids. Based on iterative saturation mutagenesis, we tuned the carbocation control of SHC for borneol production resulting in a variant reaching 90.2% borneol selectivity (99.6% ee) (Fig. 1(d)). Recently, we reported a new concept to increase the overall activity of our system. SHC spheroplasts exhibited up to 100-fold higher activity than their whole-cell counterparts [7].

With this perspective in the right economic context, we can anticipate the relevance of biotechnology to chemistry to accelerate even further [8].

Acknowledgements

We thank all contributing authors and the German Science Foundation DFG for generous funding. Grant Number: HA 1251/6-1.

References

[1] R. A. Jensen, *Annu. Rev. Microbiol.* **30**, 409–434 (1976).

[2] O. Khersonsky and D. S. Tawfik, *Annu. Rev. Biochem.* **79**, 471–505 (2010).

[3] P. O. Syren, S. Henche, A. Eichler, B. M. Nestl, and B. Hauer, *Curr. Opin. Struct. Biol.* **41**, 73–82 (2016).

[4] S. C. Hammer, A. Marjanovic, J. M. Dominicus, B. M. Nestl, and B. Hauer, *Nat. Chem. Biol.* **11**, 121–126 (2015).

[5] A. Schneider, C. Curado, T. B. Lystbaek, S. Osuna, and B. Hauer, (2022), doi: 10.26434/CHEMRXIV-2022-XWCQJ.

[6] A. Schneider, P. Jegl, and B. Hauer, *Angew. Chemie — Int. Ed.* **60**, 13251–13256 (2021).

[7] A. I. Benítez-Mateos, A. Schneider, E. Hegarty, B. Hauer, and F. Paradisi, *ChemRxiv.* (2022), doi: 10.26434/CHEMRXIV-2022-W08D0.

[8] B. Hauer, *ACS Catal.* **10**, 8418–8427 (2020).

https://doi.org/10.1142/9789811282324_0039

Discussions of Session 5 — Directed Protein Evolution for Green Chemistry

Chair: Sabine Flitsch
Auditors: U. Hennecke*, A.A. Vorobieva[†,‡]

*Organic Chemistry Research Group (ORGC), Vrije Universiteit Brussel, Pleinlaan 2, B-1050 Brussel, Belgium
[†]Structural Biology Brussels, Vrije Universiteit Brussel, Pleinlaan 2, B-1050 Brussel, Belgium
[‡]VUB-VIB Center for Structural Biology, Flanders Institute for Biotechnology, Pleinlaan 2, B-1050 Brussel, Belgium

Sabine Flitsch:

Welcome back to the discussion session. So, what we wanted to demonstrate is that we are very cross disciplinary, that we think there is a lot happening in our area and we really welcome any new scientist to join the area and join us because we think there's an enormous potential in using proteins as catalysts. With that, I open the discussion. I've been asked to remind you to put your microphones on and say your name. Otherwise, we can't write the proceedings. Maybe I will start with asking whether anyone in the audience has any question they wanted to ask the panel. Well, that's fantastic! Thank you. We already have a candidate. Please give your name.

Yamuna Krishnan:
May I just say that I just loved that session so much? A quick question for all the panelists. So, for organic chemistry, when you say catalysis, after you find the reaction that works you show broad substrate scope. Whereas in enzyme catalysis, especially these synthetic enzymes, my understanding is: the faster it becomes, the most selective or the narrower the substrate scope becomes. So, I was wondering: these are two completely different paradigms, but do you see selection ever able to break this paradigm? The enzyme has become faster, but also expanded substrate scope.

Sabine Flitsch:
That's a great question and I think quite a few of my panelists would want to comment on that. Maybe I'll just start at the very left with Stefan.

Stefan Lutz:
A great question. I hope it came across a little bit. My very last example with the polymerase, for DNA synthesis, is a perfect example. It's a matter of selecting for promiscuity to drive the broad substrate specificity, for example of enzymes. But I certainly believe that it is not given that, as you evolve, you need to narrow your spectrum.

Donald Hilvert:
In our evolutionary experiments, we usually see exactly what you predict. That is, as the enzyme becomes more active, particularly with small molecules, it becomes more selective and specific for the substrate or for the stereochemistry of the process that is carried out. However, you have many intermediates on your way up the mountain and you can take steps down. And often, the variants that have a little bit lower activity tend to be more promiscuous. Then you can re-diversify them or re-optimize them through exactly the same procedure.

Alison Narayan:
There are many different things you can screen for. It is true that those things can go hand in hand. And you can screen for an enzyme that's faster, you can screen for an enzyme that's more thermally stable, you can screen for specificity for a particular substrate, and those things don't

always have to go hand in hand. You can set up your screening in a certain way. You know, Francis Arnold, who's not here, pretty much in every talk will say: "You get what you screened for." Strategies for screening that take into consideration substrate promiscuity (if that's something you ultimately want out of your catalyst), I think, are quite powerful. And there are more and more examples of that coming out now. There's a recent paper from Andy Buller's lab at Wisconsin, that highlights this concept of protein engineering where you're not screening just based on one substrate but a panel of substrates with the hopes of engineering a catalyst that will maintain promiscuity.

Sílvia Osuna:

Yes, I think that's a very important point. Also, from our perspective, what we see is that selecting the proper scaffold for the design is key. If you start with an enzyme that is promiscuous and, computationally speaking, has many different conformations, then with our approach we try to predict mutations that stabilize certain conformations that we think are important for enhancing the activity and the selectivity. So, I think yes, that's a very important point.

Nicholas Turner:

One thing you can do is what people call substrate walking. If you imagine doing evolution on an enzyme — you can start off with one substrate, but you don't have to keep that substrate constant. You can change the substrate and that often leads to enzymes with broader substrate scope. That has been done with transaminases. We did it a number of years ago with an enzyme called monoamine oxidase. We deliberately changed — in a fairly subtle way — the substrate to try to find enzymes with broader substrate scope. Now, that has a real impact because, obviously, broad substrate scope with high activity is what you want to achieve. It also broadens the application of biocatalysis. And this is something we didn't really talk about. Biocatalysis, historically, is what I call a process technology. It's potentially a way of making specific molecules in a very, very efficient manner. But in the last few years it has gone upstream into medicinal chemistry. If you're a medicinal chemist, what you need are reagents and catalysts that have broad substrate scope because you want to make many molecules and screen them for activity. You don't know at that stage what is your active molecule. And so, you're starting to see this, particularly in industry. Because we are developing more and more panels of enzymes with broad substrate scope, they are being adopted by medicinal chemists who can screen them and create molecules for the first time using biocatalysis. I personally think that's a very exciting development. You are not just making it on a large scale, you are making a molecule that is actually not easy to make. You are accessing a new chemical space. Alison talked about that in her lecture. And that's the consequence of more enzymes with broader substrate scope.

Sabine Flitsch:
Comment to this or new question? OK, there's one new.

Peter Palese:
This is for Don Hilvert, please. When you walk up the mountain and you arrive at your best solution, is there a structural similarity to the absolute best one? Has your evolution come to a similar structure as the one which is a little bit higher up? In other words, what does evolution do?

Donald Hilvert:
To the natural enzyme you mean?

Peter Palese:
To the natural enzyme, yes.

Donald Hilvert:
In a sense, mechanistically, yes, there are similarities. In the natural enzymes — natural aldolases — there is typically a tyrosine that is used as a general acid or general base. That was absent in the original aldolase design, but it emerged during the course of evolution. In fact, this catalytic tetrad arose one residue at a time and mimics the kinds of complexity that you see in natural aldolases. Now, the substrate recognition is completely different. That does not resemble what you see in natural aldolases, because natural aldolases usually work on sugar type molecules, so very polar molecules. And there the design didn't change at all. The design was

very good at the start. In fact, the very starting design, when we characterized it structurally, was somewhat disordered in the loops that surround the active site. As we evolved it, it approached the original design model much better. The loops took on confirmations that resembled what was predicted by the original Rosetta designers.

Sabine Flitsch:

Comments on this topic? Otherwise, we switch topic. Maybe Ben Feringa next.

Ben Feringa:

Maybe a bit related to this. Thank you, by the way, for this fantastic perspective of the field. I'm really impressed! Especially I'm impressed by that where we normally need a 30-step synthesis to make a complex drug, that six enzymes can do the job very facile. But when you talk sometimes with colleagues in the field of catalysis, they go: "Oh that is great what they do. But they need a huge molecule to make a small molecule." And when you go back to Ryoji Noyori, for instance, the asymmetric catalytic hydrogenation: he takes hydrogen, pins of rhodium, a small ligand, and he has millions of turnovers, and he has hundred different substrates that hydrogenate with 99% selectivity. So why do we need an enzyme?

Nicholas Turner:

So, my response to that is: you don't have to make that enzyme by total synthesis. No, seriously, this argument has been around for a long time right now. So, let's approach this from different angles. If you try to build enzyme mimics — and this was very popular in the 80s and 90s — they are generally not very good catalysts. So, if you start to shrink an enzyme, you lose a lot of catalysis. Enzymes are large, but presumably they need to be large to be catalytically efficient. But they are so easy to produce. I mean, they really are. So, I don't mind that they're big because they are wonderful catalysts. And I hope we convinced you, that production is trivial.

Ben Feringa:

You don't have to convince me! I have to convince my colleagues!

Nicholas Turner:

So — you are right — there are small molecule catalysts that have high turnover numbers. Like Noyori: the turnover numbers of the Noyori catalysts are 10^7, so they're comparable to the turnover numbers of the best biocatalysts. But shrinking enzymes: I don't think that works, personally.

Donald Hilvert:

One might also add that the enzymes are inherently compatible with one another. So, you can run these multiple processes simultaneously, which is often hard to do with small molecule catalysts.

Ben Feringa:

I know, I agree with you. But this brings me to another comment that I often get: "Oh, but these guys have to work in water, and we like organic solvents", as most organic chemists do. Now it changed a bit. This was not discussed so much — about solvent. Maybe you can comment a bit on that. Because that interfaces to a lot of organic chemistry that we traditionally do.

Stefan Lutz:

We have lists of examples of enzymes that were evolved to work in 50% DMSO, toluene, octane, acetonitrile, methanol, ethanol. You know, it's a matter of asking the question. But I do not believe that enzyme catalysis is restricted to aqueous solvents. We have the data to show this.

Ben Feringa:

Now I want to argue again! Because I worked at Shell and, I remember, I used that same argument once and my boss said: "The worst thing we can do is to have dirty water." Because that costs a lot of energy to purify and to use it again.

Stefan Lutz:

There are enzymatic reactions that run in pure toluene.

Sabine Flitsch:

I just want to add. I think the technology has developed enormously as well. So, I think these people need to revisit it a little bit. You know, it used to be like that, and I think — hopefully — we've shown that we've really developed this technology. Professor Wüthrich?

Kurt Wüthrich:

Long ago, enzymatically active antibodies were introduced, which would perform reactions in water that organic chemists would only be able to perform in extremely dry organic solvents. I think that the Diels-Alder reaction was an outstanding example. Don (Hilvert) must know more in detail, because he was directly involved in this, if I remember correctly. Today, there was no mention of catalytic antibodies in the presentations of this session.

Donald Hilvert:

I started with catalytic antibodies and, like everybody else in the area, I dropped it because you could make catalysts for many different reactions, but they were orders and orders of magnitude less efficient than natural enzymes. The aldolase example I showed you, the aldolase was in the middle of that plot, and we've been able to go way beyond. I think the problem with the antibodies is largely technical. They were very difficult to evolve, and we spent many, many years trying to optimize them through this same evolutionary process. But in *E. coli*, they're extremely difficult to produce. And that limited us a lot.

Kurt Wüthrich:

Enzymatic antibodies nonetheless got the Wolf prize for Richard Lerner and Peter Schultz. It seems to have been some sort of success.

Donald Hilvert:

Absolutely! And, in fact, they were the inspiration for all the original computational designs. The same model reactions have been used to test the ability of the computational methods to predict active site structures. And, in fact, the original designs were less good than the typical antibodies — the corresponding antibody catalysts — but they were much more evolvable because they were embedded in scaffolds that were extremely stable, very tractable, tolerant to mutation, typically monomeric, easy to produce. So that turned out to be a huge advantage.

Martyn Poliakoff:
Following on from what you said. It was a really good session before coffee! Once you've made one of your evolved enzymes, the one that's really good, how easy is it to produce? And also, green chemistry focuses a lot on the waste. How much waste do you generate making your enzyme? Because that should be factored in. You may have a really clean reaction, but if you have made a terrible mess making your enzyme, this rather spoils the picture. Perhaps Stefan is the person to answer this?

Sabine Flitsch:
Alison first.

Alison Narayan:
I'll just start by saying: things that can be engineered for and improved upon include production of your protein. We focused today a lot on the activity and the selectivity of the enzyme. But also engineering something — or care to optimize the heterologous expressions so you can get meaningful amounts of your protein — is an important area. That is something that is possible to do. Every protein is different. For some, straight out of the gate you might be able to make buckets of your protein and it's not really a concern. Others might take a significant concerted effort to get meaningful amounts of your protein. But it is an area of science that's well developed.

Stefan Lutz:
On the expression, I think a good example is the recent development of PAXLOVID by Pfizer as an anti-COVID drug. It was identified as an enzymatic step in that synthesis. The enzyme was identified in a panel — to what Nick Turner pointed out earlier — so very quick lead identification. Within five or six months, that lead was produced in tens of metric tons for drug manufacturing. So, protein expression — once you have the evolved candidate or a lead candidate — is, I dare to say, straightforward. It's just a fermentation. As to the green chemistry, you might notice on my slide I had "greener". I didn't claim green chemistry. Because we've started doing some PMI (Process Mass Intensity)

analysis on the processes, comparing chemicals with the biocatalytic processes. Depending on where you draw the boundaries of what you take into consideration in your analysis, the analysis looks better or not so good. But, generally speaking, it is more beneficial. It is advantageous over the chemical synthesis. But the degree varies depending on where you draw the boundaries.

Martyn Poliakoff:
Thank you.

Andrew Turberfield:
It was said several times that we don't really understand how enzymes work. And Nick just said that enzymes probably have to be big; without saying why. Is there an obvious difference between trying to engineer an enzyme from a natural scaffold — which is marginally stable — and one that is synthetic and overwhelmingly stable? And is the field using that sort of comparison to try and learn how things really do work?

Donald Hilvert:
We are starting to look at this. I think one of the advantages of being able to design is that we can now compare scaffolds in a systematic fashion. And so, why is a TIM-barrel particularly engineerable? It's a beautiful fold, it allows you to bring functionality together convergently. So, it's just an ideal way of creating an environment where you can completely surround the substrate and the transition state and provide selectivity. We have now tried to take the lessons that we learned in some of these evolutionary strategies and extract the functional — the effective catalytic apparatus — and transplant it into some of the *de* novo designs that are very stable. What we found in an 8-stranded barrel, for example, is that the initial activities are much higher — by almost two orders of magnitude — compared to the original design that I showed you today. But it turned out to be much less evolvable. That may be because it's too rigid; or that the scaffold itself is just not ideally suited to do this multi-step transformation. It is maybe better for one-step reactions. That we just don't know yet. But we can now ask these questions in a systematic fashion.

Andrew Turberfield:
Thank you.

Sílvia Osuna:
If I can add something. That's why many people are also doing ancestral sequence reconstruction. They use these ancestral enzymes that are very stable and also usually promiscuous. They have this ability to adopt different conformations. And people use those scaffolds for design. And what they have seen in some cases is that, using one of these flexible scaffolds with just a reduced number of mutations, they can get much better designs. So that's another point towards that.

James Liao:
I have a question for Nick (Turner). You showed a nice example of one active site that catalyzes three reactions. Do you think that's a special case? Or that's the general principle that, eventually, one can design one active site that catalyzes multiple reactions?

Nicholas Turner:
It's definitely not a special case. I mean, there are enzymes that catalyze multiple reactions with single active sites. I think there's an enzyme called macrophomate synthase. I think it catalyzes 6 different chemical reactions in a single active site. That's possibly the world record holder. Nature has found a way of evolving enzymes that have multiple reactions within a single active site, so that's interesting. How did that happen? To me, the key, going forward is to think about what chemistries you could combine within a single active site. I mean, clearly there's going to be limitations to this. If you're going to use a single active site, the chemistries got to be related, obviously, from a mechanistic point of view. An obvious place to start is a redox reaction. Oxidation and reduction are essentially the same process, just in reverse. So, you see examples of where redox reactions are catalyzed by single active sites. Then I think you need to think about the similarity between the product of one reaction acting as the potential substrate for the next reaction. Because that could lead you into these multiple activities. Actually, if you look in the lecture of organic catalysis, this is

not unusual. Organic catalysis and biocatalysis, I think, are just sort of two sides of the same coin. In the early days, when Dave MacMillan and Ben List — and others — devised those organic catalysts, they looked at biocatalysts and I think gained some inspiration for them. Now, what we're doing is that we're looking at organic catalysts, where you can do often these multiple reactions with a single small molecule catalyst, and we're trying to use those as inspirations. Because these guys have demonstrated that you can achieve that multi-reaction sequence with a single catalyst. So, for us the question is now, can you find an enzyme that could do something similar but within the context of a protein active site?

Donald Hilvert:

Just to continue on that. The inspiration from organocatalysis is also being utilized directly for design, to try and incorporate organocatalysts in the binding pockets. This is being done by Anthony Green and Gerard Roelfes and others. They are trying to put organocatalysts into the binding pockets of a protein to gain some intrinsic reactivity that they can then shape through evolution.

Martyn Poliakoff:

It's quite a quick question. Well, I hope. Can you use the enzymes more than once or are they essentially single use catalysts? Can you recover them, recycle them, and so on?

Sabine Flitsch:

If you immobilize enzymes. We have immobilized some enzymes, not all. But if they're stable enough, you can just remove them through filtration and then put a new substrate in. We have had several examples published where we have just used them again and again if they're stable enough. So, in principle you can actually reuse them.

Martyn Poliakoff:

But since they're big molecules, as Nick (Turner) has said with pride, could you just remove them with nanofiltration, for example? Without immobilization?

Sabine Flitsch:
Yes, we do that routinely as well. You are absolutely right! They're quite easy to filter off with 10,000 molecular weight filters, or something like that. Did anyone else want to comment on that?

Bert Meijer:
First of all, I'd like to congratulate you with the progress in this field. It is immense. Not only you but the whole area, obviously. So, I synthesize often molecules that are very difficult to dissolve, or the product is actually insoluble. It's a little bit related to the solvent you are using, but also enzyme inhibition and all of this. How important is it that the substrates and the products are well soluble in the medium where you use the enzyme?

Nicholas Turner:
There are slurry-to-slurry processes that operate on the scale, so you do not need to get all of your substrate into solution. As a guideline, when people develop engineered biocatalysts, particularly in industry, they're targeting substrate concentrations typically, I would say, around at least 50, maybe 100 grams per liter. If you look at the biocatalytic processes that make it, to scale and commercialize, that's a hallmark. So, obviously, at 100 grams per liter, there may be issues around solubility; depending upon what type of substrate it is. You definitely don't need to get all of the substrate into solution. You can use co-solvents (and people do that) if you want to solubilize. Maybe that will increase the rate of the reaction — it probably will. Inhibition, that's a very important point, and we've dodged that bullet a little bit. Enzymes are prone to inhibition. Yet again, that can potentially be fixed by protein engineering. But sometimes it can't, and therefore you need sometimes strategies for *in situ* product removal. We could have had a chemical engineer on the panel talking about that issue. Some reaction products fortuitously precipitate out. The original aspartame synthesis — which was using an enzyme to do a peptide coupling — its product just continuously precipitated. So, the equilibrium of that reaction was never attained. It just came out of solution, which alleviated inhibition and drove the product formation.

Bert Meijer:
Thank you! But I still don't completely understand. You have a substrate and you make a product that has a different type of functionality because that's why you do the reaction. Then you have to know what is the distribution of that molecule in the solvent or in the precipitate, but also in the active site of the enzyme, isn't it? So, there are many reactions that will not really work because the product is more prone to be in the interior of the enzyme than the substrate. Is there something wrong in my arguments?

Nicholas Turner:
So, you're talking about inhibition of the enzyme by the product?

Bert Meijer:
Yes, that makes the catalysis so slow because the product is more prone to be in the inside of the enzyme than the substrate.

Nicholas Turner:
Yeah, that could be the case.

Bert Meijer:
How do you solve this problem, then?

Nicholas Turner:
Well, I think you probably have to either engineer the biocatalyst to be less prone to inhibition — which is possible — but without compromising its activity. Or you need to reduce the concentration of the product in solution by either removal or precipitation. You've just lowered the concentration of the product and therefore presumably you will reduce the inhibition.

Stefan Lutz:
There are really two aspects, one that actually was brought up. Cascades are a perfect solution to overcome some of the product inhibition challenges. In the Islatravir synthesis, the buildup of some of the intermediates, which otherwise could act as inhibitors for the enzymatic reactions, are simply converted (by later biocatalysis steps). The same

with cofactors. Cofactor recycling is not only a way to eliminate or reduce costs of that element within the production, but it also eliminates some aspects of product or inhibition of these cofactors. Lastly, there are plenty of engineering examples of substrate and product inhibition being eliminated as part of these directed evolution processes. So, it is an addressable problem.

John Sutherland:
Primarily addressed at Nick, but I'd like to hear what everyone has to think about it. If big enzymes are necessary and small enzymes are no good, how do we get to big enzymes in the first place?

Nicholas Turner:
I was hoping you were going to tell us this afternoon, John.

John Sutherland:
Let's play tennis Nick! I'll bat the ball back into your court.

Nicholas Turner:
That's what you call a hospital pass, isn't it? How did enzymes become so big?

John Sutherland:
If small enzymes are no good, how do you get to big enzymes in the first place, when presumably smaller proteins are initially most easily accessed? If they don't give you any advantage, what's the incentive for evolution to progress to the stage of making bigger proteins where you suddenly find the activity?

Nicholas Turner:
Ok, it's a great question.

Sabine Flitsch:
I think Professor Wüthrich has an answer to that!

Kurt Wüthrich:
I asked you this question before in private. You don't engineer sizes. You use natural scaffolds and dabble around. I think size is important to

maintain osmotic pressure in the cell. Some of the proteins are present in sufficient quantities to play a major role in maintaining needed osmotic pressure across cells in the organism.

Sabine Flitsch:
Thank you. I think Don Hilvert?

Donald Hilvert:
So, the question is: how small is small? The protein that I spoke about at the end of my talk is 94 amino acids long. It's a helical bundle, incredibly dynamic, and yet it's a very good catalyst. That protein was designed by combining peptides, initially using metal ions as an interface, and then fused and diversified. And one presumes that this would be a way to have generated the original proteins.

John Sutherland:
Just to reply. Ninety-four, with twenty amino acids in the basic set to explore that sequence space *de novo* would be very difficult for nascent biology, presumably.

Donald Hilvert:
But small peptides can also have considerable activities. People like Scott Miller have shown this. So, I don't think you need a huge protein. I think modern enzymes are large because they also need to completely surround the reaction space that they control, and large size provides you that. It also provides you a mechanism of regulating activity within the cell.

John Sutherland:
Thank you, I think that's the answer I was hoping to hear basically. So, you think there's sort of opportunity to be found in small peptide space.

Sabine Flitsch:
Maybe if we go along the table?

Daniel Nocera:
Maybe this is touching on something you mentioned, Don (Hilvert). A lot of this starts off with the active site — it is active site focused. But you used the word rate-determining step in your talk. That's why I ask this to

you first but also, I would like to know computationally. A lot of large things exist because there is allosteric regulation, away from the active site. So, if there are allosteric regulation steps, very remote from the active site, are you nailing those or is it just getting kind of averaged into the active site in directed evolution, experimentally? And then computationally, is that space too big for you to probe in these large enzymes, if you have allosteric regulation remote from an active site?

Donald Hilvert:
Just to answer quickly. We haven't focused on any proteins that are allosterically regulated in our systems. We've just been trying to look at relatively simple transformations to understand how well we might be able to catalyze them. And then, you can imagine thinking about adding regulation after the fact. But we've not actually looked at that. Sílvia might be able to give you a better example because she has this two-enzyme system.

Sílvia Osuna:
Thanks a lot for the question. So, the tool that I was mentioning in my talk, Shortest Path Map, is actually inspired by previous methods that were used to study allosteric proteins. Usually, when you have this allosteric protein, you can do these MD simulations to check for correlated movement between residues and use graph theory to try to identify communities within this protein. So, try to see: "Ok, so this region here is all the way connected to this other region there". And what we did, well, we took this kind of inspiration, but in our case, we were not that interested in communities or in regions, but rather in specific residues that we think are important for the communication. And so, with this tool, we are able to see remote — very far away — positions that are all the way connected to the active site. And I also think it's very interesting that with directed evolution we have learned that many distal mutations have a big impact. In a way, this is telling us that enzymes have some sort of allosteric regulation intrinsically. And maybe, for designing, we need to take advantage of that, and these kinds of tools can help.

Sabine Flitsch:
If you just go next.

Karthish Manthiram:

Alison, I think you mentioned the importance of cofactor regeneration. And I guess Nature is converged on a small number of cofactors that help to deliver the energy that's needed, whether it's NADH or ATP, and what not. I was just wondering, in our own industrial synthetic use of these enzymes, should we believe that those cofactors are still the best way of delivering that energy? Or should we consider other synthetic analogs or entirely other methods of satisfying those energy needs for these enzymes?

Alison Narayan:

That's an excellent question and I think that there are many different directions that are being explored. So, there's a really active area in the research of unnatural cofactors that you might use to access different chemistry. But there are also ways that you can intervene in the ways that natural enzymes power their catalysis. So, for example, with cytochrome P-450's. Typically, that cofactor might be NADPH, which is supplying electrons to a reductase. Ultimately those electrons get to the heme of the P-450. But there are ways that we can deliver electrons from organic reagents, inorganic reagents, or just straight up from an electrode to provide electrons. I think there are a lot of exciting opportunities there, and I think that when you look at a lot of the inefficiencies in the overall process of the catalytic cycle, that's an area where there is a lot of room for improvement and there are exciting developments happening there.

Sabine Flitsch:

Anyone else on cofactors?

Arne Thomas:

Very impressive session I have to say. I would like to add up on a question Ben Feringa asked. Last answer was that there are enzymes which are working in pure toluene. You can evolve enzymes who are doing this, and it is, I guess, not really a common knowledge. So, I'm impressed that this is possible. Can you give us some more details on what do I have to do to an enzyme that it maintains its structure and activity in toluene? And I think it was also mentioned that you can evolve enzymes which can work at higher temperatures, for example. This would be very interesting because, in the end, the structure of protein is hydrogen bonds and ionic interactions. So, I would have guessed that is should not be possible to go to higher temperatures. Can you give me some chemical insights on structural variations on enzymes you have to do to achieve this?

Sabine Flitsch:

When we talk about pure toluene, the close environment to the enzyme still has shells of water. And that is why toluene is quite nice, because it's not miscible with water. In fact, often, the water-miscible solvents really kill the enzyme activity. The way I look at it is that you do still have the enzyme with a small shell of water and very often you need to have a very, very precise amount of water in your samples to maximize your activity, but it can be a very, very small amount to control that. I am not aware of any enzyme that would function where there is no water around. There is still quite a lot of water around, but it's in the environment of the enzyme. About the thermal stability: obviously, there are natural enzymes that are thermostable. I personally, without any engineering: what we would do in the first instance is to look for very similar enzymes in a database of known thermophiles. And very often, if you do that, you already get quite a big win: enzymes that are stable at 70°C. Sometimes, if you express these enzymes heterologously in bacteria, you can actually boil the bacteria after you've grown them up, precipitate out all the intrinsic proteins, and you just keep your enzyme soluble, and it is stable. May be Stefan Lutz can say a little bit more about this topic.

Stefan Lutz:

Maybe to your first point: the toluene reaction. What Sabine is referring to is the primary hydration layer surrounding the protein of interest. We typically lyophilize our proteins and that reduces it down. That initial water layer is very, very difficult to eliminate because it just binds very tightly to the protein surface. So, from a preparation perspective it's feasible. It's not particular sophisticated or challenging. Water activity — adding back small amounts of water to maintain a proper layer of hydration — can be the case. As to the second part of your question, what instils solvent- or thermo-stability to these proteins? I think there is still an ongoing debate.

We have thermostable proteins that live in hot Springs or come from hydrothermal vents. People have observed certain trends, certain preferences, surface modifications but also modifications of amino acids, certain interactions — ionic for example — that can help stabilise (the protein). But there isn't a clearly defined solution. The beauty of directed evolution is, of course, that you do not need to understand the fundamentals in order to identify improved variants. All you need to do is to screen at elevated temperatures and find variants that still exhibit activity. So, you can drive to an optimization of those kind of properties. So, I am not sure that there is a satisfying answer yet as to the physico-chemical principles that Nature exploits to achieve those kinds of stabilities.

Nicholas Turner:
Just a couple of comments. Thermostability and solvent stability often go hand in hand. So, if you get one, you get the other. Which is helpful. There are increasingly available algorithms — we use them — like Protein Repair (One-Stop) Shop or things like that. You basically just type your sequence into an algorithm, and it makes predictions of where you should mutate to get improved thermostability or solvent stability. What we typically do is try a few of them. It's a little hit and miss, but often they work really well. So, it just looks at the sequence. And if you have a structure, it looks at where there might be mobile loops and it says: "Try putting mutations in there." And often they would work really well in our experience.

Kurt Wüthrich:
I'm somewhat surprised about this discussion. I am aware of thermostability but when we now talk about applications, shouldn't we aim for enzymes that have optimal activity at low temperature? Just to save energy? I think to the best of my knowledge this is being searched for in industry.

Nicholas Turner:
But if you go to higher temperature, you will get higher rates and higher solubility.

Kurt Wüthrich:
Well, you should now work trying to get comparable activity at lower temperature. Of course, you will have reduced turnover rates because of slower diffusion. But I think you should find compromises to minimize energy consumption with enzymatic reactions.

Nicholas Turner:
Fair point! Solubility is always going to be greater at higher temperature. And I take the point that, if biocatalysis is really going to make an impact, then it should work at room temperature with equivalent rates. I see your point.

Kurt Wüthrich:
You should now work on getting stable proteins at low temperature. If we consider all the problems with climate change and so on, you should definitely work towards enzymes that produce at low temperature. I am not talking about 5°C but in the range of 20 to 40°C.

Nicholas Turner:
That raises the wider question: why do people use biocatalysis? They look at the whole process. If they have a chemical process and they want to replace it, they look at the cost of the starting materials, they look at the cost of the catalyst, they look at how many steps they need to do, the energy input, etc. It's not my field but that's what I assume industrial chemists do. So, they take the whole picture into account. Maybe in some areas biocatalysis wins, but in other areas it doesn't. It's the overall economics of the process that is going to determine whether you switch from a chemical process to a biocatalytic process. Biocatalysis often gets you into a completely different synthetic regime, with different starting materials. The thing I like about biocatalysis is that, when it works, it doesn't give you something slightly better, it gives you something actually quite different. I think that's the appeal of biocatalysis, from a process chemistry point-of-view. It can make the process look really quite different.

Ben Feringa:
This is very intriguing, because of what Kurt says about temperature. I see that often people say: "We have these enzymes that are so active,

there are so many turnovers." But the real challenge, from my perspective, is also to integrate different catalysts, also synthetic catalysts. There is also a challenge to slow down the enzymes to make them compatible with the synthetic catalyst, because a lot of synthetic catalysts cannot keep up with the enzymes. And when they mismatch, you don't have the proper concentration, and so on. Is that also not a challenge in this respect?

Donald Hilvert:

There are many thermostable enzymes that are known to lose activity when you drop the temperature, natural thermostable enzymes. However, many enzymes follow a simple Arrhenius relationship. So, if you make something very active at 95°C you will get a corresponding rate acceleration at lower temperature. That said, if you have an enzyme that is stable at 95°C, it also has a potentially much longer lifetime. I think that's a big advantage, practically, to select for very stable enzymes. To answer Ben (Feringa)'s question: all you do is change the concentration. If you have an enzyme that's too fast, you use 10 times less.

Stefan Lutz:

I think, the problem you described becomes very obvious in cascade reactions, where you don't have control over the concentration. And there, there is an active effort to fine-tune and to coevolve enzymes that are part of a cascade. For example, Islatravir was mentioned several times. Those enzymes were coevolved in order to align and to maximize the flux through the entire cascade, in some cases requiring much less evolution because the native enzyme had a superior activity, others needed additional rounds of evolution to align and optimize overall flux. Maybe a quick revisit of the point that Professor Kurt Wüthrich raised about temperature. I fully agree with you, to desire to operate under greener conditions. An example where elevated temperature was a desirable feature was in the DNA synthesis, where, as you are polymerising the nucleotide, secondary structure can greatly influence the access for the enzyme and affect the synthesis. Being able to thermostabilize the protein, running the reaction at elevated temperatures to eliminate secondary structure, was a beneficial feature there. But generally speaking, I would agree with your vision of driving towards lower temperatures.

Gerald Joyce:
I want to come back to something at the very beginning of the discussion, that Andrew Turberfield started, which is: "Where to begin the game?" You kind of put two options out there, either starting with a natural occurring enzyme or starting with one of these designed scaffolds. And Don (Hilvert): you made the comment that, by starting with the design scaffold rather than an evolved enzyme from Nature, you can avoid the evolutionary baggage of an enzyme from Nature. But I would also point out that such enzymes are civilized, they are not feral, you know. I think there's more variants on this. So, one could start with a predicted ancestral enzyme, one could start with an existing enzyme that mechanistically performs a similar reaction but not the one you are interested in, one can start with an existing enzyme that through catalytic promiscuity has a bit of the activity you are interested in. So, there are all these possibilities. What's frustrating to me about the field — and I am guilty of this as well — is that too much of what we have done is anecdotal. You know, there are these shaggy dog stories about: "We started with this, we did this, and we got this and here is the result." What I haven't seen in the field is truly side by side competitive experiments where there are multiple starting points in play, that are given a fair and equal chance to get there. I know that's trickier because now you are committing some of your wells — or some of your population size — among different starting points. Isn't that what it's going to take for us to get the answer to the question of: "where to start?" And I hope the answer isn't going to be: "Well, it's different for every possible reaction." Because we really would like to find some general lessons about what a smart starting point is — if not the best starting point. How can we get more sophisticated about this?

Donald Hilvert:
I think what we want to do is to run evolutionary races from different starting points. And that's cumbersome when you are doing plate assays. But if you have the throughput of a fluorescent activated droplets sorter, where you can look at 100 million variants in a single experiment, then that becomes much more feasible because you can start at very low activity. We have shown that we can go all the way back down the mountain and optimize to a significant level with very low starting activity. So, I think

there you have a chance of actually exploring this. We're not going to be able to do hundreds and hundreds of these experiments. But I think we can begin to compare designs with similar starting activities in different kinds of scaffolds.

Gerald Joyce:

So, who's going to do this? In the protein folding field, they basically set up a challenge across the field for people to make their best predictions. But how does our field address this? I think it's going to be hard in the commercial setting because you've got milestones to make. How do we actually do this?

Alison Narayan:

I think we're starting to get to the point where we have the tools for these larger scale experiments. Sabine (Flitsch) talked a lot about the advances that have been made that make it easier for us to get different enzymes starting points in a given laboratory. We have systems that are going to allow us to get to higher throughput within our own labs, like Don (Hilvert) talked about in the droplet analysis, but also in continuous evolution systems like developed by Dave Liu and being used by Bryan Dickinson at U. Chicago. I think these platforms are going to enable us to more easily ask these questions of different starting points. I think that's exciting to do. In my own lab we often face this question of: "what is the best approach?" Do we take an enzyme with a 2% yield that gives us four different products and try to whittle that square peg to fit in our round hole or do we go back and search for a different starting point? I don't think that there are enough examples yet for us to know what the right rules are to follow, but it's an exciting time to start writing those rules.

Nicholas Turner:

You are not going to like this answer, but I personally think it's a real strength that we have these different starting points. Anthony Green designed a Baylis-Hillmanase and evolved it. That's the first enzyme that catalyses that reaction (Morita–Baylis–Hillman reaction). I can find a brand-new enzyme from the metagenomic library that hasn't been designed. To me, the multiple ways of starting, most of which will go through this direct evolution funnel at some point, becomes a real strength. We are not limited to one or two ways. We have many ways of starting. And, directly, evolution becomes a key conjurer into where we want to go.

Gerald Joyce:
I think that it's a great advantage, that there are lots of starting points. But my frustration is that the field — and I am guilty of it myself in RNA directed evolution — is more of a craft than a Science. So, it would be nice to have an agreed upon challenge that by many different approaches several different labs are all going to go for that same challenge, for example. It's not just the starting point where it's craft rather than Science. Think about all the arbitrary decisions that are made during the process of directed evolution — of an enzyme, in this case. Where to mutagenize, how high to mutagenize, what size to screen, what fraction of winners to select, whether to only select top cream of the crop or to sort of maintain a deeper call, how many rounds to do, when to mutagenize, whether to mutagenize randomly or targeted? The whole thing is craft work. So, I agree, and this session has made it clear that incredible progress has been made in the field. But it's time for the field to turn from craft to Science. That would be my ask.

Sabine Flitsch:
That's a very nice comment. One more answer.

Sílvia Osuna:
I think that's a great suggestion! We should organize something like the CASP competition, where people need to predict the structure. That's where AlphaFold turned out to be a very impressive solution. I like the suggestion! So maybe this could be done — starting with different scaffolds and try to see which is the one that works.

Sabine Flitsch:
We have a question over there.

Yamuna Krishnan:
This brings me actually to my question, which is to all the panelists. Given where the field is now, what do you see as the next technological

breakthrough that must happen and that will give you the next-step change in finding your new chemistries or finding the proteins that you want?

Sabine Flitsch:

I think that's a question where we all could answer. Maybe I get started. I think we need to dramatically expand the type of reactions. If you look at the reactions in organic chemistry, there is a very large number. Biology, we talked about this before, only uses a very limited number of types of reactions and I think that could be dramatically expanded. I personally am a big fan of C-H activations, for example, and we know biology is very good with C-H activations, looking at oxygenases and oxidase and so on. If you could pick any C-H bond on any organic molecule and activate that with an enzyme — which, in principle, you should be able to do and you have tools to do — that would be enormously beneficial, even if you did that with just one substrate. We have currently a project where we are looking at fatty acids. If you think of a fatty acid, every C-H bond in a fatty acid is different if you include stereoisomers. Could we have a toolbox of enzymes that could selectively activate every one of those and make more complex molecule? So, I think there are still a lot of challenges in the type of reactions we could do. I don't think we can do every organic reaction, but I think we can dramatically increase that set of reactions. Maybe if my colleagues want to say something?

Donald Hilvert:

I would just add one other class of reactions, that is radical reactions. In biology, radical SAM enzymes have come to the fore. They do absolutely amazing chemistry that is not duplicated using more standard methods. If biology can control radical reactivity, we should be able to capitalize on that. I think it's a huge opportunity for people interested in design and also just for being able to exploit the natural family of SAM-dependent enzymes. There are hundreds of thousands of these enzymes that are not characterised.

Alison Narayan:

I'll backup and agree with Sabine that something that's really important is development of more reactions, a greater breadth of reactions. In my opinion, one of the most important things to accomplish that is engaging a broader chemistry community into biocatalysis, such that, it's not only the people who are doing biocatalytic reactions and thinking about reactions they could do with an enzyme who sit in the biocatalyst community. But that enzymes are available enough to your ordinary organic chemists that they could think about developing a reaction that way. So, if someone wants to commercialize enzyme panels, that would be very helpful.

Donald Hilvert:

For our area of design, better computational methods are needed. Faster computational methods, more accurate computational methods, computational methods to take into account that proteins are highly dynamic, and these dynamic properties can shape the active site and contribute to catalytic efficiency. This is something that we just don't capture in the kinds of methods that I presented today. This is the future, I think.

Sílvia Osuna:

I totally agree with what Don (Hilvert) said. I think it's a very exciting time for the field, for the computational design of enzymes. With all these recent results with the application of machine learning, I think it will have a big impact. Thanks to AlphaFold, which has revolutionized the field in many different ways, machine learning techniques have been applied already for protein design. Of course, we need to move forward to not only protein design but also try to design functional enzymes. I think method development will be very, very relevant and definitely — that's my dream. So, for the next year I will try to come up with strategies to be able to take into account the flexibility, as well as the chemical steps, and try to design efficient enzymes.

Sabine Flitsch:

If you could computationally just design the enzyme we want. As wonderful as directed evolution is, if you can get away with it and just design it computationally, we could make it and we could develop catalysts for every reaction molecule very, very quickly. So, finding the right enzyme is still the slow step, despite what we said about evolution. There is a sort of "need for speed" — someone said — and if you wanted to do a reaction next week and you could computationally design your catalyst and you could make it in a week, you could do that particular reaction next week. At the moment, we are still talking about months because you have to do evolution.

Bert Meijer:
I was just thinking about what you said at the end. Many reaction that we know are strong because we also say that it doesn't work here, here and here. You now give the impression, to me at least, that every conversion can be done biocatalytically. I think the strength of the field would be also to say: if you are interested in this type of conversion, never think about biocatalysis.

Donald Hilvert:
Maybe not reactions with butyllithium.

Bert Meijer:
I mean conversions. I think it would be helpful for those who are not in the field to see when you can look, as an alternative, to biocatalysis or you shouldn't do it. And if you're not in the field it's difficult to decide whether there is any chance to do it.

Nicholas Turner:
I totally agree! If you look at biocatalysis, compared to organic synthesis, there are big gaps. There are relatively few ways of making carbon-carbon bonds using enzymes compared to organic synthesis. Organic chemists have invented way more methods, more general, making carbon-carbon double bonds. If you want to do that biocatalytically you have to essentially do an aldol condensation dehydration. There is no "Wittig-ase". I mean, people are encroaching on these, but there will always be certain reactions, I think, that are just not feasible biocatalytically because of the solvent. If you are going to generate a reactive intermediate in water, you know you can't. The most common reaction we get asked probably to invent is fluorination with F^+. You cannot do it in water. You can do F^-. You can fluorinate, obviously, there are fluorinating reagents in organic synthesis that operate by DAST, that uses F^+. So, people say: "can't you find an enzyme to do the same thing?" I don't see how you can generate F^+ under aqueous conditions.

Sabine Flitsch:
There were a few more questions to answer.

Kurt Wüthrich:
I think it relates to the questions and the contributions that were made during the last 2 minutes. Don (Hilvert) requires more efficient calculations. Now, the major breakthrough of the last two years is the advent of promising structure predictions on a large scale: AlphaFold. In some way, your field is related to what is quoted as "rational structure-based drug design". When I talk to my colleagues who are in medicinal chemistry, trying in a rational way to develop drugs, I am told that the structures obtained by AlphaFold are not quite good enough for their purpose. How is it in your field? Is this revolutionary in what you need of structural biology, or do you still need more precise experimentally determined structures? That would in no way invalidate AlphaFold, which can be used to greatly speed up experimental structural determinations.

Sabine Flitsch:
Maybe Sílvia can answer?

Sílvia Osuna:
Of course, when you run AlphaFold you get a structure that, in the end, is a model. And for some of the regions of these models, the level of confidence is not very high. But I think it's a very big breakthrough. In our group, what we are trying to do is to generate more than one structure from AlphaFold to try to have an estimation of the flexibility. I think that's very important because we can run AlphaFold in just seconds and get info about conformational heterogenicity in just seconds — that really makes a big difference! Maybe you are right, maybe the predictions that we get are not very accurate. But then you can further refine them. Ok, you need to do MD simulations that takes much longer. But as I showed you in my presentation, if you take output of a structure from AlphaFold that have different conformation and do very short, one-hour MD simulations, you can get a decent estimation of the conformational landscape. That is very exciting because that means that, just in hours, we could get some insights

and we don't need months of simulations. I think this can actually have an impact. Maybe the prediction is not very accurate, but you can couple it to other techniques and try to make it more accurate. And of course, the more structures are deposited, the better. You can train the model and can get more accurate results.

Sabine Flitsch:
Do you want to comment on that?

Ben Feringa:
I wanted to comment a little bit about the perception of using enzymes in synthesis, if I am allowed.

Sabine Flitsch:
Yes, of course. There is just one more comment.

Donald Hilvert:
David Baker published two papers last week using machine learning to actually design proteins. And the designs were much more successful than the original Rosetta fold. That said, designing structure is a lot easier than designing function, and that is the challenge. And it remains the challenge.

Ben Feringa:
You know I like butyllithium, but I also like enzymes and I tell you why. In the synthetic community, there is the perception: "Oh, enzymes, that's a different field, I should not touch that." I'll tell you a case. I think it's also how we instruct our students and what we confront them with. I tell you this case where we were struggling for quite a while with an asymmetric HCN addition. Just a simple addition of HCN. And then, at a certain point, I said to the student: "Go to the supermarket, buy some cashew nuts, don't get them roasted, mill them, ..." They used the enzyme for HCN addition and, within one day, they had 23 grams of pure adduct with 99% enantiomeric purity. Since that time, I never heard that complaining about enzymes.

Sabine Flitsch:
That's very kind of you. Rob, do you have a question?

Rob Knowles:
I did have a question. Coming back to a comment that Stefan made. With these really impressive cascades of reactions, it seems that the optimization and the coevolution become a lot more challenging. I wonder if you could just comment on general strategies for these coevolution types of approaches. If you're just looking at starting materials and products, it is not clear where you maybe want to focus your efforts. I was curious if you could just comment generally about that.

Stefan Lutz:
The way we practice the coevolution is largely on the screening side. As you evaluate individual library members, you are not just running individual reactions, you're combining them as part of your evaluation after each round of evolution, or after each round of diversification. The most practical solution is to assemble the individual variants — or the lead variants — into the actual cascade and assessing their collaborative performance in those settings. So, that's to us the practicing of coevolution optimization, does that make sense?

Rob Knowles:
When these things are coupled ... and you were talking about driving equilibrium by coupled reactions and pushing things, like flexing things through. It seems like it is challenging in that way, right? Because they're not just discrete things that you can individually optimize per se. That was, I guess, what I was trying to find.

Stefan Lutz:
For example, the evolution of the cascade for the Reb M (rebaudioside M), this artificial sweetener involving two glycosyltransferases and a cofactor regeneration enzyme, largely consisted of creating diversity in the individual enzymes, screening them in test reactions, taking the leads and then combining them in running the actual process in a microtiter

plate format to evaluate the interplay of these lead variants. There's no general protocol. It requires a great deal of expertise from the experimentalist to do that type of optimization.

Rob Knowles:
Thank you.

Andrew Turberfield:
If I can come back to the craft versus Science question. What I would do, and I accept, I'm probably totally, impossibly naïve, I would take a bad design in which the catalytic residues were very close in primary sequence, and with robust local secondary structure, and I would throw out the challenge of embedding that module in anything else you like, and then evolving it and see what happens. See what works best.

Donald Hilvert:
That's basically what we did. There's a paper that is in press in Protein Science with the 8-stranded beta-barrel. So, we've not done it on many more scaffolds yet, but that's in progress.

Gerald Joyce:
That is really good! That's what we call the skunk in the room, right? You actually put a bad piece in the middle... Just coming back again to how to put more rigor in the Science. There are so many options there. I think one technical advantage we have, and certainly more so in recent years, is the use of deep sequencing analysis. So, it's not just the answer to: "How many rounds it took us to get there? What is its k_{cat}?" It's the trajectory. I think, if one ran one of these bake-off kinds of competitions, if one also committed to doing deep sequence analysis round by round through that trajectory to look at enrichment trajectories within the population and sub-dominant individuals, and so on…: that, I think, might be a way to build more rigor into the process.

Donald Hilvert:
It will also provide the data for machine learning.

Matthew Kanan:

I want to come back to I think something that was mentioned earlier about the concentration for a scalable process. You said you push it to 100 grams per liter. I'm just wondering, in general, how challenging that is? If you've identified an enzyme and you have the great reaction metrics that we are used to. Then, for a substrate that you're going to produce on scale, what are the challenges going from there? Presumably, the concentrations you're working on at the discovery phase are far lower than that. And a related question: I'm not sure I fully understand all these cofactor regeneration approaches. For example, if I'm interested in an oxidation and I want to run it aerobically with NAD^+ or something. How feasible is that? So, I guess combining those two: aerobic oxidations at high concentration are hard in general, but really powerful for a lot of high-volume chemicals. What's the state-of-the-art with respect to that?

Nicholas Turner:

So, how general is it to be able to get up to 100–150 grams per liter with a biocatalyst? It's becoming more general. There are many enzymes that have been engineered to work at that concentration. It's not always achievable, but, if you take for example the imine reductase family of enzymes — which I briefly talked about — those wild type enzyme sequences are actually quite tolerant of reasonably high substrate concentration. And they have turnover numbers in the region of 10,000 for a wild type enzyme. So, that relates a bit to Rob's point: "How do you engineer these enzymes in the context of cascades?" What we do in our lab — and obviously you do this in industry — is we will set ourselves targets. For example, we need this level of activity, this level of turnover frequency, turnover numbers, substrate concentration. If we get to that, we think we have a catalyst that either we can put into a cascade, or we could potentially hand over to somebody to scale. And a lot of this is based on experience, you know, many biocatalytic processes have now been developed. So, people know what a good biocatalytic process looks like because it has certain hallmarks in terms of concentration, turnover, and so on. Twenty years ago, that wasn't the case, but now we know what we're shooting at in terms of parameters. In terms of oxidation, that is an issue. What we find when we wind those enzymes up, oxidases to do oxidation, oxygen can become limiting. And so, we need to think of ways of getting more oxygen into the solution because the enzyme activity is being limited by the availability of dissolved oxygen, which is relatively low in water at ambient pressure. Cofactor recycling, I would say... If you're doing oxidations with cofactors, I think that's essentially a solved problem because of the high turnover numbers of those enzymes. Codexis has a glucose dehydrogenase that has been engineered and that is so stable and so active that cofactor recycling sort of drops out of the equation in terms of being an economic barrier.

Matthew Kanan:
Sorry, so just to clarify… If I'm going aldehyde to acid, you can run that off oxygen trivially, even if it's not using oxygen as the chemical oxidant in the mechanism. It is using NAD^+, for example.

Nicholas Turner:
There are basically two ways of doing oxidation — maybe three — in biochemistry. You can for instance dehydrogenate, using some sort of nicotinamide dependent enzyme.

Matthew Kanan:
Yes, so that's what I'm asking for. Dehydrogenation mechanism, turning that into oxygen as the stoichiometric oxidant. How easy is that?

Nicholas Turner:
Then you need to switch to an oxidase or a peroxygenase, which is a mechanistically different enzyme. It is going to use oxygen to do the oxidation, and it is going to produce peroxide as the byproduct, possibly water. But it's a different way of oxidizing.

Matthew Kanan:
So, there's no way to recycle NAD^+ aerobically, I guess, it's my question.

Nicholas Turner:
So, you want to convert it back to NADH?

Matthew Kanan:

I'm saying, it's becoming NADH in the oxidation reaction, and then I want to take it back to NAD^+ with oxygen.

Nicholas Turner:

So, you would use an NADPH oxidase, which is an oxygen-dependent enzyme that would do exactly that transformation.

Kurt Wüthrich:

So far, without explicitly stating so, we talked only about homogeneous catalysis by enzymes. Over the years, there have been attempts to couple enzymes on supports and then run enzymatically catalyzed reactions through columns. What is the state of this approach these days?

Sabine Flitsch:

That is possible. It has become even quite easy in the laboratory because a lot of these enzymes contain tags anyway (these polyhistidine tags). You can immobilize them on commercially available resins. Some enzymes work really well and others not. It's a little bit of a black art, still. But that is entirely possible. Purolite is another really good Epoxy resin, very cheap resin. We've immobilized oxidases on Purolite. In fact, that stabilizes our oxidases. We can use them much more in organic solvents, they're more thermostable. So, I think that is quite a mature Science. Although it is a little bit of a black art. It works with some enzymes and not others, but it is entirely possible.

Kurt Wüthrich:

Now semantics. Would it be correct to talk about heterogeneous enzyme catalysis? I mean, what is the relative importance of the two approaches?

Sabine Flitsch:

In fact, I think that's what we did. We could say we turned homogeneous into heterogeneous catalysis in the paper.

John Sutherland:

When would you use fermentation instead of biotrans (biotransformation)? You mentioned penicillin. Would you make penicillin by biotrans?

Would you make it by fermentation then use biotrans to remove the sidechains? What are the issues?

Sabine Flitsch:

As you know, the penicillin synthase is a terrible enzyme. We are still struggling with that. It goes back to that "oxygenase" point as well. It's extremely sensitive to oxygen, so supplying with the right amount of oxygen is really difficult. Amazing enzyme, but that is definitely better for fermentation at the moment. Although we are still working on that a little bit. Also, I should say, its substrate — the ACV tripeptide (alpha-aminoadipyl-cysteinyl-valine) — you also have to make that. Which is some effort to make it. So, definitely, for penicillin fermentation is still much better.

Andrew Fire:

I was intrigued by the metagenomics that Nick mentioned. He was using it to find leads but I am wondering, if you look at the evolutionary trajectories experimentally, do you see equivalents in the metagenomic data for that? There's quite a bit out there to look at and see if the biological world has played with the same things that you're finding in experimental evolution.

Nicholas Turner:

I didn't quite understand the question. Is it: "When we look into metagenomic sequences do we find similar things to evolved enzymes?"

Andrew Fire:

Do you find similar changes to what are observed in the course of experimental evolution?

Nicholas Turner:

Oh, I see. That's a very good question! In fact, one thing we do, if we've got a new enzyme and we think we have identified some residues that are important in controlling selectivity, we will look into known sequences — which is easy to do — to see what the natural variation at that residue is. And if it's highly conserved, then it probably indicates that it's not something you can change. But if there is natural variation in that position, that is often a key indicator that this is an area of the protein that we can mutate and maybe expect to see changes. Because Nature has already explored that particular residue — or set of residues. Because it is actually occurring quite frequently in natural sequences. That is where, to me, metagenomic meets directed evolution. It's two different solutions to the same problem, natural variation as well as synthetic, laboratory-created variation.

Sílvia Osuna:
If I can add to that. The particular case that I showed you, the six positions that we targeted in this tryptophan synthase B. What we did afterwards was checking a number of multi-sequence alignments for TrpB sequences from other organisms that already had these six mutations. And we found one in an organism from a lake in Norway. We were happy to see that this sequence, this specific TrpB, had also standalone activity and very similar to the SPM6 variant that we designed.

Andrew Fire:
Another thing I would point out here is that the full breadth of metagenomic data that is out there is much greater than, for instance, any one company would have. All the sequences for all the microbiomes, and all the dumps, and all the oceans, is available now. And I think that is another tool that would come from massively parallel computing : it would be to look at a much broader set of metagenomic data.

Sabine Flitsch:
I think we're running out of time. Unless anyone has any very urgent question.

Stefan Lutz:
I just wanted to come back to your question. I think the metagenomic information, identifying variability in various positions within a protein sequence, is invaluable information for designing. Depending, of course, how far you divert from the natural function. The natural diversity in individual positions becomes less and less relevant then. But nevertheless, for the initial analysis, it's a critical element that can speed up the design component in a significant way.

Sabine Flitsch:

I think we've run out of time. I just want to thank the audience. As you could see, we didn't really have to ask each other any questions. You asked us a lot of questions and I'm really, really grateful for the very positive attitude with which you have approached our enthusiasm for the field. Thank you very much for those discussions. I close the session.

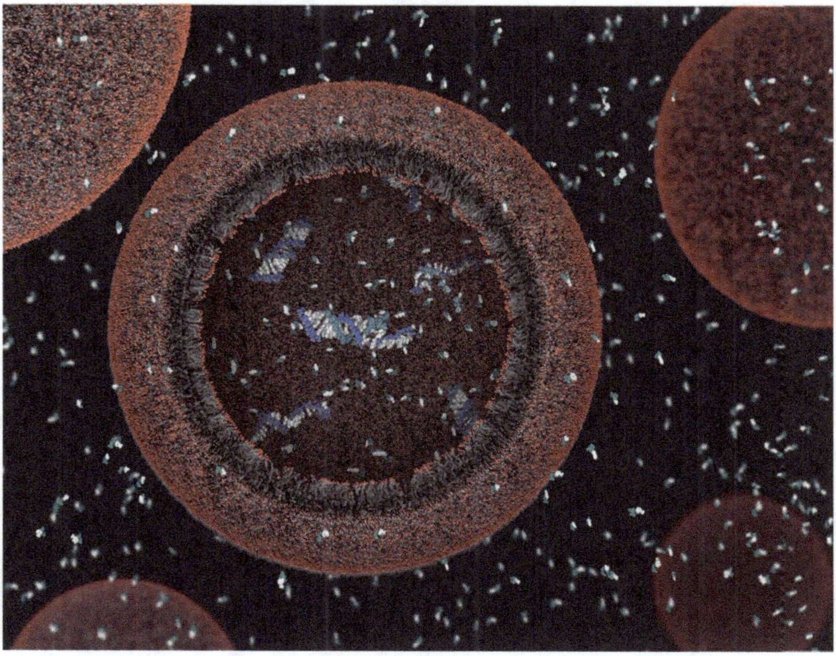

Digital Illustration of a protocell

Image courtesy of Janet Iwasa

Session 6

RNA Chemistry to Explore our Origins and Fight Viral Pandemics

https://doi.org/10.1142/9789811282324_0040

RNA Replication and the Origins of Life

Gerald F. Joyce

The Salk Institute, La Jolla, 92037 CA, USA

RNA-Based evolving systems

All known organisms rely on DNA as the genetic material and proteins as the chief agent of function, but the machinery needed to copy DNA and express proteins is far too complex to have arisen spontaneously. In the late 1960s, as the principles of molecular biology came into focus, it was first suggested that the earliest form of life instead relied on RNA as both the genetic material and the agent of function [1–3]. Special attention has been directed to what Francis Crick called the "first enzyme" of life: an RNA molecule that catalyses the replication of RNA and thus is both gene and enzyme [2]. Such a molecule could provide the basis for a living, evolving system.

There are several known examples of RNA enzymes in biology, but none have the ability to copy RNA. A larger number of RNA enzymes have been developed in the laboratory using directed molecular evolution, including those that can copy an RNA template by joining together the nucleotide building blocks of RNA (A, U, G, and C) [4–6]. Like natural evolution, directed evolution relies on processes of amplification, mutation, and selection to enrich a population with individuals that are most fit, but in directed evolution, the experimenter defines the fitness criteria.

RNA viruses also undergo processes of Darwinian evolution, resulting in the emergence of novel variants with increased fitness. Copying of the viral RNA is dependent on protein enzymes that are encoded within the viral genome, but those proteins must be synthesized by the machinery of the host cell. Thus, an RNA virus cannot be regarded as a living system in its own right. The fitness of a virus is ultimately determined by the functional copy number of its genome over time, but that fitness takes into account the material properties of the viral genome and the virally encoded proteins.

The directed evolution of RNA serves as a model of both RNA-based life and viral evolving systems. In fact, the first directed evolution experiment, carried out by Sol Spiegelman and co-workers in 1967 [7], involved the viral genomic RNA of Qβ bacteriophage, which was replicated in the test tube using Qβ replicase protein. A portion of this multi-subunit protein is encoded within the viral genome, with the remainder supplied by the host cell. The original Spiegelman experiment, and others that followed, demonstrated the evolution of variants of the Qβ genome with increased fitness. Fitness was no longer coupled to viral infectivity but simply a reflection of the increased copy number of the virus-derived RNA under the chosen set of experimental conditions.

Modern-day directed RNA evolution experiments seek to drive RNA to perform novel functions, sometimes with a practical application in mind, and also to explore the catalytic potential of RNA. RNA viruses have the benefit of encoding proteins with broad functionality, including the critical function of replicating the RNA genome. The functional superiority of proteins over RNA likely explains why there is no known example of a viral RNA that catalyses its own replication. Yet if we are to address the question of how the first living systems arose, before the advent of instructed protein synthesis, then it is important to seek RNA molecules that can function as an RNA-dependent RNA polymerase, with the ability to catalyse the replication of RNA.

Towards an RNA enzyme with RNA replicase activity

Early attempts to develop an RNA-dependent RNA polymerase focused on stringing together a few letters of RNA by adding activated nucleotides

(NTPs) to the end of a template-bound RNA primer [4, 8, 9]. Further improvements enabled several dozen nucleotides to be copied but only for unstructured, repetitive templates [5, 10]. Our laboratory entered the fray by evolving polymerases that can copy "difficult" templates to yield a functional RNA product. Selection of the polymerase was made dependent on the function of the synthesized product, requiring the synthesis of progressively more complex products. Those efforts resulted in the evolution of RNA polymerases that are faster, more accurate, and more general in copying RNA [6, 11, 12].

During the directed evolution process, the RNA enzyme underwent a dramatic structural rearrangement of its catalytic core [12]. Through the accumulation of 15 mutations within the core, an existing stem element became shortened while a new stem element was formed, together creating a pseudoknot structure that lies in close proximity to the enzyme's active site (Fig. 1). Three important attributes emerged together with this structural rearrangement. First, the catalytic rate improved by ~4,000-fold compared to the starting enzyme. Second, the polymerase gained the ability to copy templates of almost any sequence, including those with structure. Third, the polymerase gained the ability to bind the template–primer complex through high-affinity tertiary interactions, comparable to those seen with modern polymerase proteins.

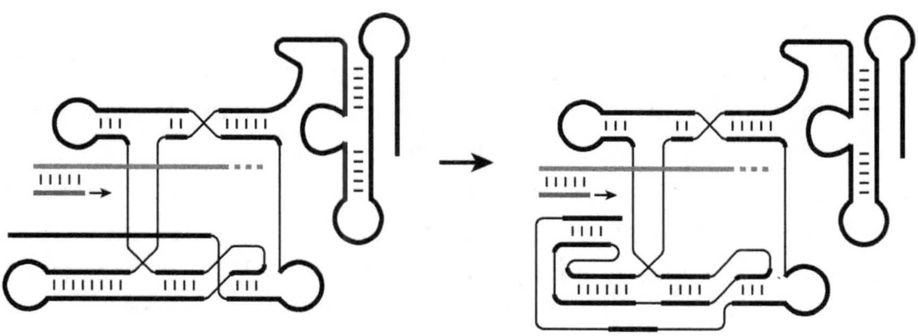

Fig. 1. Secondary structure of the RNA polymerase (black), which extends an RNA primer (red) on an RNA template (cyan). Over the course of directed evolution, the RNA enzyme underwent a tertiary structural rearrangement, whereby an existing stem element became shortened while a new stem element was formed, together creating a pseudoknot structure.

This advanced form of the RNA polymerase can copy more than 100 nucleotides in 10 minutes and can operate with an accuracy of 92–94% per nucleotide [12]. However, the polymerase itself contains 184 nucleotides and is especially difficult to copy. Furthermore, RNA replication requires copying both the template and its complement, which doubles the challenge. Thus, further improvement of the catalytic rate, copying accuracy, and sequence generality of the polymerase will be needed to recreate the first enzyme of life.

The accuracy of polymerization is critical for an RNA replicase to be able to support the self-sustained evolution of RNA. If the error rate is too high, then the copies will be riddled with mutations, exceeding the ability of selection to cull deleterious mutations [13]. Deep sequencing was used to assess the position-specific frequency of mutations for both partial- and full-length extension products on templates ranging from the most favourable to the most difficult. This analysis revealed that when the polymerase is pushed to the limits of its activity, the accuracy of synthesis declines [6, 12]. For a short, unstructured template of 11 nucleotides, the average fidelity is 97% per nucleotide, with the majority of mutations due to G•U wobble pairing. Excluding wobble mutations, the fidelity is >99%. For a longer, more structured template of 33 nucleotides, the average fidelity drops to 92% overall and 96% excluding wobble mutations. For an even longer and highly structured template of 77 nucleotides, pushing the limit of polymerase activity, the average fidelity is 84% overall and 88% excluding wobbles.

Examination of the partial-length products revealed that fidelity is lowest for the last added nucleotide and increases monotonically for positions further upstream from the last nucleotide. The longer the polymerization reaction is allowed to continue, the greater the overall yield and also the lower the fidelity of the full-length products [6]. Taken together, these facts indicate that the polymerase stalls after adding a mismatched nucleotide but over time can extend past the mismatch to incorporate the mutation within full-length products. It will not be sufficient to evolve a faster polymerase unless the polymerase also evolves either a lower frequency of mismatched NTP addition or a reduced propensity to extend mismatched termini.

We are continuing the directed evolution process to develop ever more capable forms of the polymerase, focusing especially on improving the fidelity of template copying. The polymerase now has sufficiently high activity that it can synthesize its own evolutionary ancestor, an RNA-joining enzyme that contains 97 nucleotides. By challenging the evolving population of polymerases to synthesize a functional copy of its ancestor, we are placing unprecedented selection pressure on improving polymerase fidelity because about half of the nucleotides within the synthesized product cannot be mutated without loss of activity [14]. Furthermore, the RNA being synthesized has the same catalytic domain as the polymerase itself, thus training the polymerase to synthesize an RNA of similar composition.

As a result of the most recent rounds of directed evolution, the fidelity of polymerization has improved from 84% to 89% for the synthesis of the 97-nucleotide functional product. This is the first time that the ability to synthesize longer products has been accompanied by improved fidelity. We appear to have entered the long-anticipated virtuous cycle, where the ability to synthesize longer products enables us to impose selective pressure to drive further improvement of fidelity due to the greater number of immutable nucleotides within those longer products. In turn, every improvement in fidelity enables the synthesis of ever longer functional products.

The threshold of heritable information

The propagation of heritable information requires both efficient and accurate copying of that information. It has long been recognized that there is an "error threshold" based on the relative advantage of a selectively advantageous individual compared to the population as a whole, taking into account the probability of producing error-free copies [13]. For the copying of RNA genomes, there is an inverse relationship between the per-nucleotide fidelity of polymerization and the maximum length of RNA that can be maintained through successive rounds of replication. For the RNA polymerase we have been studying, an average fidelity of >98% will be needed to achieve self-sustained Darwinian evolution. The greater the efficiency and fidelity of the polymerase, the more readily it can be evolved towards further improvements in efficiency and fidelity because one can then impose greater selection pressure to drive those improvements. This bootstrapping process is analogous to what is thought to have driven the evolution of more complex genomes during the early history of life on Earth [6, 15].

The genomes of RNA viruses typically contain 10^3–10^4 nucleotides, and the error rate of the corresponding viral RNA polymerase proteins that copy those genomes is in the range of 10^{-3}–10^{-4} [16]. Some RNA viruses, such as HIV-1 and poliovirus, operate very close to the error threshold, which facilitates their rapid evolutionary adaptation but also places them close to overstepping the error threshold and no longer able to maintain heritable information.

Considerable effort has been devoted towards pushing RNA viruses over the error threshold by exposing them to mutagens, an approach that has been termed "lethal mutagenesis" [17]. This effect must be distinguished from the way in which a mutagen can reduce copying efficiency. Lethal mutagenesis is the result of a cascade of copying errors that cannot be balanced by selection [18]. For example, the purine analogue ribavirin, in addition to inhibiting viral replication, exerts an antiviral effect through enhanced mutagenesis [19]. A more contemporary example is the cytidine analogue molnupiravir, which has been approved for the treatment of patients with symptomatic SARS-CoV-2 infection [20]. This compound promotes G-to-A mutations and is resistant to the proofreading exonuclease encoded by the virus [21].

Both self-replicating RNA enzymes and RNA viruses lie close to the edge of life. Both are able to maintain heritable genetic information and undergo Darwinian evolution. However, both lie precariously close to the error threshold, beyond which it is no longer possible to maintain that genomic information. In addition to being interesting in their own right, these systems serve as simplified models to study the fundamental processes of Darwinian evolution. These processes provide the basis for all known life, from the time of its origins and throughout its natural history.

Acknowledgements

This work was supported by grant NSSC22K0973 from NASA and grant 287624 from the Simons Foundation.

References

[1] C. Woese, *The Genetic Code*, Harper & Row, New York (1967), pp. 179–195.

[2] F. H. C. Crick, *J. Mol. Biol.* **38**, 367 (1968).

[3] L. E. Orgel, *J. Mol. Biol.* **38**, 381 (1968).

[4] W. K. Johnston, P. J. Unrau, M. S. Lawrence, M. E. Glasner, and D. P. Bartel, *Science* **292**, 1319 (2001).

[5] A. Wochner, J. Attwater, A. Coulson, and P. Holliger, *Science* **332**, 209 (2011).

[6] K. F. Tjhung, M. N. Shokhirev, D. P. Horning, and G. F. Joyce, *Proc. Natl. Acad. Sci. USA* **117**, 2906 (2020).

[7] D. R. Mills, R. L. Peterson, and S. Spiegelman, *Proc. Natl. Acad. Sci. USA* **58**, 217 (1967).

[8] K. E. McGinness and G. F. Joyce, *Chem. Biol.* **9**, 585 (2002).

[9] H. S. Zaher and P. J. Unrau, *RNA* **13**, 1017 (2007).

[10] J. Attwater, A. Wochner, and P. Holliger, *Nat. Chem.* **5**, 1011 (2013).

[11] D. P. Horning and G. F. Joyce, *Proc. Natl. Acad. Sci. USA* **113**, 9786 (2016).

[12] X. Portillo, Y.-T. Huang, R. R. Breaker, D. P. Horning, and G. F. Joyce, *eLife* **10**, e71557 (2021).

[13] M. Eigen, *Naturwiss.* **58**, 465 (1971).

[14] E. H. Ekland and D. P. Bartel, *Nucleic Acids Res.* **23**, 3231 (1995).

[15] G. F. Joyce and J. W. Szostak, *Cold Spring Harb. Perspect. Biol.* **10**, a034801 (2018).

[16] E. C. Holmes, *The Evolution of RNA Viruses*, Oxford University Press, New York (2009).

[17] L. A. Loeb, J. M. Essigmann, F. Kazazi, J. Zhang, K. D. Rose, and J. I. Mullins, *Proc. Natl. Acad. Sci. USA* **96**, 1492 (1999).

[18] J. Summers and S. Litwin, *J. Virol.* **80**, 20 (2006).

[19] S. Crotty, C. E. Cameron, and R. Andino, *Proc. Natl. Acad. Sci. USA* **98**, 6895 (2001).

[20] A. J. Bernal, M. M. Gomes da Silva, D. B. Musungaie, E. Kovalchuk, A. Gonzalez, *et al.*, *N. Engl. J. Med.* **386**, 509 (2022).

[21] C. J. Gordon, E. P. Tchesnokov, R. F. Schinazi, and M. Götte, *J. Biol. Chem.* **297**, 100770 (2021).

https://doi.org/10.1142/9789811282324_0041

Systems Chemistry Assembly of RNA and Peptides

John Sutherland

MRC Laboratory of Molecular Biology,
Francis Crick Avenue, Cambridge CB2 0QH, UK

Introduction

The discovery that nucleotides and amino acids can both be made by cyanosulfidic chemistry [1] suggests that the assembly of these building blocks into higher-order structures should be carried out in mixtures if we want to simulate how biomolecules could have come into being on early Earth. Following this logic, we have discovered RNA aminoacylation chemistry that could have led to the emergence of translation.

My view of the present state of research on prebiotic RNA chemistry

I think that RNA should not be viewed in isolation but as part of a double act with peptides. This is because nucleotides and amino acids are most plausibly made prebiotically by cyanosulfidic chemistry and because the roles of RNA and proteins in extant biology are so intertwined. I think the genetic code and translation are rooted in an early phase of synergistic macromolecular assembly from nucleotides and amino acids.

My recent research contributions to prebiotic RNA chemistry

We have focused on two aspects of RNA chemistry recently. First, we have connected the cyanosulfidic chemistry synthesis of nucleotides [2] to a very specific geochemical scenario on early Earth (Fig. 1).

Second, we have discovered that the selectivity of aminoacylation of tRNA acceptor stem-overhang mimics by interstrand aminoacyl-transfer [3] is dependent on the sequence of the terminal three base pairs of the stem (Fig. 2). This discovery suggests how tRNAs could have been selectively aminoacylated as a prelude to the development of translation according to the genetic code.

Outlook to future developments of research on prebiotic RNA chemistry

I think that clues from experimental chemistry will allow us to describe an ever more detailed geochemical scenario that drove multi-step organic synthesis of building blocks destined to become biological on early Earth. I further think that translation and the synthesis of loosely coded peptides comprising about half the canonical amino acids must have been a very early process. Recapitulation in the laboratory should one day allow us

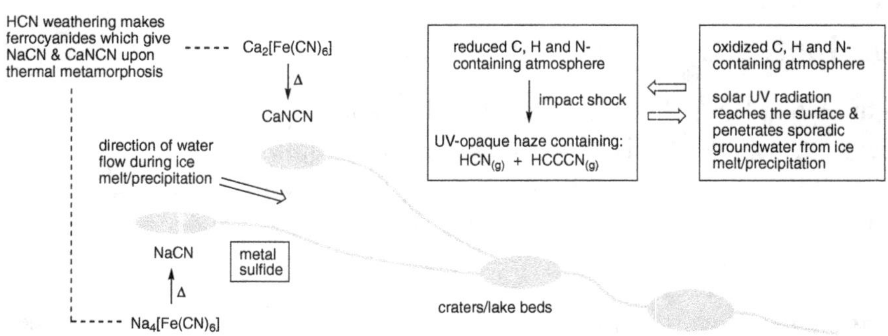

Fig. 1. Setting the stage for cyanosulfidic synthesis of nucleotides and amino acids. The chemistry would play out during and after sporadic flow of groundwater.

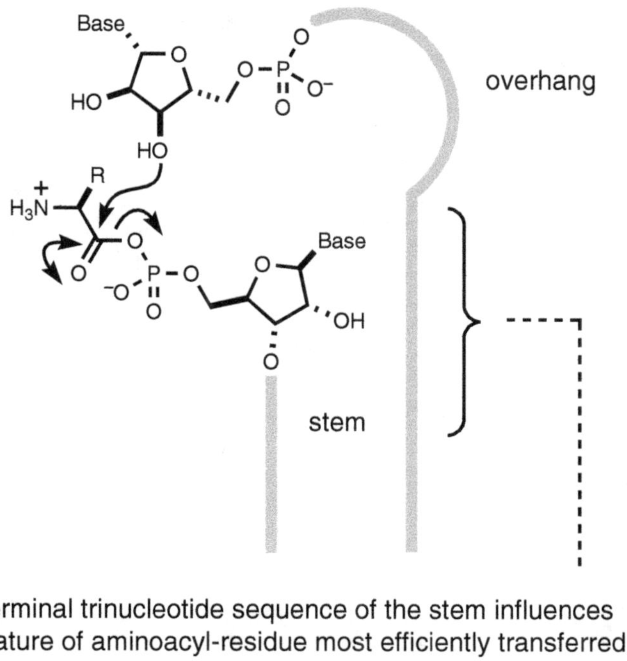

Fig. 2. Interstrand aminoacyl transfer in an RNA acceptor stem-overhang mimic.

to create living systems from feedstock molecules, such as hydrogen cyanide, hydrogen sulfide, cyanoacetylene, and inorganic phosphate.

Acknowledgements

I would like to thank the Medical Research Council (MC_UP_A024_1009) and the Simons Foundation (290362) for their long-term support of our research.

References

[1] B. H. Patel, C. Percivalle, D. J. Ritson, C. D. Duffy, and J. D. Sutherland, *Nat. Chem.* **7**, 301 (2015).

[2] J. Xu, M. Tsanakopoulou, C. J. Magnani, R. Szabla, J. E. Šponer, *et al.*, *Nat. Chem.* **9**, 303 (2017).

[3] L.-F. Wu, M. Su, Z. Liu, S. J. Bjork, and J. D. Sutherland, *J. Am. Chem. Soc.* **143**, 11836 (2021).

https://doi.org/10.1142/9789811282324_0042

From Prebiotic Chemistry to the Beginnings of Biology

Jack W. Szostak

Department of Chemistry, University of Chicago, 5735 S Ellis Ave., Chicago, 60637 IL, USA

Current status of research on the origin of life

How, given an appropriate geophysical environment, inputs of energy and prebiotically synthesized chemicals, did life actually begin? Progress in understanding the conversion of simple starting materials such as cyanide into the chemical building blocks of biology has been summarized in the accompanying article by John Sutherland. Here I focus on our efforts to understand how those prebiotically available chemicals spontaneously assembled into the first primitive cells, commonly referred to as protocells. Since the beginning of biology corresponds to the beginning of Darwinian evolution, those protocells must have contained a genetic polymer capable of encoding advantageous functions in its sequence. Moreover, the chemistry of the cell and its environment must have been sufficient to drive the replication of this material, which would have generated the variation that is the substrate for evolution. In addition, this genetic material must have been segregated into spatially localized units so that any advantage provided by a particular sequence could accrue to itself and not to other unrelated sequences.

It has become clear in recent years that the most likely primordial genetic material was in fact something very much like modern RNA. This conclusion was by no means obvious even a few years ago. Going back somewhat further, Jerry Joyce and the late Leslie Orgel concluded, on the basis of difficulties encountered in explaining the prebiotic synthesis of ribonucleotides and the non-enzymatic copying of RNA sequences, that RNA might have been preceded by some simpler ancestral genetic polymer. This outlook sparked a remarkable outburst of creative chemistry in the form of a search for alternative genetic polymers that might have been easier to synthesize and/or replicate than RNA. The highlight of this approach was the tour de force synthesis by Eschenmoser of diverse families of nucleic acids all capable of self-association into antiparallel Watson–Crick base-paired duplexes. Given the many seemingly plausible alternatives to RNA, the question arose as to why, in the end, did life settle on RNA and not something else.

Our recent contributions

Much of our recent work has focused on obtaining a mechanistic understanding of the chemistry of non-enzymatic template-directed RNA copying. Somewhat surprisingly, our efforts in this regard have led not only to a framework for the chemical copying of arbitrary RNA sequences and a model for replication but also to a potential explanation for why RNA won out over all potential competitors. We started by addressing some of the peculiar aspects of the kinetics of template-directed primer extension that had first been noted by Orgel. The most puzzling of these observations was the very slow rate of primer extension when only a single activated monomer can bind next to the primer, and the dramatic acceleration of the reaction when a second activated monomer binds downstream of the first monomer [1]. This catalytic effect had been unexplained for some 25 years until it was rediscovered by Noam Prywes in my lab [2] and then explained by Travis Walton [3] as the consequence of the generation of a highly reactive covalent intermediate which forms by the reaction of two activated monomers with each other. A series of structural studies led by

Wen Zhang in my lab culminated in the observation by time-resolved crystallography of all steps in the primer extension reaction, from monomer binding to intermediate formation to the formation of a new phosphodiester bond [4]. Subsequent studies showed that primer extension via reaction with imidazolium-bridged dinucleotides was both faster and of higher fidelity than reaction with activated monomers [5]. Our most recent studies have shown that monomers bridged to short oligonucleotides lead to even faster reactions and, importantly, allow all four canonical nucleotides to be copied and incorporated at similar rates. We now believe that the formation of imidazolium-bridged intermediates is the key to the efficient and accurate non-enzymatic copying of RNA templates.

While the above mechanistic insights provide a potential explanation for how primordial RNA sequences could have been copied prior to the evolution of polymerase ribozymes, we still have to ask why RNA and not some other genetic polymer. To address this question, we have begun to evaluate the copying kinetics of alternative genetic polymers. The two that we have studied in most detail are arabino- and threo-nucleic acids since the corresponding mononucleotides are likely to form during the synthesis of ribonucleotides. We see that threo-nucleotides exhibit a slower formation of the bridged intermediate, while both threo- and arabino-nucleotide bridged intermediates exhibit slower primer extension. Furthermore, once a threo- or arabino-nucleotide is added to a primer, subsequent extension is also slower. In contrast, non-canonical nucleotides in the template can be copied over by primer extension with ribonucleotides with only modestly reduced rates. Other aspects of template heterogeneity such as 2′–5′ linkages and pyrophosphate linkages can also be copied over to generate a canonical RNA product. Based on these experiments, it appears that RNA is intrinsically better able to take part in non-enzymatic copying chemistry, and thus over repeated rounds of copying chemistry would have won out over competing nucleic acids. At this time, it appears that potential alternatives to ribonucleotides are either harder to make, more susceptible to degradation, or less able to take part in copying chemistry, leading to potential alternatives to RNA being filtered out at one or more stages.

Returning to the subject of non-enzymatic RNA replication, there are several problems that make replication more difficult than simple template copying. In a prebiotic situation, there is no way to supply the defined primers needed for the replication of a linear genome, while the replication of a circular genome would encounter severe topological difficulties. To overcome these and other issues, we have proposed a model for primordial RNA replication that we call the virtual circular genome or VCG model [6]. In this model, a circular genomic sequence is represented by a collection of oligonucleotides that map to a circular sequence, but no actual circular molecules need to exist. Replication is then driven in a distributed manner by oligonucleotide elongation by primer extension, with longer oligos acting as templates and shorter oligos acting as primers, downstream helpers, and invaders for catalysis of strand displacement. We are actively engaged in experimental tests of this model.

At this point, we can see at least the outlines of an overall prebiotic, non-enzymatic process leading to the synthesis and replication of a primordial RNA genome. But a protocell is not just replicating RNA — that RNA must be encapsulated within a replicating compartment boundary. By analogy with modern cells, that boundary is likely to be a bilayer lipid membrane. However, primordial cell membranes must have been quite different than modern cellular membranes. Critically, they must allow the passive transfer of nutrients such as nucleotides from the external environment to the cell interior; in addition, they must be able to grow and divide solely in response to the chemistry and physics of the environment.

For 20 years now, we have studied the properties of vesicles assembled from simple fatty acids, ranging from oleic acid/oleate as a convenient model system to decanoic acid/decanoate as a more prebiotically plausible system. Such vesicles are permeable to activated nucleotides and even to di- and tri-nucleotides so that in principle RNA replication could take place internally but be fed from an external source of material. Vesicle growth can be driven by the addition of alkaline micelles to vesicles in a solution at lower pH. Ting Zhu in my lab showed that fast growth transforms initially spherical multilamellar vesicles into filamentous vesicles that are easily broken apart by mild shear forces into smaller daughter vesicles [7]. Subsequent slow growth allows the daughter vesicles to increase in size so that the cycle can repeat. Thus, a fluctuating environmental supply of fatty acids can drive repeated cycles of growth and division. More recently, Anna Wang found conditions that lead to the spontaneous assembly of fatty acids into giant unilamellar vesicles. Fast addition of alkaline micelles again leads to a rapid increase in surface area, after which shape fluctuations lead to spontaneous division [8].

An alternative pathway for vesicle growth involves competition between vesicles. Irene Chen initially showed that osmotically swollen vesicles grow following an influx of iso-osmotic vesicles because the tense membrane of the swollen vesicles can relax by absorbing fatty acid molecules from the empty vesicles [9]. This pathway relies upon the dynamic behaviour of single-chain amphiphiles, which exchange rapidly between vesicles. Subsequently, Itay Budin showed that fatty acid vesicles that contained a fraction of two chain lipids could also grow by absorbing fatty acids from pure fatty acid vesicles, with growth into filamentous forms again leading to facile division [10]. However, in this case, the daughter vesicles do not have the same composition as the parental vesicles, so continued cycles would require *in situ* synthesis of two-chain lipids. Another route to competitive growth was found by Kate Adamala, who showed that the internal synthesis of a hydrophobic peptide could drive the competitive growth of fatty acid vesicles [11]. In this case, continued cycles of growth and division would require the replication of an internal catalyst of peptide synthesis, presumably a ribozyme. Overall, fatty acid vesicles seem like ideal models for protocell membranes.

There is, however, a big problem: the high concentrations of Mg^{++} that are required for RNA copying chemistry rapidly disrupt and destroy fatty acid membranes. Finding solutions to this mutual incompatibility is critical to the field and is increasingly the focus of our attention. The first partial solution to this problem was identified by Kate Adamala, who showed that chelation of Mg^{2+} by citrate allowed RNA copying to proceed while protecting membranes. This allowed RNA synthesis to occur within model protocells [12], but unfortunately, the model of growth following the addition of alkaline micelles was no longer effective. We are currently investigating the potential for competitive growth in the presence of Mg^{2+}-citrate, which could provide a solution to the compatibility problem noted above. However, such a solution would not be prebiotically realistic, and so our search for other solutions to this problem continues.

Our work with mixed fatty acid–phospholipid vesicles points to a potential solution, with interesting consequences. Such vesicles are stable in the presence of moderate levels of Mg^{2+}, and in the presence of even higher levels of Mg^{2+}-citrate, they exhibit enhanced permeability to activated nucleotides, which allows RNA template copying chemistry to proceed well internally. We know they can grow by absorbing fatty acids from fatty acid vesicles, but in order to maintain a constant membrane composition, we would need a non-enzymatic pathway by which fatty acids could be transformed into phospholipids. An interesting possibility is that the same chemistry that activates phosphates for RNA synthesis could also activate carboxylates for esterification and phospholipid synthesis [13]. Another attractive possibility involves vesicle membranes that contain single-chain cyclophospholipids, which also confer tolerance to Mg^{2+} and which would be easier to synthesize from fatty acids. An important direction for future research is therefore to explore synthetic pathways that would allow the maintenance of a steady-state membrane composition during multiple cycles of vesicle growth and division.

Outlook for the future

In summary, the twin goals of non-enzymatic RNA replication and sustained vesicle growth and division are in sight, and I expect to see experimental demonstrations of these key processes in the near future. The integration of these two processes remains challenging, but several approaches to solving the compatibility problem are under investigation, and I am optimistic that one or more solutions to this problem will become apparent in the coming years. At that stage, a fully functional protocell system will open the doors to new modes of evolutionary exploration as we see how adaptation to a range of physical and chemical environments occurs in a reconstituted laboratory-scale version of the RNA world.

Acknowledgements

I thank the many brilliant students and postdoctoral fellows who have carried out the work described above over the past two decades. I also thank the Simons Collaboration on the Origins of Life, the National Science Foundation, and the Howard Hughes Medical Institute for supporting this work.

References

[1] T. Wu and L. E. Orgel, *J. Am. Chem. Soc.* **114**, 5496 (1992).

[2] N. Prywes, J. C. Blain, F. Del Frate, and J. W. Szostak, *eLife* **5**, e17756 (2016).

[3] T. Walton and J. W. Szostak, *J. Am. Chem. Soc.* **138**, 11996 (2016).

[4] W. Zhang, T. Walton, L. Li, and J. W. Szostak, *eLife* **7**, e36422 (2018).

[5] D. Duzdevich, C. E. Carr, D. Ding, S. J. Zhang, T. S. Walton, J. W. Szostak, *Nuc. Acids Res.* **49**, 3681 (2021).

[6] L. Zhou, D. Ding, J. W. Szostak, *RNA* **27**, 1 (2021).

[7] T. F. Zhu and J. W. Szostak, *J. Am. Chem. Soc.* **131**, 5705 (2009).

[8] J. Kindt, J. W. Szostak, and A. Wang, *ACS Nano.* **14**, 14627 (2020).

[9] I. A. Chen, R. W. Roberts, and J. W. Szostak, *Science* **305**, 1474 (2004).

[10] I. Budin and J. W. Szostak, *Proc. Natl. Acad. Sci. USA.* **108**, 5249 (2011).

[11] K. Adamala and J. W. Szostak, *Nat. Chem.* **5**, 495 (2013).

[12] K. Adamala and J. W. Szostak, *Science* **342**, 1098 (2013).

[13] C. Bonfio, D. A. Russell, N. J. Green, A. Mariani, and J. D. Sutherland, *Chem. Sci.* **11**(39), 10688 (2020).

https://doi.org/10.1142/9789811282324_0043

Ribozymes and RNA Modifications

Claudia Höbartner

Institute of Organic Chemistry,
Julius-Maximilians-Universität Würzburg,
Würzburg, Germany

Introduction

Nearly all classes of coding and non-coding RNAs undergo post-transcriptional modification, and methylated nucleotides belong to the evolutionary conserved features of RNA. Recent studies revealed a dynamic RNA modification landscape in mRNA and hint at important roles of methylated RNA in regulating cellular functions [1]. Synthetic RNA modifications, including methylated nucleotides, are also important for emerging RNA-based medicine. For example, the modified nucleoside 1-methylpseudouridine (m^1Y) increased the effectiveness of COVID-19 mRNA vaccines [2], and 2′-*O*-methyl or 2′-*O*-methoxyethyl nucleotides are key components of therapeutic antisense oligonucleotides [3], while other modified nucleotides interfere with viral replication by preventing efficient translocation or introducing massive copying errors, as recently shown for SARS-CoV-2 RNA polymerase and the repurposed antiviral nucleosides prodrugs remdesivir [4] and molnupiravir [5].

In an era before modern life emerged, RNA was thought to function both as a genetic material and as a catalyst. *In vitro* directed molecular evolution experiments discovered ribozymes that function as RNA-dependent RNA polymerases and can copy themselves or their ancestors [6, 7]. RNA may have had additional catalytic roles during evolution, and nucleotide-containing co-factors used by contemporary metabolic enzymes may constitute molecular remnants of ancient ribozymes [8, 9]. The discovery of riboswitches as regulatory RNA elements that bind small molecules including various co-enzymes [10] nourished the thoughts of more widespread catalytic competences of RNA [11, 12]. Indeed, *in vitro* selection endeavours have revealed aptamers and ribozymes that bind and utilize nucleotide co-factors, including redox co-factors and acetyl co-enzyme A [13, 14].

Besides nucleotide-derived cofactors, the involvement of modified nucleosides in RNA catalysis is interesting to consider. Spontaneous nucleobase or ribose modification/methylation may have occurred in reactive environments and the presence of non-canonical nucleotides in RNA could benefit or burden the evolution of RNA catalysts. Methylation of nucleic acids is a widespread modification found in all domains of life, and most methyl groups are installed by methyltransferase enzymes that use *S*-adenosyl-methionine (SAM) as the methyl donor. Methylated nucleosides and methyl group transfer reactions may have even occurred on the primitive Earth [15]. Directed molecular evolution of methyltransferase ribozymes that catalyse the installation or selective removal of methyl groups from RNA nucleotides may shed light on the potential catalytic abilities of primordial ribozymes and could reveal distinctive properties of now universally conserved methylated nucleotides in tRNA and rRNA.

Ribozymes that install RNA modifications

Earlier experimental attempts to address the question of whether RNA can catalyse site-specific RNA methylation using SAM resulted in the enrichment of co-factor-binding aptamers without methyltransferase activity [16]. *In vitro* selection by enrichment for catalytic activity identified self-alkylating ribozymes using reactive iodo- or chloroacetyl derivatives [17–19], and later electrophilic epoxides [20] resulted in N^7-alkylation of guanine with the RNA. Based on the design of these experiments, these ribozymes could not be used as methyltransferases. The required methyl halogenides would be too reactive reagents, and epoxides would transfer at least two carbon atoms to the guanine. Searching for alternatives, we hypothesized that a methyltransferase ribozyme should use a co-factor that makes specific non-covalent contacts to the RNA, including H-bonding, π-stacking, and others. Our selection strategy was designed for the ribozyme to form a binding site for the "leaving group", and we chose O^6-methylguanine (m^6G) as a potential methyl group donor. Transfer of the methyl group from m^6G to the target RNA would result in the release of guanine. Guanine is well established as an RNA ligand in natural riboswitches and *in vitro* selected aptamers [21]. A possible evolutionary relationship between ribozymes and riboswitches has previously been discussed [11, 12]; metabolite-binding riboswitches may resemble inactivated ribozymes that lost their catalytic activity during evolution from the RNA world.

Following the hypothesis outlined above, we have discovered the first methyltransferase ribozyme (MTR1). This ribozyme catalyses a site-specific intermolecular methyl transfer to install 1-methyladenosine (m^1A) at a defined position in a target RNA by utilizing O^6-methylguanine as the methyl group donor [22]. We showed that the ribozyme can be engineered to methylate natural RNA sequences, including tRNAs, which contain m^1A at conserved positions. The ribozyme showed accelerated transfer rates at pH 6 and could be split into two fragments that assemble on the target RNA into a functional ribozyme. We solved the crystal structure of MTR1 and investigated the mechanism of RNA-catalysed methyl transfer [23]. The structure revealed the products of the reaction, i.e., the post-catalytic state, with m^1A and the free guanine bound in close proximity in the active site. The ligand is fully contacted by hydrogen bonds with the RNA, and the base pairing pattern is highly reminiscent of guanine binding observed in natural guanine riboswitches [24].

The MTR1 ribozyme utilizes general acid catalysis to enable the methyl group transfer. Structure probing in solution as well as activity assays of structure-guided mutants provided strong support for the mechanism, in which a protonated cytidine is involved in binding and activation of the co-factor. The mechanistic analyses revealed two key nucleotides that act in concert to accelerate the methylation reaction. We found a synergistic effect of two methylated ribose residues (at C12 and U42), which enhanced the reaction rates by at least 120-fold over the rates obtained with unmodified RNA under *in vitro* selection conditions. This finding supports the speculation that modified nucleotides may have enhanced early RNA catalysis.

In analogy to MTR1, it seems feasible to evolve additional methyltransferase ribozymes and explore potential preferences for the generation of various methylated nucleotides in RNA-catalysed reactions. It may also be rewarding to perform *in vitro* evolution experiments starting from native RNAs (riboswitches) that are already known to bind methyltransferase co-factors or derivatives thereof, such as SAM, cobalamine, or tetrahydrofolate. Indeed, in 2021, two additional RNA-catalysed RNA methylation reactions were described. Micura and co-workers demonstrated that a natural preQ$_1$ riboswitch RNA can bind the synthetic co-factor m^{6}preQ$_1$ and mediate the transfer of the methyl group to a

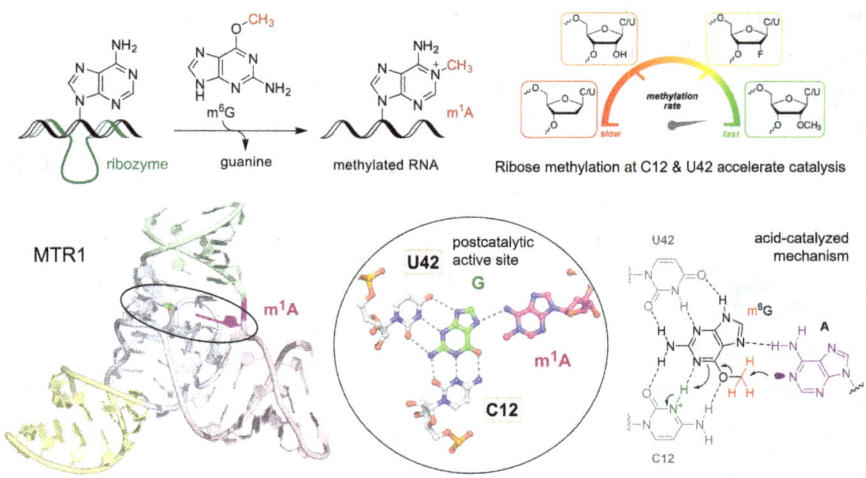

Fig. 1. The methyltransferase ribozyme MTR1 catalyses site-specific RNA methylation using m^{6}G to install m^{1}A in RNA. The architecture of the active site is reminiscent of natural purine riboswitches. The catalytic mechanism involves a protonated cytidine. 2′-OMe nucleotides in the active site synergistically accelerate methyl transfer.

cytidine in the binding site, thus generating 3-methylcytidine (m^3C) [25]. Murchie and co-workers found a ribozyme that uses SAM and Cu^{2+} to generate 7-methylguanosine (m^7G) in the ligand-binding site [26]. The active sequence identified by *in vitro* selection in the laboratory was then also found in genomic sequences from all domains of life, suggesting that it may have ancient origins. Recently, we found another SAM-utilizing ribozyme that alkylates the minor groove side of adenosine [27] and we expect that more RNA-modifying ribozymes will be discovered in the future.

Perspectives for the future

More than 70 different methylated nucleosides are known in natural RNA, and it is an open question which fraction of methylation sites can be accessible by RNA-catalysed RNA methylation. While some *N*- and *O*-alkylation sites seem more easily addressed than others, the whole family of C5-methylated pyrimidine nucleosides requires more sophisticated reaction mechanisms and likely additional co-factors. Nevertheless, the recent advances in the field of methyltransferase ribozymes and the insights into catalytic mechanisms beyond RNA-catalysed phosphotransfer reactions suggest that other co-factor-utilizing ribozymes may be found in the laboratory and possibly in natural RNAs to catalyse more diverse reactions than currently known. The added benefit of modified nucleotides shaping the active sites of ribozymes is worthy of further exploration. This may lead to the first XNAzymes catalysing reactions other than cleavage or ligation of RNA. The key challenges for finding new ribozymes by directed molecular *in vitro* evolution experiments lay in the design of the selection strategies for enrichment and amplification. Novel high-throughput analyses of ribozyme activities directly from sequencing data may accelerate the discovery rate. A bold speculation on possible co-factor-assisted RNA-catalysed reactions could address radical reactions, for example, to explore the potential transition from RNA to DNA, which would require cleavage of a carbon–oxygen bond, mimicking the enzymatic activity of ribonucleotide reductase in nature or Barton–McCombie deoxygenation in the organic chemistry laboratory. In another direction, we are looking for RNA-catalysed site-specific nucleobase deamination, akin to enzymatic RNA editing that converts adenosine to inosine. Fundamental studies in these and other directions will continue to explore the potential of ribozymes in shaping early RNA-based life and its evolution. In addition, the development of nucleic acid catalysts as research tools in RNA biology and as potential future RNA-based therapeutics is an exciting and challenging research line ahead.

Acknowledgements

This work was supported by the European Research Council (ERC consolidator grant 682586), the Deutsche Forschungsgemeinschaft (DFG SPP1784 and HO4436/3-1), and the University of Würzburg.

References

[1] S. Zaccara, R. J. Ries, and S. R. Jaffrey, *Natl. Rev. Mol. Cell. Biol.* **20**, 608 (2019).
[2] K. D. Nance and J. L. Meier, *ACS Cent. Sci.* **7**, 748 (2021).
[3] M. Egli and M. Manoharan, *Acc. Chem. Res.* **52**, 1036 (2019).
[4] G. Kokic, H. S. Hillen, D. Tegunov, C. Dienemann, F. Seitz, *et al.*, *Nat. Commun.* **12**, 279 (2021).
[5] F. Kabinger, C. Stiller, J. Schmitzova, C. Dienemann, G. Kokic, *et al.*, *Nat. Struct. Mol. Biol.* **28**, 740 (2021).
[6] K. F. Tjhung, M. N. Shokhirev, D. P. Horning, and G. F. Joyce, *Proc. Natl. Acad. Sci. USA* **117**, 2906 (2020).
[7] J. Attwater, A. Wochner, and P. Holliger, *Nat. Chem.* **5**, 1011 (2013).
[8] H. B. White, 3rd, *J. Mol. Evol.* **7**, 101 (1976).
[9] A. Kirschning, *Angew. Chem. Int. Ed.* **60**, 6242 (2021).
[10] R. R. Breaker, *Biochemistry* **61**, 137 (2022).
[11] R. R. Breaker, *ACS Chem. Biol.* **15**, 2020 (2020).
[12] T. J. Wilson and D. M. J. Lilley, *Wiley Interdiscip. Rev. RNA* **12**, e1651 (2021).
[13] X. Chen, N. Li, and A. D. Ellington, *Chem. Biodivers.* **4**, 633 (2007).
[14] S. Tsukiji, K. Ramaswamy, and H. Suga, *Pure. Appl. Chem.* **76**, 1525 (2004).
[15] C. Schneider, S. Becker, H. Okamura, A. Crisp, T. Amatov, *et al.*, *Angew. Chem Int. Ed.* **57**, 5943 (2018).

[16] D. H. Burke and L. Gold, *Nucl. Acids Res.* **25**, 2020 (1997).

[17] C. Wilson and J. W. Szostak, *Nature* **374**, 777 (1995).

[18] A. K. Sharma, J. J. Plant, A. E. Rangel, K. N. Meek, and A. J. Anamisis, *et al.*, *ACS Chem. Biol.* **9**, 1680 (2014).

[19] S. Ameta and A. Jäschke, *Chem. Sci.* **4**, 957 (2013).

[20] R. I. McDonald, J. P. Guilinger, S. Mukherji, E. A. Curtis, W. I. Lee, and D. R. Liu, *Nat. Chem. Biol.* **10**, 1049 (2014).

[21] R. T. Batey, *Q. Rev. Biophys.* **45**, 345 (2012).

[22] C. P. M. Scheitl, M. Ghaem Maghami, A. K. Lenz, and C. Höbartner, *Nature* **587**, 663 (2020).

[23] C. P. M. Scheitl, M. Mieczkowski, H. Schindelin, and C. Höbartner, *Nat. Chem. Biol.* **18**, 547 (2022).

[24] A. Serganov, Y. Yuan, O. Pikovskaya, A. Polonskaia, L. Malinina, *et al.*, *Chem. Biol.* **11**, 1729 (2004).

[25] L. Flemmich, S. Heel, S. Moreno, K. Breuker, and R. Micura, *Nat. Commun.* **12**, 3877 (2021).

[26] H. Y. Jiang, Y. Q. Gao, L. Zhang, D. R. Chen, J. H. Gan, and A. I. H. Murchie, *Nat. Catal.* **4**, 872 (2021).

[27] T. Okuda, A. K. Lenz, F. Seitz, J. Vogel, C. Höbartner, *Nat. Chem.* (2023). 10.1038/s41557-023-01320-z.

https://doi.org/10.1142/9789811282324_0044

Pandemic RNA Viruses and Vaccines

Peter Palese

Department of Microbiology,
Icahn School of Medicine at Mount Sinai,
New York, 10029 NY, USA

Virus: "A piece of bad news wrapped up in a protein"
Sir Peter Medawar

Introduction

That viruses are different from bacteria has been known for about one hundred years. That some viruses contain DNA and some RNA as their genetic material has become clear only in the second half of the 20th century. With this biological knowledge, we became aware that some of the worst infectious diseases in humans are caused by viruses. Smallpox, caused by a DNA virus, is high on the list of harmful diseases to humans; it was responsible for untold number of deaths for centuries. But also among RNA viruses, there are major disease-causing pathogens, especially those associated with pandemic, global epidemics.

Fortunately, vaccines against viruses have been successfully developed. In fact, several RNA viruses have been eliminated by vaccination or close to. New platforms for the development of vaccines have dramatically improved the health in many countries globally. But much remains to be done to provide vaccines and vaccinations to those who need them.

Pandemic influenza and pandemic corona viruses

In the last 100 years, members of the influenza virus family have caused four pandemics. The 1918/1919 pandemic influenza virus is estimated to have been responsible for the deaths of up to 100 million people (Fig. 1). While the more recent influenza viruses (pandemic and epidemic) have been less virulent than the one of 100 years ago, we are concerned about the emergence of new pandemic influenza virus strains, which arise from "genetic mixing" of human and animal influenza strains [1]. My laboratory was involved in the reconstruction of the pandemic 1918 influenza virus using a technology, reverse genetics, developed at Mount Sinai [2–4]. We were able to identify what made the 1918 virus so virulent and we showed that presently available antivirals and influenza vaccines are highly effective against that early pandemic strain.

Vaccines against influenza were developed as far back as the 1940s, barely ten years after the discovery of the virus. The early vaccines consisted of whole virus treated with chemicals to make them non-infectious. This inactivated material was then injected to induce a protective immune response in the patient. Though not perfect, these vaccines have lowered the number of deaths and hospitalizations, and they have shown to reduce

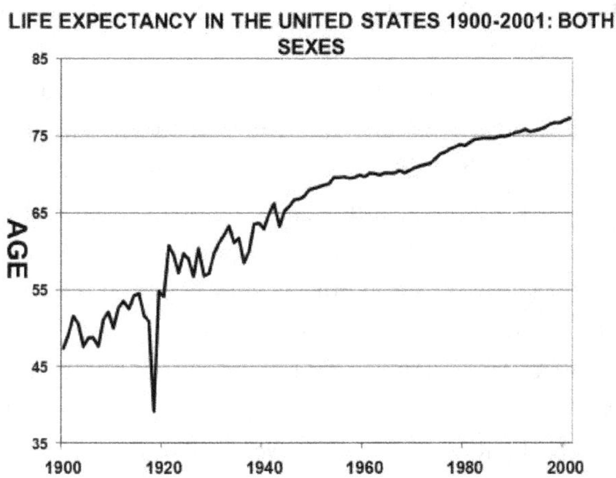

Fig. 1. The 1918 influenza pandemic caused a reduction in average life expectancy of 11 years.

the severity of disease in patients who were immunized but became ill. More recently, live attenuated influenza virus vaccines and synthetic protein vaccines have been introduced to better protect children and the elderly, respectively.

Unfortunately, influenza viruses come in different flavours defined by the major surface glycoproteins of the virus, the hemagglutinins. We now know of 18 such subtypes of viruses infecting different species, but we have only seen three of these flavours in humans: H1, H2, and H3 (Fig. 2). Each time a virus with a new subtype emerges in humans, we experience a pandemic. In addition to pandemic changes, the virus also undergoes annual variations which force the annual reformulation of the vaccines and annual influenza virus (re)-vaccinations. Many efforts are underway to develop a universal influenza virus vaccine which does not have to be administered annually and would last for 1–20 years or even longer. The same technology, reverse genetics, which allowed us to resurrect the 1918

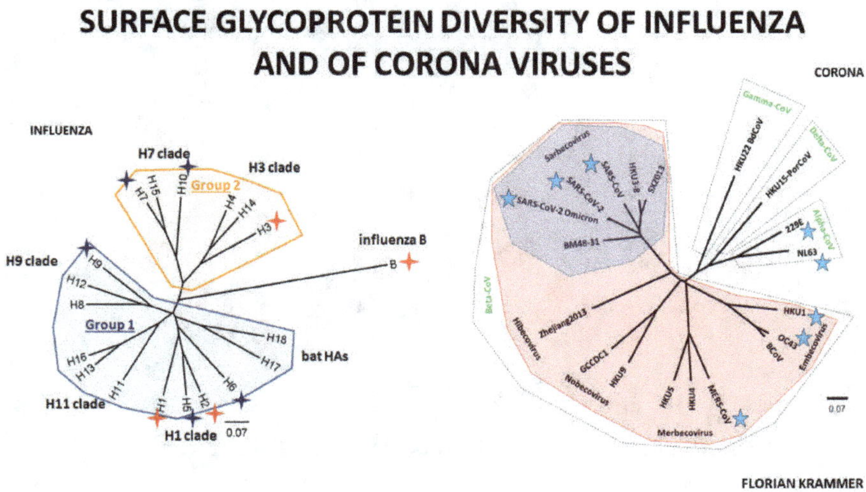

Fig. 2. Left. Dendrogram of the 18 hemagglutinin subtypes of influenza viruses. Red asterisks identify the human H1, H2, H3, and B hemagglutinins. Blue asterisks highlight avian hemagglutinins of viruses having been observed to infect humans. Bar (0.07) identifies 7% amino acid differences. Right. Dendrogram of the S spike proteins of the alpha, beta, gamma, and delta corona viruses. Blue asterisks identify S spike proteins of coronaviruses isolated from humans. Bar (0.07) identifies 7% amino acid differences.

virus is now used to develop better influenza virus vaccines for humans as well as for animals. We take advantage of having identified a conserved region in the stalk of the viral hemagglutinin. A vaccine that directs the human immune response towards this conserved domain in the hemagglutinin is presently in phase 1/2 trials (Fig. 3). In terms of improving veterinary vaccines against avian respiratory pathogens, we again use reverse genetics [5]. Several of these veterinary vaccines have been commercialized.

Highly surprising was the emergence of SARS-CoV-2 at the end of 2019. This RNA-containing virus is also a respiratory pathogen and it has been declared by the WHO a pandemic virus on March 11, 2020. Like the influenza virus in 1918/1919, SARS-CoV-2 has also contributed to a loss of average life expectancy. In the last two years, 2020 and 2021, the

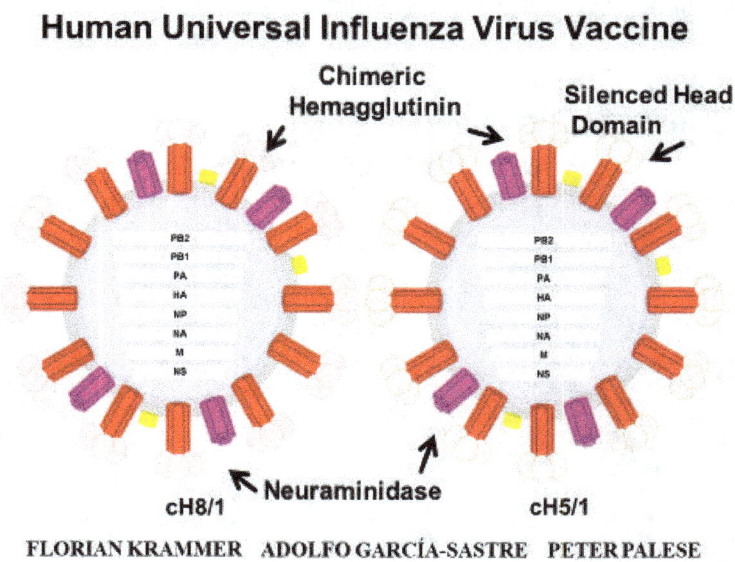

Fig. 3. Vaccination with influenza viruses expressing chimeric hemagglutinins to boost preexisting antibody responses against the conserved hemagglutinin stalk (red) domain. The head domains (light orange) of the chimeric hemagglutinins (cH8/1) and (cH5/1) are silenced. By sequential vaccination with vaccine strains expressing different heads, but the same hemagglutinin stalk domain (red), the immune system preferentially targets the epitopes that remain constant (red). The neuraminidase is also conserved (purple) and induces protective immune responses [6, 7].

average life expectancy in the US dropped by 1.9 and 0.9 years, respectively. The impact on lives and economies has seriously affected people in the US (Fig. 5) and around the world. Like influenza virus, SARS-CoV-2 comes in different flavours with respect to its spike surface proteins. The genetic (antigenic) diversity of SARS-CoV-2 is even broader than that of influenza viruses (Fig. 2). Major efforts have been directed towards the development of protective vaccines and they were effective beyond all expectations. Foremost are the mRNA platforms which use lipid nanoparticles (LNP) to engulf mRNA molecules (Fig. 4) [8].

Extraordinary development work was done with "warp" speed and the resulting vaccines have saved millions of lives. The LNP platform is a real game changer which cannot be under-appreciated. The mRNA vaccines

Fig. 4. Each dot represents one death in the US caused by SARS-CoV-2 (January 2020–May 13, 2022).

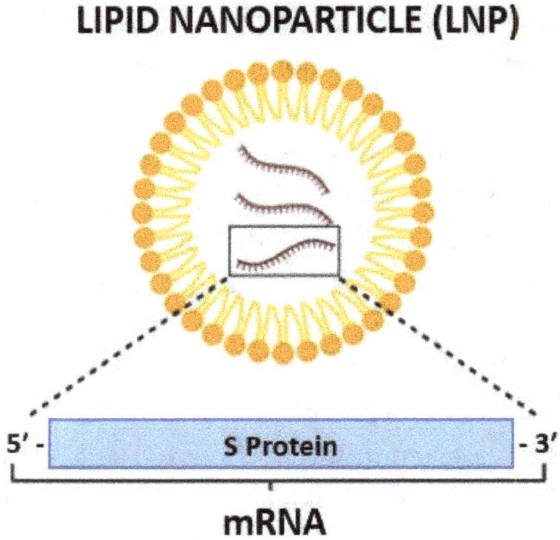

Fig. 5. LNP encapsulated mRNA expressing the S Protein of SARS-CoV-2.

are wonderfully effective and are extraordinarily safe. In addition to the LNP-mRNA platform, COVID-19 vaccines have been produced based on viral vectors or by employing conventional methodologies, such as inactivated whole virus preparations. Extensive trials have shown that many of these vaccines are effective and protective against SARS-CoV-2.

While the success of COVID-19 vaccines has been exceptional, breakthrough infections after (multiple) vaccinations continue to be a problem; also, present vaccines do not efficiently prevent transmission from one patient to the other. Our efforts at Mount Sinai are directed to develop a viral vector-based vaccine, which can induce mucosal immune responses in the respiratory tract (Fig. 6). Such a vaccine should further reduce infections in vaccinated patients and it should minimize transmission of the virus [9–11]. As the S spike protein of a corona virus defines its antigenicity, it is the extraordinary variation of the viral surface protein, which makes it likely that novel pandemic SARS-CoV will emerge in the future. Since corona viruses are also zoonotic (can jump from animals to humans) like influenza viruses, the emergence of four influenza pandemics over the last 100 years may foretell additional catastrophes caused by corona viruses.

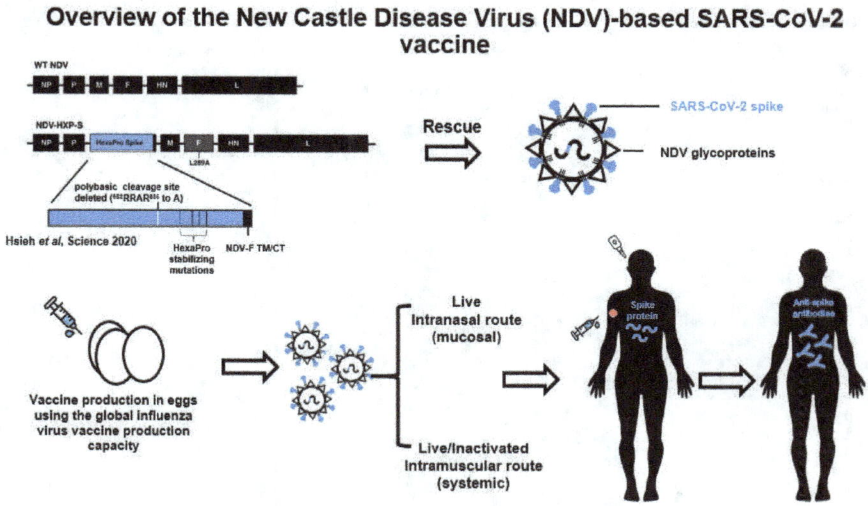

Fig. 6. NDV-based SARS-CoV-2 vaccine which is cheaply produced in embryonated eggs (like the influenza virus vaccine). Also, when given intranasally, it induces mucosal immunity preventing reinfection with the virus and reduces transmission from person to person [9–11].

Poliomyelitis, measles, mumps, rubella, hepatitis A, and rotavirus vaccines

Classical vaccine development and manufacturing have been extremely successful following in the footsteps of the Salk (killed poliomyelitis vaccine) and Sabin (live attenuated poliomyelitis vaccine) platforms. In fact, six diseases (poliomyelitis, measles, mumps, rubella, hepatitis A, and rotavirus gastroenteritis) have seen a dramatic decrease in high-income countries because of the development of these vaccines. In some instances (poliomyelitis, measles, mumps, and rubella), the diseases have mostly disappeared in the US. Figure 7 shows the success of the eradication of poliomyelitis as a result of vaccinations in all the states.

The future: Pandemic RNA viruses and vaccines

Influenza and corona viruses have been the two virus families which have caused pandemics in the last 100 years. Both families are characterized by

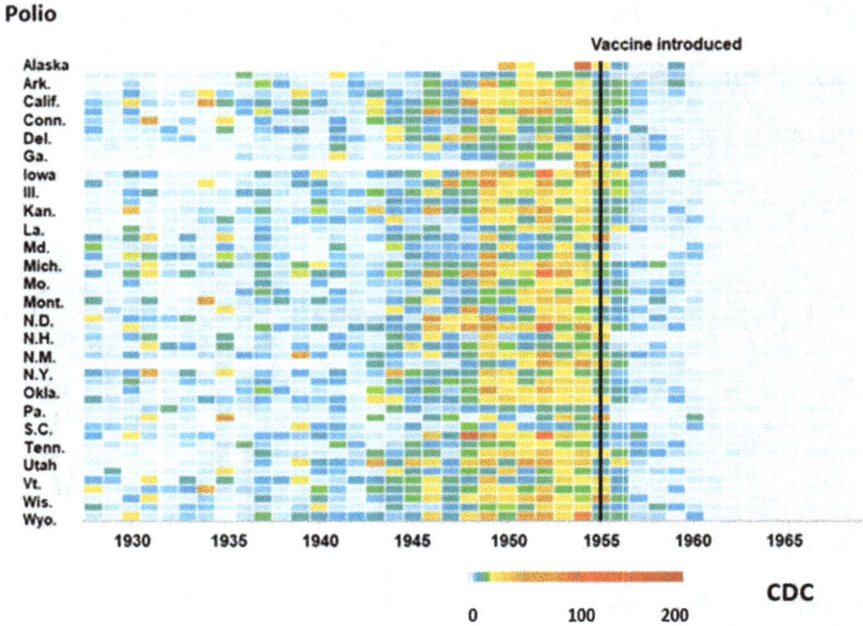

Fig. 7. Cases of poliomyelitis.

coding for viruses which have on their surface highly variable glycoproteins. Such variable strains are circulating in many different animal species and occasionally they jump into humans. These events are unpredictable but are likely to occur in the future as more humans on the globe are interacting with more animals. To prepare for such eventualities (or rather certainties), we need increased surveillance in humans as well as in animals for novel pathogens. If history is any guide, RNA viruses which are transmitted via the respiratory tract are high on the list of these threats. In addition, we need novel methods of pathogen detection and of diagnostic tests and overall substantial resources to do molecular research on host-pathogen interactions. Only the continued advances in molecular biology, immunology, broadly basic research, and big data analysis (including artificial intelligence) will prepare us against future pandemics. The development of LNP-RNA was a major accomplishment and has made all the difference with this most recent pandemic virus, but it is not a panacea. It is unlikely that this platform (with our present knowledge) will help develop an effective HIV/AIDS or hepatitis C virus vaccine. We need certainly more highly visionary research in order to develop new and improved vaccines against present strains and future outbreaks of pathogens which have acquired resistance to present medical interventions. In order to overcome these new challenges brought about by global warming, increases in the world population, bio-safety risks, and disturbances in the natural habitat of many animals. It will also be important to have public health systems which can respond to these special demands.

Acknowledgements

It is thankfully acknowledged that work in the author's laboratory was mostly supported by the National Institutes of Health and institutional funds.

References

[1] F. Krammer and P. Palese, In *Fields Virology*, P. M. Howley and D. R. Knipe, (eds.), Vol. 1, Chapter 14, p. 596 (2021).

[2] T. M. Tumpey, C. F. Basler, P. V. Aguilar, H. Zeng, A. Solorzano, *et al.*, *Science* **310**, 77 (2005).

[3] G. Neumann, *Cold Spring Harb. Perspect. Med.* **11**, 4 (2021).

[4] P. Palese, H. Zheng, O. G. Engelhardt, S. Pleschka, and A. Garcia-Sastre, *PNAS* **93**, 11354 (1996).

[5] M. S. Park, J. Steel, A. Garcia-Sastre, D. Swayne, and P. Palese, *PNAS* **103**, 8203 (2006).

[6] N. Folschweiller, C. Vanden Abeel, and P. Van Damme *et al.*, *Lancet Infect. Dis.* **22**, 1062 (2020).

[7] G. R. Nachbagauer and P. Palese, *Annu. Rev. Med.* **71**, 315 (2020).

[8] N. Chaudhary, D. Weissman, and K. A. Whitehead, *Nat. Rev. Drug Discov.* **20**, 817 (2021).

[9] A. Duc Dang, T. Dinh Vu, and H. Hai Vu *et al.*, *Vaccine* **40**, 3621 (2022).

[10] P. Pitisuttithum, V. Luviza, and S. Lawpoolsri *et al.*, *EClinicalMedicine* **45**, 101323 (2022).

[11] S. Ponce-De-Leon, M. Torres, and L. E. Soto-Ramirez *et al.*, *medRxIv* PMID: 35169806.

https://doi.org/10.1142/9789811282324_0045

Discussions of Session 6 — RNA Chemistry to Explore our Origins and Fight Viral Pandemics

Chair: Gerald Joyce
Auditors: C. Gueydan*, C. Martens†

**Université libre de Bruxelles, Laboratoire de Biologie moléculaire du gène, Departement of Molecular Biology, 12 rue Pr. Jeener et Brachet, 6041 Gosselies, Belgium*
†Structure and Function of Biological Membranes — Chemistry Department, Université libre de Bruxelles, Campus Plaine, Boulevard du Triomphe, 1050 Brussels, Belgium

Gerald Joyce:
Welcome back, everybody. We covered a lot of ground in that last session. I'm Jerry Joyce. I'm going to start with a kind of apocryphal story about evolution, and then open it up to general questions and also comments from our panel, of course. I think we see that we're very grateful to Darwinian evolution. We wouldn't be here, our biosphere wouldn't be here, all the innovation of our biosphere wouldn't be here. But it is, of course, a double edged sword. That especially with regard to pathogen evolution, especially with regard to viral pathogen evolution, the pace at which it occurs is too fast a pace for the human timescale; even

for our defenses, both our biological defenses and our technological defenses sometimes. So that's something I think, as a society, we need to recognize — that pace.

The apocryphal story I want to tell was triggered by Peter's slide (Peter Palese). First of all, seeing the kids in the iron lung with polio, and then that graph, which I've never seen before about how the vaccines just shut it down. As you know, polio is still with us. I made that remark very briefly, but it's been in the news lately. Why is polio still with us? We've had the vaccine since the early 1950s. It's an extremely efficacious vaccine. The reason is, I'm not sure everyone knows the story (and I am from the Salk Institute), that it was the Salk vaccine, a killed viral vaccine — formaldehyde killed whole viral particles. That was the original vaccine. But the Sabin vaccine, like the Newcastle disease vaccine that you're working on now, Peter, is a live vaccine. The benefit of a live vaccine, of course, is that it's the gift that keeps on giving. It's transmitted to household contacts, and to neighbors and so on. So you don't have to inoculate everyone with the vaccine, just enough people that then the vaccine strain transmits vaccination-based immunity. But, of course, evolution happens. And because the vaccine, the live vaccine strain is still used, especially in the Global South, there's a lot of viral load of that vaccine strain, and the vaccine strain evolves, and it has evolved back into a pathogenic strain. And so the polio virus cases that we see now are not a direct recapitulation of the original pathogenic poliovirus, but a different route to pathogenicity. So now, of course, everyone is going back to the Salk vaccine, which is a killed vaccine, and doesn't have that problem, but also doesn't have the benefit of the propagation. Again, a classic story of the two-edged sword of evolution. But just a remark, though, to Peter is… and he didn't have time to go through this in detail, but he is working on a Newcastle disease based vaccine that encodes the spike protein of SARS-CoV2. In that case, it is a live vaccine, but of a non pathogenic virus. It's not some softened form of SARS-CoV2. It's just carrying the epitope for the spike protein in an innocuous virus. It's not completely implausible that some variant of Newcastle might turn pathogenic, but that's a long way away. How many people found that last session reassuring? Yes, so I think that's some of the questions that come to mind to myself. You know, RNA is on the edge of life, both historically and with regard to viral evolution. And then the more extreme versions that you heard in Andy's talk (Andrew Fire) where he can probably tell you, if you bring him your empty cup, whether you had milk or cream in your coffee or not. Because he can let his little replicons replicate, and they will pick up a little bit of the DNA from the milk that you used in your cup. So there the dual axis of both the benefit and the glory of evolution, and the challenge of it. That's not really a question, but I'm just trying to capture the mood. So, Karla, go ahead.

Karla Kirkegaard:

I have a question for Jack (Szostak). You know, there's always this problem, with the primordial world, how do you bound the RNA to give it a membrane, and you've showed nice membrane images. But it's been really exciting this phase separation, that happens with simple proteins. So you guys know about phase separation, when you have simple proteins, and so called natively denatured, what they do is form these phase separations, and as soon as I learned about that, I thought, ah, that's it! That's what RNA did! It made simple amino acids, strings of them and encapsulated itself.

Jack Szostak:

So there's been quite a lot of work along those lines. And we've looked at it a little bit in the phase separated systems that we've looked at, which tend to be arginine rich peptides forming a coacervate. With negatively charged RNA, what we see is that RNA molecules exchange rapidly between droplets. And also, of course, the droplets have a tendency to ripen and coalesce. So I think over time, they're not a good kind of compartmentalization that would allow for evolution. Where I think they might come in really useful, is forming within a membrane compartment, localized microenvironments that might be good for some kinds of chemistry, maybe even RNA replication.

Gerald Joyce:

So along those same lines, we saw that there's not enough time to complete the replication process before the exchange between the phase-separated regions, right? Is it so? It's really a kinetic problem? Is that how you see it?

Jack Szostak:
Yes, yes, I mean, you know, in principle, if there's a way of forming droplets where the RNAs are permanently inside, they can't exchange except maybe very rarely but they can still grow and divide, then that would work fine. But there's no system like that so far.

Gerald Joyce:
Please, any comments or questions from the audience? Anything's fair game.

Yamuna Krishnan
A question for John. When I saw your results where you had an RNA, but also had a sort of affinity for a certain amino acid. I started thinking immediately as we were talking about charging tRNAs, and things like this, which of the RNAs came first : mRNA, tRNA rRNA? Because I've only had to have this ribosome activity, and I still haven't sorted it out in my head. But when you talked about the second genetic code, I wanted to ask, does this mean that the second genetic code came first?

John Sutherland:
So, good question. Paul Schimmel always had it, that tRNA is basically a dimeric product. And the acceptor stem basically arose before the anticodon stem. And there is some quite, pretty good supportive evidence for the idea of tRNA being a dimer. Which bit comes first is difficult to say, but Christian de Duve attitude was that if you actually have a chemical selection to attach an amino acid, then a frozen accident makes more sense. So my feeling is that you would actually have the stereo chemically coded amino isolation first that's just occurring because of fundamental chemistry. It's not something that biology is trying to do for a purpose. But if that then becomes spliced or joined to an anticodon stem, which has a family box anti-codon, it can then participate in transpeptidation. And the idea is really that the selection for that might be based more upon the fact that the transpeptidation liberates an RNA terminus for replication chemistry. And the peptide is just a byproduct. So you can produce uncoded or partially coded peptides as a byproduct of engumming the ends of RNA that has been gummed up because of this competing amino isolation chemistry that's taking place.

Jack Szostak:

Okay, can I give her an alternative hypothesis? The way we've been thinking about is not contradictory to this, but maybe builds on it in a different way. So we were thinking that you couldn't evolve coded translation unless specifically aminoacylated RNAs already existed. And so, you saw a way to bias that with this second genetic code. But it has to be linked, right, everything has to be linked, the codon, the stem specificity. So we're thinking: maybe a way around that is if aminoacylated RNAs did something else first. And what we found is that if you have little bits of RNA, you can help them to assemble into larger structures, including ribozymes, if you make those out of aminoacylated RNAs. And so that would provide the potential for the side chains of amino acids, which are now in the backbone of RNA to actually take part in, for example, ribozyme catalysis, that would provide a selective pressure for the evolution of ribozymes that are RNA and amino acid specific aminoacyl-RNA synthetases, that I think would set the stage for coded translation. That's our hypothesis.

Gerald Joyce:

Just to relate that to what we talked about this morning, that also gives a role, a selectively advantageous role for small peptides, even individual amino acids, but also small peptides, before you could have more complex evolvable proteins. The other one I would put on the table is: we don't actually know whether the first RNA aminoacylation events, or for that matter amino acid activation events, were chemical reactions or RNA-catalyzed reactions. Either one is possible. One could make arguments both ways.

Donald Hilvert:

I also had a question about the evolution of translation. It went by kind of fast. It wasn't clear to me whether you were suggesting that the first amino acids were those in the fourfold degenerate codon boxes? Because if so, there's only one that really is functionally interesting. And that's the arginine. Everything else is quite hydrophobic.

John Sutherland:

The idea basically can be seen two ways round. You can say, look, the amino acids we find easiest to make fit the fourfold degenerate codons. Or you can take the energetics argument and say, therefore it makes sense that the amino acids that were first designed were first designed to fourfold degenerate codons. But yes, there is functionality in serine, and threonine. There's N terminal amino group functionality, and there's arginine. But the real idea is that there's no negative charge, there's no aromatics, but then RNA can contribute negative charge and aromatics. So the idea really is it's like the two wounded First World War soldiers who can walk by leaning on each other, but neither one can walk on their own. So the idea is short peptidilated RNA, either conjugates where that covalent attachment is still maintained or aggregates where they just associated.

Gerald Joyce:
Maybe a question for Claudia (Höbartner). How much decoration and functional decoration of RNA can occur? Not just as markers, but as catalytic enhancers, as you showed with RNA methylation?

Claudia Höbartner:
Well, I think there are two things that we can consider. One is that already modified building blocks could take part in the evolution. So they can properly be also formed spontaneously and then be incorporated and the information can be retained. Or cofactors can separately evolve and then be used by evolving ribosomes to decorate other RNAs.

Karla Kirkegaard:
I asked you briefly in the break. And I want to go back to it about the effect of modification on RNA-RNA interactions on base pairing, because I'm interested in what the enzymologists here think. It seems to me that with RNA catalysis, the rate limiting step is often product release. And so I'm wondering about the effect of modification on that affinity.

Claudia Höbartner:
Yes, very good point. And, in fact, it looks like that would go the opposite direction, because when we modify antisense oligonucleotides to hybridize to mRNAs, then ribose methylation enhances that affinity rather than doing the opposite. So I think that whole question of how we get strand separation is also something that goes back to the replication problem. So it helps to copy one strand. But in order to make the next copy, we need

to separate it again. And I don't know if I have an answer to how we would overcome this problem by modifications.

Andrew Fire:
Actually, I'm not sure that what you're saying is completely incompatible. Because if you enhance the interaction with a short duplex, that actually could have advantages. You are able to then have a short duplex that's more stable, but is able to come apart more easily than it would be for a longer duplex. So there could be some something there.

Sabine Flitsch
Can I just ask about the temperature these reactions should happen? Because it's not possible that, you know, these were elevated temperatures, and then maybe base pairing was not as strong as you might imagine? And maybe you need the methylation to enhance the base pairing and recognition.

Gerald Joyce:
Yes, I would be curious. Others, please comment. We don't have a really firm handle on what the temperature of the planet was at 4.0 billion years ago; even if you don't think in terms of the global temperature, but of different niches and what the temperature might be there. Yes, in principle, you would have the destabilizing effect of higher temperature. But then of course, the propensity of RNA to spontaneously undergo cleavage would be enhanced, especially in the presence of divalent metals. So I don't know. What do people think? What do we actually think the temperature was like, at the time of origins? And what should it be to make things more beneficial?

Karla Kirkegaard:
Maybe you just got strand separation once a day.

Jack Szostak:
You wouldn't want to use a day night cycle for strand separation, if we cook the RNA for hours, there will be nothing left. We've been thinking, I think maybe we were biased initially, but because of PCR, we were

thinking along the lines of very short, high temperature spikes to get short bits of RNA apart and shuffle configurations. That's what I'm leaning towards now and John (Sutherland) may comment on this. Are there other kinds of fluctuations that could be more gentle to RNA, like solvent fluctuations, or salt fluctuations, even pH fluctuations? So we're just starting to explore those alternatives.

Gerald Joyce:
This is a big problem. The separation of sizeable RNAs is quite difficult, takes extreme conditions. Yes, you could cycle temperature, hopefully a short spike. You could bring in hot formamide or something. But a question for Andy (Fire): There might be different ways to skin the cat beside the kind of simplified version I showed. Plus strand makes minus, they come apart, each one does it again. Jack (Szostak) talked about the virtual circle genome. So the idea would be that each little fragment is small enough that it could come apart without extreme conditions, and then collectively they carry information. Another possibility would be a rolling circle type replication system like occurs in certain viral systems, or some other reciprocal strand displacement type system like is used in some diagnostic assays. So the question for Andy (Fire): What about something like the mechanism inspired by the viroid where there's alternative confirmations that basically get a way to hairpin back for both the plus and minus strands.

Andrew Fire:
I think that's a good model. I think that was some of the models that came out early. Other people in the Q beta field were part of the driving factor. And the monsters that would form in those replication systems was their ability to self denature in some way. Multi forms have been able to do that in multiple stages where the activation energy for each stage is relatively small. It's attractive, but I'm sure it could be done in many different ways.

Gerald Joyce:
What he's referring to is test tube evolution experiments where Q beta replicase just like Andy (Fire) explained using T7 RNA polymerase, is the replicator. And the replicatee is the Q beta bacteriophage genomic RNA

that gets copied and copied and copied and evolves in the test tube. But there's a special trick there, which is, as the newly synthesized strand is produced by the polymerase, it exits in a way that is separate from the template, and then begins to fold on its own. So that's another trick, you know, but how to do that perbiotically is a little tricky. It takes that structure to make sure the newly synthesized strand peels off into the distance. So I think there's lots of options, but it is still an unsolved problem in the field.

Ben Feringa:

Talking about templating. I read a number of articles in the past that people said inorganic templating, inorganic surfaces played a crucial role in prebiotic chemistry. I didn't hear you about that. Is that out of the picture completely? Or what is the role there?

John Sutherland:

It's not out of the picture. No, I think the problem is searching that space. Basically, there are lots of potentially available enabling surfaces. There's been a lot of work done, for example, on clays, by Jim Ferris on oligomerisation of nucleotides on clays. And there are some results from that which are quite encouraging. And Jack's got some data on clays, basically. And the formation of vesicles.

Ben Feringa:

On clays you got a lot of organization!

John Sutherland:

Yes, you get the organization. But the organization when you polymerize, it's sometimes more difficult to get things off the surface. So it's the same issue basically.

Karla Kirkegaard:

Actually, for some RNA viruses on those membranes on which they replicate, you get a lattice of replication proteins, like in the case of poliovirus, like a polymerase lattice. And so that's what we think. I think there is no long double stranded RNA intermediate, even though people always draw it. I think there's no evidence for it. I think you see it after you extract the RNA with detergent. What I think is that it's splayed out on those surfaces, and kept pretty much single stranded by the protein interactions.

Martyn Poliakoff:
I'm not sure that temperature cycling is a problem. If you think a bit about the geysers in Yellowstone National Park, and you imagine some slow, flowing stream that gets doused with hot water every hour or something that you could easily imagine cycling of temperature. Because the important thing is solely you only need this to be done locally. You don't need whole lakes suddenly to heat up or cool down. And if it was associated with the stream, the reaction products would be nicely washed down and distributed across the landscape.

Jack Szostak:
Yes, we were trying to think of geological scenarios like that, where we could get a pulse of high temperature, but you need to be able to repeat it, it has to be a cyclic process, but if the products get washed away downstream, then it's not so great. We have to explore all these things, see what will work the best.

Karla Kirkegaard:
I think when you're talking about evolution, you don't have to have things endlessly logically repeatable. And we're talking about very rare events, in general, and so sometimes looking at that rare event, and then adding, you know, arguing post hoc, something very very rare doesn't work…

Jack Szostak:
Well, we were talking about how to get basically a cell cycle going, right? Where it may take multiple cycles of high temperature, cool phase, in order to get one replication cycle. So you don't want to have too much degradation or dilution.

Gerald Joyce:
We're also mixing up two different eras here. Once you have evolution, Darwinian evolution, you can invent ways to tame your two strands. But before you have that, it has to be a physicochemical way. And although

temperature cycling, a repeating geyser or some hotsprings might do it, there's also the problem of strand reannealing. Just separating two strands, you need that. But then, if those two strands come together before either strand can be copied, you lose. Of course, you don't have the help of evolution at the beginning.

Thomas Ebbesen

There's a lot of work on vents, including complete chemical cycles that you can drive in inorganic materials, and none of you talked about that. The problem with a light system using sunlight is that as the molecules go bigger, they absorb more, their absorption cross section increase, catabolism is faster than metabolism. And it's an old question regarding evolution of chemicals on the surface of the earth. Vents seem to not have that problem since they're very deep in the ocean.

John Sutherland:

So, the best to date experimental data on vent based chemistry is the synthesis of one micromolar formate from CO_2. And it's a long way from one micromolar formate to building blocks for ribonucleotides and proteins. Second comment is, you have a problem with accumulation of materials and separation and purification, no crystallization in vents. No basic way of concentrating, you've got dilution to effectively homeopathic levels, okay, and you don't have the energy associated with UV photons to drive many, many reactions which are required.

Thomas Ebbesen:

Okay, UV photons are very destructive too, so that you have to take into account if you do a model on that.

John Sutherland:

Granted, they are destructive, but they also provide a selection mechanism as well.

Jack Szostak:

I can just add one thing, you know: when you have a genome that's mega bases, or more, then it doesn't take much UV to be harmful. But if the genome is 50 bases, you can tolerate a lot more.

Thomas Ebbesen:

I'm talking about before the genome. I'm talking about just the chemistry of the chemical soup. I mean, there are articles reviews written on this 30–40 years ago already about this issue of the balance of the cross section. So… But this is not the topic.

John Sutherland:

Just one last comment. There are plenty of things written about it but there's absolutely no supporting experimental evidence.

Nick Turner:

I want to go back to these very first peptides that might have been produced. And this is a question I guess, for John and Jack. Are you thinking about trying to make a peptide that will have catalytic activity? I can think of some candidates, I'm sure you've thought of these. So di-peptides diketopiperazine can catalyze asymmetric Strecker reactions. So you've got cyanide, you've got imines that make amino acids that could help. Is that the way you're thinking? Does that help to advance your argument that you can make for the first time a catalytic peptide?

Jack Szostak:

That's an extremely important point, you can't evolve something like the ribosome unless you have both these substrates. And the product has to be useful in some way. And so we've been thinking more and more about that. You gave some examples, John gave me a possible example for "Gly-Gly", which would be amazing because there's certainly lots of glycine around. So that's what we need to sort of close the loop. To have a selection for the evolution of coded synthesis, we need a function for the first peptide.

John Sutherland:

The example I gave Jack was simply that when you have divalent metal ions that catalyze the hydrolysis of RNA, for example, Cu^{++}, passivated, by binding dipeptides by tridentate binding, you can stop copper from destroying RNA hydrolytically by making a dipeptide that chelates it.

Rob Knowles:

Just to follow up a little bit on what Nick was just speaking about. I know there's no clear answer to this, but I was curious to hear all your opinions on sort of the chirality questions and how those sort of arose? And what the sort of current thinking is on that stereochemical question?

John Sutherland:

The idea basically is that if you can sort out the stereochemistry of RNA, then when you do the amino acid transfer chemistry, you get the relative stereochemistry. So you can go from receiving amino acid pools to get the correct relationships. There are chemical relationships between amino acids and RNA. So the trick then is how do you sort out the RNA? And all I would say at this stage is there is an intermediate in the synthesis which is crystalline, and it crystallizes as a conglomerate, which gives you the chance basically, if you do this on the surface of the planet where you can crystallize, it gives you the chance to use conglomerate resolution techniques basically, or imagine how you could resolve by conglomerates.

Henry Snaith:

So I'm apologizing in advance, I'm going to ask a layperson's question. This seems like sort of fair understanding of what could have happened. But obviously, you're a little away from having the complete picture of the evolution of life. Can you make any estimation of the probability of this happening? Or what do you need to know before you can estimate the probability of life having occurred within the last 4 billion years on Earth?

Jack Szostak:

The simple answer is, we're here. So the probability is one.

Henry Snaith

I was fearing that was going to be the answer!

Gerald Joyce:

I think this is a place where there's been a lot of incautious statements, particularly from astronomers. We don't know the probability, we really don't. So you know, we know now how incredibly abundant extrasolar planets are, right? Certainly with the James Webb telescope, we're going to start seeing smaller, more potentially Earth-like rocky type planets, within the so-called habitable zone. Where it gets murky is in the denominator, when calculating probabilities, where it's n-minus-one. And since our n is one, this denominator makes the equation undefined. We don't actually know, given a terrestrial-like rocky planet within the habitable zone, what the chances are of starting life. I mean, we can talk about it, but we don't actually have a number.

Jack Szostak:

Let me give a more serious answer, which is we don't know. So one of the many things we don't know is how many different environments have to be present and involved in mixing, transport, in the right order. Getting all the bits and pieces to work together perfectly by chance might be incredibly rare. And maybe that's why it could have taken hundreds of millions of years to get the right combination of events. Or it could turn out to be incredibly easy. And life's popping up and dying out all over the place. And that's really what we don't know.

Henry Snaith:

And are there key points? There is large uncertainty in terms of the whole evolution from the first RNA through to complex life or multicellular life. Are there key points where there were pauses? I guess there's no way of looking to see what happened.

Jack Szostak:

There are a whole bunch of things that we don't know, right? Once life, say RNA world type life, got started, how long did it take to evolve protein synthesis? Metabolism, protein, membrane transport, all those things? Once you get to something as complex as LUCA, which is basically a modern bacterium, it would be very hard to drive that to extinction, even by huge impacts. But we have no idea how long that took.

Gerald Joyce:

I agree completely, Jack. So once you get to something like the *last universal common ancestor*, or what we call LUCA, game on! If there was an RNA world prior to that, which seems highly plausible, it couldn't have been a very long interval between that and LUCA. Best estimate when there was a last universal common ancestor is 3.6 to 3.7 billion years ago. And then, on the earth, it wasn't really habitable for aqueous life until 4.2, maybe 4.3, billion years ago. So it depends on what you call « long ». That's still hundreds of millions of years, which is many, many generations. But in the lifetime of a planet around an M star, you've got enough time. To me the bigger uncertainty is the interval before evolution starts. Because evolution finds a way, right? Some would say, if you don't have a magnetosphere, you're not going to have a suitable planet because the cosmic radiation from the parent star is going to make it impossible. Some would say, if you don't have plate tectonics, you're not going to have a way because you don't have a fluxing environment. We just don't know the answer to all these things. What will change the equation is when n equals 2, right? And so, obviously, there's a lot of effort to find evidence of life, not extant life, but extinct life; evidence of extinct life on Mars, or on one of the satellites of Jupiter or Saturn, maybe for the younger people within their lifetime. And then, of course, whether there could possibly be a signal of life on an extrasolar planet. And then finally, SETI, if there's actually a signal that we pick up.

Ben Feringa:

So that was exactly what I wanted to ask. This new satellite has now been announced. It gives us beautiful pictures, you know, and I get the impression that my colleagues in Physics know precisely what they are looking for. What should we look for, as chemists?

Gerald Joyce:

Obviously, with JWST (James Webb Space Telescope), they're going to be able to get atmospheric data on extrasolar planets. So what should they look for? There are different arguments. I'll put a couple of them out there, and others feel free to chime in. The slogan was: search for water! Because that's necessary for life. That's now considered not to be a defining enough criterion. Another mantra that one hears is: search for non-equilibrium chemistry. That there's some process that is operating out of equilibrium, which would be tied to a living thing. The more recent counterargument to that is: Life drives itself extinct if it stays out of equilibrium for too long because it consumes the resources of the planet. So a long lived life form would be one that actually can operate very close to equilibrium. Then people point to specific chemicals like ozone, so that there would be an oxygenic atmosphere associated with ozone. Phosphine was one that was in the news recently because of the phosphine signal, at a very low level, that was found in the upper atmosphere of Venus. And although it's true that one of the sources of phosphine on Earth is biogenic, and there isn't a fully known chemical mechanism for getting that level of phosphine on Venus, there are plausible mechanisms. So that's kind of my view of the laundry list. None of them are smoking guns. I think what the extrasolar planetary astronomers are going to look for is some of those things. My prediction is that there'll be some initial overclaims and a lot of publicity, and then some counter discussion. And then hopefully, rationality comes back.

Ben Feringa:
The reason I asked this question, you know, origin of life, to me, prebiotic chemistry looks like a chemical problem and molecular problem. And I don't want to leave it to my colleagues or Physics all the time, you know.

John Sutherland:
Ben, I agree with you. I've collaborated with Didier Queloz. And one of the things we've done is basically say "Look, we know we need a certain amount of light to do this chemistry". So he can then look at his exoplanets and say they are or are not close enough to their star to receive enough light to do that chemistry, to outdo competing chemistry that occurs in the dark, for example. So you can constrain which exoplanets are most interesting to look for. But I kind of completely agree with what Gerry (Joyce) said : there are going to be false claims. And it's going to mess the field up. I would love to see cyanide. So when Schumacher-Levy impacted Jupiter, cyanide in Jupiter's atmosphere spiked too. In principle, James Webb could spot a meteorite impact on an exoplanet: a spike in cyanide in the atmosphere. And that would then tell you that what we're invoking — cyanide and atmospheres of rocky planets — is not unachievable, for example.

Jack Szostak:
So there are a couple of problems. What I would most like to see is evidence of prebiotic chemistry on an exoplanet. What I hear when I talk to astronomers who are thinking about this is is that it's super difficult not just because of how rare the relevant gases might be, but because a young planet around a young star is in a very dusty environment, which makes getting good spectra almost impossible. And then there's contamination. As you know, we can see tons of cyanide in the dusty rings of a proto planetary nebula. So I think finding evidence of this chemistry on a young planet is going to be really tough unfortunately.

Kurt Wüthrich:
In the timeframe that we are considering, there have been extreme climate changes on Earth. How do you take account of these when looking at the evolution? Let me just give you a hint. With proteins, we know that there is heat denaturation. It's perhaps less well known that there is also cold denaturation of proteins. Now, is there a reason why an RNA world would have survived more elegantly during all these climate changes than a protein world? Should we look at this when considering the most recent, ongoing change of climate?

Gerald Joyce:
Okay, I'm going to start, but I want someone else to help here! So, again, let's separate the eras. The era of the RNA world, I would view as a more fragile era. The era since LUCA, the last universal common ancestor, has proven to be a very robust era with regard to changes in climate, including far more dramatic changes than we've seen in the course of human existence on this planet. If one looks at so-called extremophiles, where microorganisms or even eukaryotic organisms can live, the range of temperatures are from –10 Celsius to 200 Celsius — although not replicating at 200°C but surviving — replicating at 120°C, and extreme differences in salinity and pH. So again, evolution finds a way. Once evolution is going and robust, the proteins and all the machinery adapt to those conditions. That doesn't mean that one can't imagine a kind of climate change that would cause a global extinction. A very large impactor hitting this planet, which

fortunately we seem to have escaped in the last 4.3 billion years, could sterilize the planet. A 250 km impactor would boil off the oceans and melt the crust of the planet. So too are moon forming type events; I don't think one of those are going to happen ever again. Or a gamma ray burst. One can imagine really extreme events. But over the course of the last 3.8 billion years or so, evolution has found a way. During the RNA world, I imagine that to be a more fragile time and, who knows, there may have been a start, and a wipeout and another start and another wipeout, that would help the denominator factor, if it really happened more than once. Others please?

Claudia Höbartner:

Regarding the comparison to cold denaturation of proteins, I'm not aware of any comparable situation in RNA, since, even under ice conditions the evolution of ribozymes has been done and can occur. And so maybe a preservation, in this way, may also help. And the other aspect, in direction of heat denaturation and disability of RNA under hot conditions in the presence of harsh environments, it's a place where I feel that modifications could play a role to prevent the hydrolysis of the strand.

Gerald Joyce:

I want to come back to the point of life's biosignatures. And we were talking about chirality and the origins of chirality. There was a time not that long ago when people felt a strong biosignature of life would be chiral asymmetry — strong enantiomeric excess of organic compounds or a complex mix of organic compounds. That would not be a strong biosignature, because we know that in things like the carbonaceous chondrite meteorites there can be a substantial enantiomeric excess of amino acids, including non-biological amino acids. We know these aren't contaminations, with locales within the meteorite having up to 18% enantiomeric excess. So if you find enantiomeric excess, it isn't proof of life. Then this touches on the question of how to break chiral symmetry in the first place. Because if that was necessary to start life, as I think many people believe, then by definition, it's not a biosignature because it had to be broken before you start life. Then there's also racemization, but that's not one that I think you can hang your hat on as to chiral asymmetry or chiral excess.

Ben Feringa:
I always learned that circular polarized light in outer star systems is huge, massive circular polarized light, and this signature of chirality in meteorites probably come from there. At least that's the theory I always learned. Prof. Eschenmoser said chirality is the signature of life. That was his firm statement, but I'm happy to hear what you're thinking.

Gerald Joyce:
Maybe others can comment on circularly polarized light, or that the pole of a neutron star has been hypothesized to be sufficient? But that's abiogenic, right? That's not a biosignature.

Jack Szostak:
The circularly polarized light was kind of the common explanation that was talked about for decades and it's experimental fact that you can make that work. But the coupling is very weak, right, it's very hard to get to even the levels that we are seeing in meteoritic samples without essentially destroying almost everything. So I think a revolution over the last decade is now we have many physical mechanisms that can drive homochirality. And so the question now is which of these physical mechanisms actually was responsible?

Chad Mirkin:
More of a generalist outsider looking in type question. This is a pursuit of something that you never really know if you're right, you have a definitive possibility, right? And it's probably one of the biggest questions you could ask. And I think of big questions driving new developments. If you step back and look at what this pursuit has done in terms of new developments, have there been major advances in chemistry, major advances in techniques? Are there new advances required that would accelerate your ability to get to more definitive positions?

John Sutherland:
Chad, I completely agree, 10–20 years ago, I'd have completely agree with you that we may never know. But I'm increasingly of the opinion that we will know in really intricate detail exactly what must have happened,

because things are coming together now. When you start putting networks together, and you start sort of suggesting geochemical scenarios, the geochemical scenarios then feed back into the chemistry. If you then take the input they're putting into the chemistry, see how it plays out in the chemistry, the chemistry gets better, you can sort of bootstrap between the scenario and the chemistry. And I think if you get that, you saw the origin of chirality, you start to see amino acylation of RNA and you start to see peptidation. I think you start to get something that's so good, it wouldn't constitute legal proof. But I'd certainly die happy if I could see all of that.

Chad Mirkin:
What about other techniques? In other words, is there something missing that would accelerate? Has there ever been something that we've learned that doesn't answer the question, but drives other developments in chemistry? What's come out of this pursuit is what I'm trying to get at?

Jack Szostak:
I think maybe what has come out more than anything else is how we think of the problems, that the things that have held us back for so long are just preconceptions about how things work. And, you know, just one example: There are some synthetic organic chemists who would say that this is completely impossible to happen on the surface of a planet all by itself. Then you start to think about the geochemical scenarios, the fact that key intermediates crystallize out beautifully, that you could accumulate reservoirs, that you can capture low levels of cyanide in iron complexes and accumulate huge reservoirs of starting materials. It's how you think about these problems and how you connect simple known reactions into pathways that give us what we need. That's the advance.

Andrew Fire:
Briefly to that, one can draw a line between the origin of life work that was done and the study of the RNA polymerase within very simple systems, and then getting those RNA polymerase to do things like making vaccine RNA and to do it cleanly and knowing what to look for in those preps. And so I do think that there's been a pretty strong return on investment in the field through just that little part of it, you know, the work that

came through that as earlier work on on RNA polymerases, but certainly, it was extraordinarily valuable.

Chad Mirkin:
How about other drivers? Are there analytical techniques that need to be developed that would accelerate? I haven't heard that yet. Is there something missing that slows you down?

Andrew Fire:
We're not doing the origin of life stuff.

Jack Szostak:
The only thing that's slowing us down is having time to think about the problems!

Gerald Joyce:
Yes, the origin of life research field, both prebiotic and early studies of RNA-world-like systems, have benefited from developments elsewhere. The advances of analytical chemistry have helped a lot. Advances in mass spectroscopy, especially, have helped the field a lot, because many of these reactions involve complex sets of products. And so that's probably one I would point to. Thank you to the analytical chemistry field for helping our field!

John Sutherland:
I agree with Jerry, I think systems chemistry was made possible by advances in analytical chemistry. In the past, you'd see people publishing a synthesis of something we thought was prebiotic by making A to B: one product. Life's not one product, you've got to operate in mixtures. So embracing the concept of systems chemistry, I think, has been key to a lot of us. And that was enabled by technology development. Now, analytical technology gets even better, we can play with bigger systems and understand what's going on in these systems, that will be a huge advantage.

Donald Hilvert:
So half of the session was focused on viroids and viruses. Where are viruses in the history of life? Are they intermediates between the prebiotic

soup and cells? Are they what's leftover from primitive cells that got rid of everything they didn't need, so that they could adopt a parasitic lifestyle? Something in between?

Andrew Fire:

So, one of the real experts in this is Eugene Koonin at the NIH. He has found a number of characteristic viral protein families that are not usually seen in a cellular component. And that argues that some fraction of viral components diverge very, very early from LUCA, last common ancestor, and maybe before that. So there's a model that viruses predate or at least date from the earliest cells. As soon as we had working members of society, aka cells, we had parasites of them aka viruses, depending on how you call that, at the very least. And so viruses are probably extremely old. And perhaps at the very root of the tree is the model there. And I want to pine on, I think there's debates about that. But I think there are strong arguments that viruses are indeed quite old, even the ones we have now.

Donald Hilvert:

Maybe a short follow up question to you. It wasn't clear to me what viroids are. Do they actually have a protein shell? Or are they just self replicating RNA?

Andrew Fire:

A plant viroïd is a self replicating RNA that uses the cell machinery to replicate itself. There probably are proteins that are involved in that and the transfer from cell to cell almost certainly involves another element of the viroïd, sort of stealing something from the plant to do that. Either cell-cell junctions or other cellular components. And the hepatitis Delta does encode a single protein, probably that's involved in replication and then uses another virus to get its way from cell to cell. So there are proteins that are employed by the viruses, but they are more canny about doing other entities proteins than one might expect.

Karla Kirkegaard:

It seems to me that when you have the pieces we've been talking about, you have nucleotides that can condense, you have them encoding a couple of amino acids, amino acids participating in the condensation. It's almost a virus already.

Sabine Flitsch:

I was just curious about sulfur. When that comes in? For example, recently, I came across this whole area of the thioesther world where people really claim that there was chemistry which involved sort of thioesters. What do you think about that?

John Sutherland:

The thioester world was proposed by Christian de Duve and it's a very attractive idea, because, as you well know, thioesters play a key role in connecting nucleotide base chemistry with isolation based chemistry. And sulfur is involved in some of the schemes we've drawn and some of the schemes other people draw. There are other problems with sulfur. I think that if you look at the genetic code, it looks like the self containing amino acids were fairly late additions. So if sulfur was readily available, and files were stable to the conditions, because they're so functionally useful, one could argue that they should have been incorporated earlier and it looks like they were incorporated late.

Yamuna, Krishnan:

I was just wondering about what Chad said about, you know, what's missing, that could help accelerate your progress. And I was just wondering whether you could speak about two things that I found missing so far in your talks. Radicals, radical chemistry, and how do you make lipids? And if we did know how to make lipids, then would that clarify your geochemical scenario?

John Sutherland:

The answer to the lipid problem was presented in a talk the other day, which is that Fischer Tropsch and methenation are two extremes of the same equation, basically. And if you talk to the geochemist about what likely would have prevailed, they say that methenation is far more likely than Fischer Tropsch. So this mechanism to provide large amounts of hydrocarbons is no longer as favored as it used to be. So you're right, identifying how best to make lipids is key. There are some syntheses out there, but I don't think they're any good, including my own. That's a key issue. I think how to get lipids that give you the encapsulation that Jack told us is needed. And I think we all agree it's needed.

Jack Szostak:
I think John didn't answer your question about radical chemistry. But radical chemistry is central to the whole cyanosulfidic photoredox system. So it played an important role in making all of the building blocks.

Bert Meijer
I'll ask a question as a supramolecular chemist. And as you see, it's all supramolecular chemistry, all of these molecules together. And in supramolecular chemistry, it's very prone to have the effect of cooperativity, in which suddenly things become very strong. And you need a certain concentration, or a certain length or whatever, in order to get the cooperativity to work, and sometimes it becomes even stronger. How is that issue here? I don't hear it actually, at all. Not in the original life, but also not if you look to the viruses, if they're very small, maybe there's less cooperativity than when they're bigger. For us, cooperativity is an enormously important aspect of all of our supramolecular chemistry. I was curious to know how you think about this.

Karla Kirkegaard:
I think that's one of the fascinations of these macromolecular assemblies, and especially one of the really interesting things about capsides and also the large aggregations of polymerase etc, that we see, for example, in these molecules that hyper stabilized capsid. What sometimes people say about them, that they don't like them as antivirals, is because they're needed stoichiometrically. What we've shown is that they're not, you can look at the binding curves of the small molecules to capsides, and you get inhibition way before you saturate all the subunits. So, there's a lot of cooperativity and assembling something like a capside that you can take advantage of in that way. So yes, I mean, cooperativity drives a lot of these things.

Jack Szostak:
Yes, I think there are many places where there are cooperative interactions in this. The assembly of RNA goes on, on a template to be cooperative, because of stacking interactions. The assembly of membrane systems or the condensation of peptides into phase-separated droplets, there's a high degree of cooperativity in all of these phenomena.

Bert Meijer
Yes, that I understand. But I mean, in the beginning, you wouldn't have the cooperativity. And then there's a certain point in which the cooperativity does work, and then it goes very fast. And I was just wondering whether the experiments you are now all doing are in the pre-cooperative regime or already in the cooperative regime.

Karla Kirkegaard:
Well, you guys keep making your replicating RNAs longer and longer. What have you seen?

Jack Szostak:
We've been looking at the assembly of ribozymes from fragments that can certainly be cooperative. So we definitely see cooperativity in these assembly processes. That's just one example.

Bert Meijer:
But if the primer is short, then there is no cooperativity and then it's not a primer. I'm just trying to figure out at what point it becomes cooperative and what point it is. We call it isodesmic, but then the association constants are actually independent on everything. And as soon as it becomes cooperative, you add up and it goes. I was just wondering.

Jack Szostak:
Actually, let me pass the question to John and ask if you have two hairpins assembling on a template? Is that cooperative or anti cooperative?

John Sutherland:
I think it's cooperative. Well, I think the question is basically "does one help the binding of the other?". And we haven't actually established that.

I think the messenger RNA in accommodating two tRNAs on the ribosome is kinked. And that kink is probably entropically stabilized by the 30S subunit. And so we're probably up against an entropy issue when we try and bind the two. So it's probably not cooperative. But there could be interaction between the two elsewhere, in which case it would be cooperative. So I think my answer is that we should find out.

Bert Meijer:
I think it's an important issue, because the entropy issue is a very complicated one. As we do interactions of individual ones that are weak, and you take a whole string, then it starts to come together to cluster. And that, in your view, should be entropically unfavorable, but due to the fact that the rest can do this, it becomes entropically driven. So I can imagine that that plays a role here as well. If you have these individual ones, on the template that they like to come together, in order to have more mobility on the one that is not called.

John Sutherland:
I think the first complex in which trans-peptidation took place would have to be cooperative. And I think that cooperativity would be the key to it. There's been this lovely description of the ribosomes as an entropy trap. And I think that's what you want to try and recapitulate in a simpler system, something that sort of is an entropy trap, just by proximity induced by cooperative behavior.

Bert Meijer:
Okay, thank you so much. I just didn't hear the word cooperativity at all in all of the talk. So I thought it's an important issue.

Gerald Joyce:
I'm glad for this. I think this is very stimulating. I've never heard this concept discussed in the many, many discussions about the origin of life problem. We're all kind of winging an answer here. But I think it's one we should take forward as food for thought. Certainly by the time you get to genomic systems, especially when genomic length is at a premium, as I talked about earlier, you don't have enough fidelity to have a big

genome. There can be a tremendous benefit from the redundancy that makes use of multiple subunits that are then assembled in a simple manner. I think about something like tobacco mosaic virus, which doesn't have a big genome, but has proteins that very beautifully make the entire scaffold of what is then the assembled viral particles. So I can see it by the time you have a genomic system. But I've never heard it seriously discussed until now in terms of prebiotic. Thanks for that.

Karthish Manthiram:
I guess I was just wondering if there's any possibility whatsoever that, from the era of the RNA world, any physical evidence whatsoever could have at all been preserved? Is it improbable, is it unlikely that, on a very flux set of narrow, narrow, narrow conditions, that something could exist out there?

Gerald Joyce:
I'm going to start on this one, because I love this topic. When are we going to have a second example? Are we going to find it on an extrasolar planet? We're not going to get a smoking gun, let's face it. Are we going to find it on Mars? There'll be some false alarms, I'm afraid. Maybe we'll see evidence of extinct life; maybe we'll see cross contamination; maybe we'll see evidence of our same life form that once upon a time was on Mars as well. Maybe it'll just be synthesized in the lab. But then there's the one you're saying, which is: maybe we'll find it in some isolated niche on this planet. And I won't name the name of a certain individual in Boulder, who has…, I shouldn't even said that! I have seen a draft manuscript that never was officially sent for review, where some bioprospecting in very unusual environments gave potentially some evidence of an extent RNA-based organism. But of course, extraordinary claims require extraordinary proof. And this is a very careful scientist, who has not found that extraordinary proof. I think the general argument is that if there was a RNA-based life form, still wiggling around on this planet, it would not just be food for thought… it'd be food; it'd be gone! So it would have to either be in a quite isolated niche from our biosphere or have some mechanism that keeps it at arm's length. For example, if it's based on a different set of nucleotides, or something kind of far out like that. I don't know,

I think it's a great question and it's one I always fantasize about, but there's no hard answer for it. Others?

Daniel Nocera

So I'm going to preface : first I'm an inorganic chemist. And then secondly, I started proton-coupled electron transfer (PCET) at a kinetic level. And radicals are important. I'm wondering, John. So with sulfur, hydrogen bonds are great. And then I'm thinking about Rob's work who does photo catalysis. And that has shown that phosphate is great as an H dot donor. How much have you looked at that? And I'm sure you have things like just heterogeneous chemistry being basically radical reservoirs. So for instance, sulphide minerals, where I would be actually getting my H dot phosphate. Actually, Rob has even induced chirality from each dot phosphate transfer. So I'm wondering about that. And then I was thinking about the vesicles. Could all that be happening inside of a vesicle? Just an inorganic rock vesicle? Just how much has the inorganic heterogeneous surface chemistry come into play?

John Sutherland:

I think that this is a really interesting question. And the protonation of hydrated electrons by phosphate dates back to the 60s, I believe. But you needed fairly high concentrations to get effective concentrations that would start interfering with either proton coupled electron transfer, or direct electron addition, unless you have something like hydrosulfide. So the second part about the inorganic or heterogeneous aspects: we find that copper with hydrogen sulfide and cyanide is a very effective combination. The copper is doing some sort of redox cycling, it's almost certainly forming copper thiocyanate, because the oxidative byproduct is called SS bonds. And those SS bonds are cleaved by cyanide to generate thiocyanate. And then copper thiocyanate is a p type semiconductor. So if you get nanoparticulate, semiconductor particles, an awful lot of your photoredox is probably going by heterogeneous chemistry.

Omar Yaghi:

John, imagine there are clearly lots of different reactions that are producing products at different rates and some of them might be needed for certain other reactions. So could you comment about the potential presence of reservoirs to hold certain products until they're needed? You discuss clays as potentially surfaces for reactions and templating reactions and things like that. Could you talk about the reservoir question? Storage of certain important molecules? And then my other very naive question: Could you simulate those early scenarios in a closed system and where you can perturb the system? And maybe use it as a.., I don't want to say, … as a control experiment? But in the lab?

John Sutherland:
Quick answer on the reservoir concept. I think, as you summarized, it's really important to be able to accumulate just as a taster. Cyanide from the atmosphere can be concentrated by formation of Ferrocyanide and then we can have long term precipitation of ferrocyanide salts. The cyanide can then be recovered by thermal metamorphosis. So you can end up with very constant but sort of dry powder containing sodium or potassium cyanide in quantity, in bulk. So, I think reservoirs are important and coupling the reservoir concept to the environment is very important.

Omar Yaghi:
How about for organics, not just cyanide, but for organics?

John Sutherland:
So for organics, basically, we invoke the crystallization. In these crystallization steps, those materials can crystallize and then you can have a time lag. So, you could imagine a reservoir of dry organic crystals being produced and lasting for a long time, as long as they're not UV unstable. And the one I talked about (the intermediate in the nucleotide synthesis) is completely UV stable, so it could accumulate and persist for a long time. But I think this is something we need, that is to have accumulation via reservoirs and persistence for long enough to accumulate enough material.

Gerald Joyce:
So, we've talked about clays and other minerals as a way to sequester materials for surface catalysis. There's a more radical proposal that's been

made, with very little experimental support, that says that something more dramatic could be happening in clays, which is Darwinian evolution. When you were talking about multivariate MOFs (Metal Organic Frameworks) the other day, and thinking about differential occupancy along a linear string, it has been suggested that on a two-dimensional surface of a clay, there can be informational representation of solutes that are deposited as features of the clay that have an informational sense to them. Now that information doesn't have value unless it's propagated and preserved, if it's going to be a Darwinian system. But it has been proposed — this is Graham Cairns-Smith, back in the 70s, who was the first to get explicit about this — that the successive layers of the growing clay would, either in a complementary manner or in an identical manner, recapitulate that informational layout on the surface, and propagate it with sufficient fidelity that Darwinian information can be maintained over time. And furthermore, it would adapt, because any surface pattern that was especially advantageous would propagate better than one that was less advantageous, and so on, and so on, and so on. And I just put that one out there to the audience. It's, of course, logically consistent, but there is very little experimental evidence for that. And this, I guess, comes back to John's question: Is there a whole other place to look than where we are now? We've been so much under the lamppost of "life as we know it" that we need a completely different set of techniques that would be necessary, in this case, to interrogate what is the very detailed distribution of features or charge or facets. Could it be something on the more meso scale, and can that propagate and evolve? Omar?

Omar Yaghi:
I was just gonna say that thanks to your answers now Bert' supramolecular chemistry and my chemistry have their origins in the origin of life. So next time, maybe you should invite us as speakers in your session? What about the control experiment or under control conditions? Laboratory conditions are trying to simulate the original parameters of the origin of life? Maybe that's too naïve?

John Sutherland:
There are some toxicity issues. You'd have to have a closed vessel, basically, to contain the cyanide and the hydrogen sulfide. So we've considered it, but we've not found a deserving postdoc yet!

Omar Yaghi:
But I mean, you can work with toxic materials, right, safely?

Jack Szostak:
Yes! So John didn't go into it. But people are trying to use flow systems to sort of mimic the model reservoirs and streams and starting small with just a few steps... But I think, on the long term, the goal would be to try to connect as many steps as possible in a way that's consistent with a reasonable geological scenario.

Gerald Joyce:
Okay, I'm going to put Andy (Turberfield) on the spot here. Since we're talking, getting a little further out, about what would still constitute a living system, but wouldn't be anything like ours. Can you see a system, this would be an artificial life type system, based on, say, DNA origami type structures, that contains information, propagates that information , and begins to evolve in a Darwinian sense?

Andrew Turberfield:
I mean, information comes for free if you're playing with ready formed DNA. I guess the question is the self replication, and especially the selection pressure. So, as you know, there are lovely experiments from Erik Winfree and Rebecca Schulman on propagation of DNA crystals, self-assembled crystals carrying information. They certainly showed replication, generation to generation, they showed variation. I think they just about showed selection pressure, because they selected for fragmentation, which was there for nucleation of new replicating strands. So, again, those experiments are the nearest I can come to an off the cuff answer.

Gerald Joyce:
It is a little too obscure a reference. But these are obviously just artificial constructs. Yet as long as the information is represented, and can be replicated with reasonable fidelity, and that there's variation, then there's the opportunity for selection. In that example, and in what we all study too, which is more related to the historical origins of life, there's another problem besides just the fidelity problem that I talked about. There's the replication problem. Just doing the reactions. Then there's the fidelity problem.

But there's also the opportunity space, right? So it's not enough for a chemical system to begin to undergo Darwinian evolution. It has to be able to go somewhere. If it only can just flip a bit up or down. No offense, but that's kind of where the Schulman system is. It can make a left right choice, that's not going to be robust to significant changes in the environment, for example. So, you wouldn't want to say that it's universal, like a Von Neumann machine, a universal replicator constructor that can do all possible logical things, or all possible chemical things. But it needs to have some opportunity space to explore, some more or less open-ended space for evolution. And that has obviously happened at least once. I haven't seen another example of a system that does that. And I guess to take it one further step: I always wonder, is there a whole other way to do this that isn't Darwinian? Is there a way to propagate information and persist despite the changing environment, to have a kind of memory for what has worked and apply that to the future? But it's not the Darwinian paradigm, okay? That doesn't mean things like autocatalytic cycles, which are just existential events of chemistry, but some other way in which a system can sustain itself over time, despite a fluctuating environment. So I would put those two pieces out there; it's got to actually have room to roam. And maybe there's a whole other paradigm for how to do it that none of us have conceived. So Andy, I know I cut you off, you're going to say something?

Andrew Fire:
Merely that I now know what a straw man feels like!

Gerald Joyce:
We're coming to the end. But Jack, maybe, and then Peter, one last comment, and then we'll close.

Jack Szostak:
Yes, I just wanted to bring up the fact that there are several labs, ours, Chris Hunter's and several others who are trying to design genetic polymers that can do templated replication, and that could be the basis of totally different Darwinian system. So it would be artificial, intelligently designed. It's a huge challenge for chemistry to make anything that's

anywhere near as good as RNA and DNA. But I think that makes it a very attractive challenge.

Peter Palese:

Yes, I just wanted to say we talked a lot about prebiotic situations, viruses, even though we have very successful RNA viruses, I think they really come after cells have been formed in the way we know them now. Because, as Gerald pointed out, viruses are not living in this same sense as bacteria, for example. So they really need components of a well functioning cell. And therefore, I think viruses have come later, whether they are DNA or whether they are RNA. And I think, in that sense, I was the last of the speakers, but I think they are really much newer, and much more recent, in the evolution of life.

Ben Feringa:

One question about viruses and vaccines. We get, as scientists, often confronted with the question, why do we need vaccines? And what is this with this viruses? And of course, then I mentioned polio, you know, and I mentioned Spanish flu that we heard today. But do you have an advice for us? Being scientists, what message should we give to our politicians or the general public? I'm not even talking about politician, the general public in our general lectures, when we get these questions? The reason I get confronted with this is when I talk about my nano machines, they say, "Oh, this is so dangerous". And then I say no! More dangerous then nanomachines are viruses! Then we get this discussion about viruses and then you have these disbelievers. So do you have an advice for us?

Peter Palese:

Pray!

Karla Kirkegaard:

I think one of the things that's happened during this pandemic is that finally there has been an appreciation that you do it for other people, as well as yourself. I think, what I used to tell people was: sure if you don't want to vaccinate your child, you want your little numb kid not to get stuck with a needle. Fine! Just have them stay home the rest of their lives.

But you know, it's not just about selfishness. It's about other people. It's about other people's grandmothers and stuff. And I think that really helps people. Some kinds of people sometimes be more accepting of vaccines.

Ben Feringa:
Yes. But you know, I had these cases where there was a big guy with a lot of tattoos and said, I don't want to have a needle in my body!

Karla Kirkegaard:
Well, one of my favorite things that happened during the pandemic was a bunch of guys with tattoos were outside a bar. And I was visiting my friend in the Olympic Peninsula. And I walked by, and they were talking, and one of them said "you know, I'm gonna just do what Tony says, I believe Tony". And that was one of the high points of my pandemic.

Daniel Nocera:
Could I just ask one quick question to Peter? You're doing a nasal spray, and I just saw AstraZeneca failed on their trial, instead of pharmacological problem, that seems pretty straightforward. Just from way on the outside, I was just wondering why that would have failed so quickly.

Peter Palese:
The adenovirus has the S protein incorporated in the genome. So when it comes into the nasal cavity, upper respiratory tract, the virus, the carrier goes in, and the adenovirus itself does not expose the S protein, it is only genetic information. And so the adenovirus doesn't replicate probably well enough to really express the S protein in infected cells. So the titer is a little bit too low. But, if you give it probably three or four times it will probably be okay. So there's nothing really wrong about it. It's not dangerous, I think it's just that it does not reach the expression of the S protein to levels where we make enough antibodies.

Karla Kirkegaard:
I want to say another reason why I love vaccines, and some of it has to do with the genetics that I was talking about. So if you have, for example, an antiviral, and you're worried about resistance. There's like one mutation

that you need, usually to get that. But then you think about our antibody response, and we make so many antibodies that's like built in multi drug therapy. First of all, there's that. And then there's the other thing that I talked about, which is dominance, basically. That if there's a vaccine resistant, an antibody resistant virus that comes out, it's coming up in a cell that's full of other antigens that they're going to put on the surface. So it's was shown by Esteban Domingo, who's a famous RNA geneticist, a long time ago, their resistance to monoclonal antibodies is exactly suppression of that is suppressed, because there are so many different epitopes in the cell at same time, that the new resistant one doesn't express its phenotype, he called that « phenotypic masking ». So, between those two phenomena of phenotypic maskings, you're not getting selection out of the cell for something that's antibody resistant, and that there are multiple antibodies that recognize different parts of it will never get antivirals as good as vaccines. It's just that it's the best example of Preventative Medicine. And so sometimes people don't have the foresight to do it.

Peter Palese:
Just half a sentence. Per dollar or per euro, vaccines are really a fantastic bargain, just to reiterate what Kierkegaard said.

Gerald Joyce:
All right. We're going to close the session at this point. Thank you, everybody.

https://doi.org/10.1142/9789811282324_bmatter

Index

www.ingramcontent.com/pod-product-compliance
Lightning Source LLC
LaVergne TN
LVHW010325220926
848484LV00004BB/839
* 9 7 8 9 8 1 1 2 8 2 3 1 7 *